Pharmaceutical Suspensions

Alok K. Kulshreshtha · Onkar N. Singh
G. Michael Wall

Editors

Pharmaceutical Suspensions

From Formulation Development
to Manufacturing

 Springer

Editors
Alok K. Kulshreshtha
Alcon Research Ltd
Forth Worth, TX
USA
alok.kulshreshtha@alconlabs.com

Onkar N. Singh
Alcon Research Ltd,
Forth Worth, TX
USA
onkar.singh@alconlabs.com

G. Michael Wall
Alcon Research Ltd
Forth Worth, TX
USA
michael.wall@alconlabs.com

ISBN 978-1-4419-1086-8   e-ISBN 978-1-4419-1087-5
DOI 10.1007/978-1-4419-1087-5
Springer New York Dordrecht Heidelberg London

Library of Congress Control Number: 2009937643

Printed on acid-free paper

Springer is part of Springer Science+Business Media (www.springer.com)

Preface

The suspension dosage form has long been used for poorly soluble active ingredients for various therapeutic indications. Development of stable suspensions over the shelf life of the drug product continues to be a challenge on many fronts.

A good understanding of the fundamentals of disperse systems is essential in the development of a suitable pharmaceutical suspension. The development of a suspension dosage form follows a very complicated path. The selection of the proper excipients (surfactants, viscosity imparting agents etc.) is important. The particle size distribution in the finished drug product dosage form is a critical parameter that significantly impacts the bioavailability and pharmacokinetics of the product. Appropriate analytical methodologies and instruments (chromatographs, viscometers, particle size analyzers, etc.) must be utilized to properly characterize the suspension formulation. The development process continues with a successful scale-up of the manufacturing process. Regulatory agencies around the world require clinical trials to establish the safety and efficacy of the drug product. All of this development work should culminate into a regulatory filing in accordance with the regulatory guidelines. *Pharmaceutical Suspensions, From Formulation Development to Manufacturing*, in its organization, follows the development approach used widely in the pharmaceutical industry. The primary focus of this book is on the classical disperse system – poorly soluble active pharmaceutical ingredients suspended in a suitable vehicle.

After discussing various disperse systems, the theory of disperse systems and commonly used excipients, the remaining chapters in this textbook systematically explain the development of pharmaceutical suspensions, from the pre-formulation stage to clinical development, regulatory submission and commercial manufacturing. Additionally, the emerging area of nano-suspensions as applied to the pharmaceutical field is also discussed. Each of the chapters in *Pharmaceutical Suspensions* was written independently by scientists who are skilled in their specific areas. Contributing authors represent a cross-sections of scholars from academic, industrial and governmental affiliations.

Pharmaceutical Suspensions, From Formulation Development to Manufacturing is organized in a total of ten chapters: Chapter 1 introduces various pharmaceutical disperse systems in-depth. Chapter 2 presents the general principles of suspension dosage form and Chapter 3 discusses commonly used excipients in pharmaceutical suspensions. Chapter 4 systematically highlights steps involved in pharmaceutical development of suspension dosage forms. Chapter 5 focuses on preclinical development of suspension formulations. Analytical tools needed to characterize pharmaceutical suspensions dosage forms are discussed in Chapter 6. The clinical development aspects of suspension drug products are discussed in Chapter 7. Chapter 8 highlights scale up and technology transfer of the development of pharmaceutical suspensions. Chapter 9 reviews the science and regulatory perspectives of pharmaceutical suspensions. Finally, Chapter 10 deals with the pharmaceutical applications of nano-suspensions as nanomedicine, an emerging technology area.

Pharmaceutical Suspensions, From Formulation Development to Manufacturing should serve as a good resource for pharmaceutical scientists, process scientists and chemical engineers involved in the areas of research and development of pharmaceutical suspension dosage forms, and for new and sustaining scientists of the pharmaceutical and chemical fields. The fundamental aspects together with the practical case studies should also make this a useful source for undergraduate and graduate education.

Forth Worth, TX   Alok K. Kulshreshtha, Ph.D.
                                                   Onkar N. Singh. Ph.D., M.B.A.
                                                   G. Michael Wall, Ph.D.

Acknowledgements

We wish to acknowledge our families for their support and encouragement. We also wish to express our sincere gratitude to all the contributors who made this textbook possible. Their patience and perseverance throughout this process is greatly appreciated.

Contents

Contributors

Yusuf Ali, Ph.D.
Vice President, Santen Incorporated, Napa, CA, USA

Ashwin Basarkar, Ph.D.
Pharmaceutical Sciences Department, North Dakota State University,
Fargo, ND, USA

Diane Burgess, Ph.D.
Professor of Pharmaceutics, University of Connecticut, School of Pharmacy,
Storrs, CT, USA

Martin J. Coffey, Ph.D.
Principal Scientist, Pharmaceutical Product Development, Bausch and Lomb, Inc.,
Rochester, NY, USA

Sudhakar Garad, Ph.D.
Group Head, Novartis Institute for Biomedical Research, Cambridge, MA, USA

Abhay Gupta, Ph.D.
Pharmacologist, Division of Product Quality Research, Office of Testing
and Research, Center for Drug Evaluation and Research, Food and Drug
Administration, Silver Spring, MD, USA

Yatindra Joshi, Ph.D.
Vice President, Generic R&D, Teva Pharmaceuticals, North Wales, PA, USA
(former Vice President, Novartis Pharmaceuticals, NJ, USA)

Akio Kimura
Head of Pharmaceutical Development, Santen Incorporated, Osaka, Japan

Mansoor A. Khan, Ph.D.
Director, Division of Product Quality Research (DPQR), Office of Testing
and Research (OTR), Center for Drug Evaluation and Research,
Food and Drug Administration, Rockville, MD, USA

Alok K. Kulshreshtha, Ph.D.
Associate Director, Pharmaceutical Technology, Alcon Research Ltd.,
Fort Worth, TX, USA

Chandrasekar Manoharan, Ph.D.
Pharmaceutical Sciences Department, North Dakota State University,
Fargo, ND, USA

R. Christian Moreton, Ph.D.
FinnBrit Consulting, Waltham, MA, USA

Mohammed T.H. Nutan, Ph.D.
Irma Lerma Rangel College of Pharmacy, Texas A & M University,
Kingsville, TX, USA

Riccardo Panicucci, Ph.D.
Global Head, Novartis Institute for Biomedical Research, Cambridge, MA, USA

Yashwant Pathak, Ph.D.
Assistant Dean of Academic Affairs, Professor and Chair,
Pharmaceutical Sciences, College of Pharmacy, Sullivan University,
Louisville, KY, USA

Alan F. Rawle, Ph.D.
Divisional Manager, Malvern Instruments, Inc., Westborough, MA, USA

Indra K. Reddy, Ph.D.
Professor and Dean, Irma Lerma Rangel College of Pharmacy,
Texas A & M University, Kingsville, TX, USA

Vilayat A. Sayeed, Ph.D.
Division of Chemistry, Office of Generic Drugs, Center for Drug Evaluation
and Research, Food and Drug Administration, Rockville, MD, USA

Jagdish Singh, Ph.D.
Professor and Chairman, Pharmaceutical Sciences Department,
North Dakota State University, Fargo, ND, USA

Onkar N. Singh, Ph.D., MBA Assistant Director, Development,
Alcon Research Ltd., Fort Worth, TX, USA

Deepak Thassu, Ph.D., MBA
Vice President of Pharmaceutical Development, PharmaNova Inc.,
Victor, NY, USA

Praveen Tyle, Ph.D.
Senior Vice President and Global Head, R&D and BD&L, OTC,
Novartis Consumer Health Care, Parsippany, NJ, USA

Sudhir Verma, Ph.D.
University of Connecticut, School of Pharmacy, Storrs, CT, USA

G. Michael Wall, Ph.D.
Senior Director, Alcon Research Ltd., Fort Worth, TX, USA

Jianling Wang, Ph.D.
Senior Research Investigator, Novartis Institute for Biomedical Research,
Cambridge, MA, USA

Terry K. Wiernas, Ph.D., MBA
Vice President, Alcon Inc., Fort Worth, TX, USA

About the Editors

Alok K. Kulshreshtha, Ph.D., Associate Director, Pharmaceutical Technology at Alcon Research Ltd., Fort Worth, TX, has 20 years of experience in the development of sterile ophthalmic, contact lens care, and dry eye products. During this tenure in pharmaceutical industry, Dr. Kulshreshtha has worked on several suspension drug products in different stages of development. Dr. Kulshreshtha's interest in the suspension area started almost 30 years ago when he began his doctoral dissertation on suspension rheology. Over the years, Dr. Kulshreshtha has worked on many new technologies for producing sterile suspensions. Dr. Kulshreshtha is the inventor/co-inventor of several patents and has authored many peer-reviewed publications. Dr. Kulshreshtha is a member of the American Association of Pharmaceutical Scientists (AAPS) and has been reviewer for AAPS abstracts screening for posters and symposia. Dr. Kulshreshtha holds a Ph.D. and M.S. in Chemical Engineering from Purdue University, and a Bachelor's degree in Chemical Engineering from the Indian Institute of Technology, Kanpur, India.

Onkar N. Singh, Ph.D., M.B.A., Assistant Director, Development at Alcon Research Ltd., Fort Worth, TX, has over 16 years of experience in pharmaceutical product research and development of nasal, sterile ophthalmic, sterile otic, and parentral products. Dr. Singh is inventor/co-inventor of many patents related to formulations, process and formulation technology and has authored several peer-reviewed articles. Prior to joining Alcon Inc., Dr. Singh was Manager at Access Pharmaceuticals, Dallas, TX, leading efforts in the area of site-specific targeted drug delivery systems for diagnostic and therapeutic areas of cancer therapy. At Access Pharmaceuticals, Dr. Singh worked on nanoparticles, liposomes, emulsion formulations, protein pharmaceuticals and gene delivery technology. Dr. Singh served as Secretary/Treasurer of the "Ocular Focus Group of AAPS" from 2002 to 2005 and has been a reviewer for AAPS abstracts, posters and symposia. Dr. Singh is a member of the American Association of Pharmaceutical Scientists (AAPS) and currently serves as an editorial board member of *American Pharmaceutical Review Journal*. Dr. Singh holds a B.S. in Pharmacy, M.S. in

Pharmaceutics from Banaras Hindu University, India, Ph.D. in Pharmaceutics from University of Illinois at Chicago and M.B.A. from University of Texas at Arlington, TX.

G. Michael Wall, Ph.D., Senior Director of Pharmaceutical Products Development at Alcon Research, Ltd., Fort Worth, TX, has over 21 years of pharmaceutical industrial experience in the development of intranasal, sterile ophthalmic and otic drug products. Dr. Wall is author or co-author of two patents and over 100 publications including peer-reviewed articles, abstracts, book chapters, book supplements and one book. Dr. Wall has led international, multidisciplinary teams in the development and approval of Ciprodex® Sterile Otic Suspension and Patanase® Nasal Spray, as well as other products in countries around the world. Dr. Wall is a Scientific Fellow of the American Academy of Otolaryngology – Head and Neck Surgery. He has also served the American Chemical Society as a member for 30 years. Dr. Wall received a B.S. degree in chemistry from Auburn University, a M.S. degree in medicinal chemistry from Auburn University School of Pharmacy, a Ph.D. in medicinal chemistry from the University of Mississippi School of Pharmacy, and completed the Advanced Management Program (168th session) of the Harvard Business School.

Chapter 1
Various Pharmaceutical Disperse Systems

Chandrasekar Manoharan, Ashwin Basarkar and Jagdish Singh

Abstract This chapter discusses various types of dispersed systems, including applications of coarse and colloidal dispersions as pharmaceutical dosage and delivery systems. Applications of colloidal dispersions as controlled drug delivery systems are also discussed in this chapter.

Introduction

Dispersed systems consist of at least two phases: the substance that is dispersed known as the dispersed (or) internal phase, and a continuous (or) external phase. Based on the particle size of the dispersed phase, dispersions are generally classified as molecular dispersions, colloidal dispersions, and coarse dispersions. Molecular dispersions have dispersed particles lower than 1.0 nm in size. Colloidal dispersions have particle sizes between 1 nm and 1 μm. Microemulsions, nanoparticles, microspheres are some of the examples of colloidal dispersions. Coarse dispersions have particle size greater than 1 μm, which includes suspensions and emulsions. The scope of this chapter is to elaborate coarse and colloidal dispersions as pharmaceutical dosage and delivery systems. Applications of colloidal dispersions as controlled drug delivery systems have been discussed in the final section of this chapter.

1.2   Coarse Dispersions

1.2.1   Suspensions

Suspensions are a class of dispersed system in which a finely divided solid is dispersed uniformly in a liquid dispersion medium (Nash 1988). Suspensions can be

C. Manoharan, A. Basarkar and J. Singh (✉)
Department of Pharmaceutical Sciences, College of Pharmacy, Nursing, and Allied Sciences,
North Dakota State University, Fargo, ND 58105, USA
e-mail: Jagdish.Singh@ndsu.edu

A.K. Kulshreshtha et al. (eds.), *Pharmaceutical Suspensions: From Formulation
Development to Manufacturing*, DOI 10.1007/978-1-4419-1087-5_1,
© AAPS 2010

classified as coarse or colloidal dispersion, depending on the size of particles. Typically, the suspensions with particle size greater than ~1 μm are classified as coarse suspension, while those below 1 μm are classified as colloidal suspension. When the particles constituting the internal phase of the suspension are therapeutically active, the suspension is known as pharmaceutical suspension. Depending on their intended route of delivery, pharmaceutical suspensions can be broadly classified as parenteral suspension, topical suspensions, and oral suspensions (Martin et al. 1983).

The following can be the reasons for the formulation of a pharmaceutical suspension:

– The drug is insoluble in the delivery vehicle.
– To mask the bitter taste of the drug.
– To increase drug stability.
– To achieve controlled/sustained drug release.

Physical characteristics of a suspension depend on their intended route of delivery. Oral suspensions generally have high viscosity and may contain high amounts of dispersed solid. A parenteral suspension on the other hand usually has low viscosity and contains less than 5% solids.

Ideally, the internal phase should be dispersed uniformly within the dispersion medium and should not sediment during storage. This, however, is practically not possible because of the thermodynamic instability of the suspension. Particles in the suspension possess a surface free energy that makes the system unstable leading to particle settling. The free energy of the system depends on the total surface area and the interfacial tension between the liquid medium and the solid particles. Thus, in order to minimize the free energy, the system tends to decrease the surface area, which is achieved by formation of agglomerates. This may lead to flocculation or aggregation, depending on the attractive and repulsive forces within the system. In a flocculated suspension, the particles are loosely connected with each other to form floccules. The particles are connected by physical adsorption of macromolecules or by long-range van der Waals forces of attraction. A flocculated suspension settles rapidly, but can be easily redispersed upon gentle agitation. This property is highly desirable in a pharmaceutical suspension to ensure uniform dosing. A deflocculated suspension on the other hand stays dispersed for a longer time, however, when the sedimentation occurs; it leads to formation of a close-packed arrangement resulting in caking. Subsequent redispersion of this type of emulsion is difficult as the energy barrier is much higher compared with a flocculated suspension.

In summary, a flocculated suspension sediments faster and is easy to redisperse, whereas a deflocculated suspension sediments slowly and is difficult to redisperse.

The rate of sedimentation of particles can be determined by Stokes' law:

$$V = \frac{d^2(\rho_1 - \rho_2)g}{18\eta_0} \tag{1.1}$$

where V is the terminal velocity of sedimentation (cm/s), d is the diameter of the particle (cm), ρ_1 and ρ_2 are the densities of the suspended particles and the medium, respectively, g is the acceleration due to gravity, and η_0 is the viscosity of the medium.

1.2.1.1 Important Considerations in Formulation of Suspension

Formulation of a pharmaceutical suspension requires a knowledge of the properties of both the dispersed phase and the dispersion medium. The material for the formulation of suspensions should be carefully selected keeping in mind the route of administration, intended application, and possible adverse effects. The following are the most important factors to be considered during the formulation of pharmaceutical suspensions:

1. Nature of suspended material: The interfacial properties of the suspended material are an important consideration during the formulation of a suspension. Particles that have low interfacial tension are easily wetted by water and therefore can be suspended easily. Particles of materials with high interfacial tension, however, are not easily wetted. The suspension of such materials is normally achieved by using surfactants. Surfactants increase wettability of the particles by reducing their surface tension.
2. Size of suspended particles: Reduction of particle size leads to a decrease in the rate of sedimentation of the suspended particles as explained by Stoke's law. Reduction in the size of particles can be achieved by processes such as milling, sieving, and grinding. Particle size also affects rate and extent of absorption, dissolution, and biodistribution of the drug. However, reducing particle size beyond a certain limit may lead to formation of a compact cake upon sedimentation.
3. Viscosity of the dispersion medium: Greater viscosity of dispersion medium offers the advantage of slower sedimentation; however, it may compromise other desirable properties such as syringability for parenteral suspensions, spreadability for topical suspensions, ease of administration for oral suspensions. The property of shear thinning is highly desirable so that the suspension is highly viscous during storage when minimal shear is present so that the sedimentation is slow and has low viscosity after agitation (high shear) to facilitate ease of pourability from the bottle.

1.2.1.2 Sustained Release Suspensions

Emphasis has been placed lately on achieving sustained release of drugs by using pharmaceutical suspensions. Sustained release of drugs can be achieved after injection through intramuscular or subcutaneous route because of limited availability of dissolution medium. The particle size of suspension can be tailored to fall in colloidal or coarse range, depending on a variety of factors such as type of drug, formulation, site of action, and route of delivery. Depending on the aqueous solubility, the drug can be delivered in conjugation with polymers or metal salts to sustain the release (Morales et al. 2004; Gietz et al. 2000). Polymers can efficiently alter drug release kinetics through encapsulation or conjugation of the drug. A variety of microparticulate or nanoparticulate delivery vectors can be prepared that can efficiently prolong the release of drug. These systems will be discussed later in the chapter. Addition of metal salts cause aggregation of drug molecules by formation of metal-hydroxy and -oxy polymers (Masuoka et al. 1993). This strategy has traditionally been used for precipitation of insulin and corticotrophin and more recently to achieve sustained release of recombinant hirudin (Gietz et al. 1998).

1.2.2 Emulsions

An emulsion is a dispersion of at least two immiscible liquids, one of which is dispersed as droplets in the other liquid, and stabilized by an emulsifying agent. Two basic types of emulsions are the oil-in-water (O/W) and water-in-oil (W/O) emulsion. However, depending upon the need, more complex systems referred to as "double emulsions" or "multiple emulsions" can be made. These emulsions have an emulsion as the dispersed phase in a continuous phase and they can be either water-in-oil-in-water ($W_1/O/W_2$) or oil-in-water-in-oil ($O_1/W/O_2$) (Fig. 1.1).

The size of the dispersed droplet generally ranges from 1 to 100 µm, although some can be as small as 0.5 µm or as large as 500 µm. Emulsions are subdivided arbitrarily such as macro, mini, and microemulsions, based on the droplet size. In macroemulsions, the droplet size usually exceeds 10 µm. In the case of miniemulsions, the droplets are in the size range of 0.1–10 µm, and in microemulsions the droplets are below 100 nm. Due to a small droplet size of the dispersed phase, the total interfacial area in the emulsion is very large. Since the creation of interfacial area incurs a positive free energy, the emulsions are thermodynamically unstable and the droplets have the tendency to coalesce. Therefore, the presence of an energy barrier for stabilizing the droplets is required. Surfactants reduce the interfacial tension between the immiscible phases; provide a barrier around the droplets as they form; and prevent coalescence of the droplets. Surfactants are mostly used to stabilize emulsions and they are called as emulsifiers or emulsifying agents. Based on the constituents and the intended application, emulsions may be administered by oral, topical, and parenteral routes.

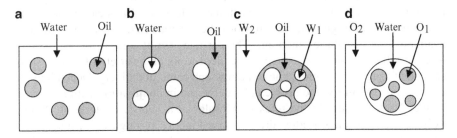

Fig. 1.1 Different types of emulsions. Simple emulsions: (**a**) O/W and (**b**) W/O, Double or multiple emulsions: (**c**) $W_1/O/W_2$ and (**d**) $O_1/W/O_2$

1.2.2.1   Mechanism of Emulsification

When two immiscible liquids are in contact with each other, the molecules at the interface experience an imbalance of perpendicular forces. The net forces at the interface are called interfacial tension and they tend to minimize the surface area of individual liquids. In emulsions, the process of dispersion of one liquid in the other results in an increase in surface area between the dispersed droplets and dispersion medium, and surface free energy, which can be expressed as follows:

$$\Delta W = \gamma \Delta A \qquad (1.2)$$

where ΔW is the increase in free energy at the interface, γ is the interfacial tension, and ΔA is the increase in surface area. In order to reduce the total surface area and the free energy, the dispersed droplets tend to coalesce and tend to separate, making emulsions thermodynamically unstable. One possible approach to preventing coalescence and stabilizing the dispersed droplets is to reduce the interfacial tension. By reducing the interfacial tension, both the surface free energy and surface area are reduced, which leads to a stable emulsion. Emulsifying agents are employed to reduce the interfacial tension by forming a barrier between two immiscible liquids. The emulsifying agents can be ionic, non-ionic or zwitterionic surfactants, and proteins or amphiphilic polymers. Finely divided solids such as bentonite and veegum can also act as emulsifying agents. Surfactants are most widely used to stabilize pharmaceutical emulsions because of their well-marked emulsifying properties. Surfactant molecules are amphiphilic, i.e., have both polar and nonpolar groups, and tend to be oriented between the two phases, with the polar group in the polar phase and the nonpolar group in the nonpolar phase. The polar group is often an ammonium, carboxylate, sulfate, or sulfonate group and mostly contains a succinate or sorbitan group. The non-polar group is generally a linear hydrocarbon chain, and in some cases it is branched and may contain phenolic or other aromatic groups. The amphiphilc nature of surfactants can be expressed in terms of an empirical scale of so-called hydrophile–lipophile balance (HLB) system, established by Griffin (1949). The HLB system provides a scale of hydrophilicity (0–20) and the relationship between HLB values and the expected activity from surfactants is given in Table 1.1. A good emulsifying agent should have a limited solubility in both the oil and water phases of the system. Surfactants having HLB values from 3 to 6 are generally lipophilic and produce W/O emulsions, and those agents with

Table 1.1   HLB ranges of surfactants

HLB (overlapping words) range	Application
1–3	Antifoaming
3–6	W/O emulsifier
7–9	Wetting agent
8–18	O/W emulsifier
13–15	Detergent
15–18	Solubilizer

Table 1.2 HLB values of typical emulsifying agents

Class	Agent	HLB
Anionic	Triethanolamine oleate	12.0
	Sodium oleate	18.0
	Sodium dodecyl sulfate	40.0
Cationic	Cetrimonium bromide	23.3
Nonionic	Sorbitan monolaurate (Span 20)	4.3
	Sorbitan monooleate (Span 20)	8.6
	Polyoxyethylene sorbitan monolaurate (Tween 20)	16.7
	Polyoxyethylene sorbitan monooleate (Tween 80)	15.0
	Glyceryl monostearate	3.8

HLB values from 8 to 18 produce O/W emulsions. Typical surfactants with their HLB values used as emulsifying agents are listed in Table 1.2.

Several theories exist that describe how emulsifying agents promote emulsification and maintain the stability of the resulting emulsion. The most prevalent theories are the oriented-wedge theory, surface-tension theory, and the interfacial film theory. Indepth discussions of these theories are beyond the scope of this chapter and can be found elsewhere in literature (Becher 1977; Ansel et al. 1995). However, a general way in which emulsions are produced and stabilized has been discussed in this chapter. Emulsions do not form spontaneously when liquids are mixed and hence an input of energy is required to break up the liquids into small droplets. As the energy is applied, the interface between the oil phase and water phase is deformed resulting in the formation of droplets. Surfactant molecules get rapidly adsorbed at the interface formed between the droplets and lower the interfacial tension. After the formation of emulsions, surfactants prevent coalescence of newly formed droplets by providing a strong short-ranged interfacial repulsion (Myers 1992). By lowering the interfacial tension, surfactants also reduce the energy needed to break up the large droplets into smaller ones.

1.2.2.2 Emulsion Stability

An emulsion is thermodynamically unstable, meaning that the dispersed droplets will tend to coalesce to minimize the interfacial area and break into two separate equilibrium phases with time. Three major phenomena, namely flocculation, creaming, and coalescence can take place before the emulsion separates or breaks into two phases (Fig. 1.2). The moving droplets due to Brownian motion can either adhere or repel, depending upon the Vander Waals attraction and repulsion forces that exist between the droplets. If the repulsion forces are weak, the attractive forces will pull them into contact and flocculation takes place. In flocculation, the droplets become attached to each other but are still separated by a thin film. When more droplets are involved, they aggregate and form three-dimensional clusters. At this point, the size of the droplets is not changed and the emulsifying agent is located at the surface of the individual droplets. Based on the density of the dispersed phase

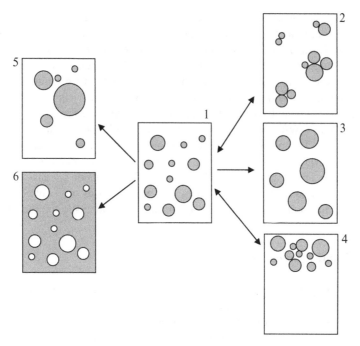

Fig. 1.2 Schematic representation of different types of instability processes in an emulsion: (1) freshly prepared emulsion, (2) flocculation, (3) coalescence, (4) creaming, (5) Ostwald ripening, and (6) phase inversion

and dispersion medium, the aggregated droplets may concentrate in one specific part of the emulsion. Creaming occurs when the aggregated droplets rise through the medium or sink to the bottom (sedimentation). Creaming depends upon the radius of the droplets, the relative difference in the densities of the two phases, and the viscosity of the continuous phase. The rate of creaming can be assessed by Stokes' equation (1.1). It is apparent that the rate of creaming is increased by increased droplet size, a larger density difference between the two phases, and a decreased viscosity of the continuous phase. Creaming can be minimized by reducing the droplet size to a fine state, keeping the difference in the densities of the two phases as small as possible and increasing the viscosity of the continuous phase. Creaming is reversible to some extent, because the dispersed droplets are still surrounded by the protective film and behave as a single drop. Coalescence occurs when two or more droplets fuse together to form a single larger droplet, which leads to the complete separation of the two immiscible phases. Contrary to creaming, the thin liquid film between the droplets is ruptured and therefore, coalescence process is irreversible. This phenomenon is called cracking of emulsions. Altering the viscosity and/or forming a strong interfacial film, using particulate solids, can stabilize coalescence. Coalescence process requires the droplets to be in close proximity. However, a different phenomenon called Ostwald ripening occurs even when the droplets are not in direct contact. Ostwald ripening is a process that involves the growth of large particles at the expense of smaller ones because of high solubility of the smaller droplets and molecular diffusion through the continuous phase. A certain solubility of the dispersed in the continuous phase is required for Ostwald ripening to take place and is driven by the difference in Laplace pressure between droplets having different radii (Capek 2004). It is possible to stabilize emulsions against Ostwald ripening by adding components of high molecular weight that reduce the rate of diffusion of molecules within the dispersed species. Emulsions can invert from an O/W to a W/O emulsion or vice versa during homogenization or sterilization procedures. This phenomenon is called as phase inversion and can be regarded as a form of instability. The temperature at which phase inversion occurs is called phase inversion temperature (PIT). The stability of an emulsion can also be affected by microbial contamination and oxidative decomposition of oils. This can be prevented by adding suitable preservative agents and antioxidants to the formulation.

1.2.2.3   Emulsification Methods

In an emulsion preparation, the liquid that forms the dispersed phase and the liquid that forms the continuous phase are influenced by the volume ratio of the liquids, the kind of emulsifying agent used, and its concentration in strong connection with the PIT. Emulsions may be prepared by high-pressure homogenizers, ultrasound homogenizers, colloid mills, ball and roller mills, rotor/stator systems such as stirred vessels, and electrical and condensation devices. The principle and operation of these various equipments can be found in the literature (Walstra 1983; Becher 1977).

Membrane emulsification is a relatively new technique for producing emulsions and has received increased attention over the past few years (Charcosset et al. 2004). In the membrane emulsification process, the dispersed phase is pressed through the membrane pores to form droplets that are then carried away by a continuous phase flowing across the membrane surface. The conventional direct membrane emulsification process is shown schematically in Fig. 1.3.

Various other types of membrane emulsification processes and their applications have been reviewed by Vladisavljevic and Williams (2005). Membrane emulsification is a suitable technique for the production of single and multiple emulsions. Compared with other emulsification processes, membrane emulsification processes allow the production of emulsions at lower energy output. W/O/W emulsions of a water soluble anticancer drug epirubicin prepared by membrane emulsification have been used successfully for the treatment of liver cancer by hepatic arterial injection chemotherapy (Higashi et al. 1995).

1.2.2.4   Emulsion Characterization

The type of emulsion prepared can be determined by simple tests such as dilution test, dye solubility test, and electrical conductivity test. The dilution test is based on the principle that an emulsion can be diluted with its continuous phase. If water is

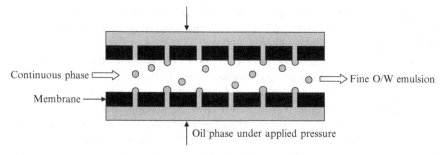

Fig. 1.3 Schematic diagram of membrane emulsification process. The dispersed phase (i.e. oil) is forced through the membrane by applying pressure. The droplets formed at the membrane surface are detached by the continuous phase flowing across the membrane

added to an O/W emulsion, the emulsion is diluted indicating that water is the continuous phase. In contrast, if water is added to a W/O emulsion, the emulsion is not diluted and the separation is apparent. The dye solubility test is based on the principle that water-soluble dye will be dispersed in aqueous phase and oil-soluble will be dispersed in the oil phase of an emulsion. For example, if a water-soluble dye is added to an emulsion, and if the dye diffuses uniformly throughout the emulsion, water is the continuous phase and it will be an O/W emulsion. The principle underlying the electrical conductivity test is that water conducts an electric current and oils do not.

The droplet size and size distribution are important parameters when characterizing emulsions. Flocculation, creaming, coalescence, and Ostwald ripening can be evaluated by monitoring the changes in the droplet size distribution of the emulsion. A diverse range of techniques applied to determine the droplet size includes optical microscopy, dynamic light scattering, cryo-scanning electron microscopy, transmission electron microscopy, and nuclear magnetic resonance. Recently, differential scanning calorimetry has been used to determine the droplet sizes (Clausse et al. 2005). In this technique, a nondiluted emulsion is submitted to a regular cooling and heating cycle between temperatures that include freezing and melting of the dispersed droplets. The droplet size can be determined from the thermograms. Information regarding emulsion type, purity, and stability can also be obtained using differential scanning calorimetry.

1.2.2.5 Emulsion Applications

Emulsions have been used as drug carriers for more than a century. Today, several types of emulsions exist and are used for a variety of applications from solubilizing drugs to controlled release. Emulsions can be delivered by oral, topical, and parenteral routes. When administered by oral route, O/W emulsions can efficiently improve the oral absorption and bioavailability of poorly water-soluble drugs. Improved absorption was seen for drugs such as griseofulvin, theophylline, and phenytoin (Carrigan and Bates 1973; Gagnon and Dawson 1968; Diamond 1970).

Oral administration of a W/O emulsion containing ovalbumin, a model antigen has been shown to be more efficient in enhancing the immunogenic response than that of ovalbumin in saline (Masuda et al. 2003).

For parenteral drug delivery, both W/O and O/W emulsions have been investigated but O/W emulsions are predominantly used. W/O emulsions are easy to prepare; have good physical stability; and are easily injectable because of their low viscosity (Bjerregaard et al. 1999a). They have the potential for sustained release of hydrophilic drugs, since the surfactant layer acts as a release barrier for drugs present in the aqueous phase (Davis et al. 1985). In addition, the release properties from a W/O emulsion can be controlled within certain limits by adjusting parameters such as droplet size, osmotic gradients, and volume fraction of the dispersed phase (Bjerregaard et al. 1999b). In vivo sustained release of aprotoin, a 58 amino acid polypeptide, from W/O emulsion has been demonstrated in mice and rabbits (Bjerregaard et al. 2001a, b). O/W or lipid emulsions are used for parenteral nutrition therapy, as well as for therapeutic agents. In lipid emulsions, the oil phase is typically a glyceride. Oils mostly used in lipid emulsions are soybean oil, cottonseed oil, safflower oil, and medium-chain triglycerides. Egg phospholipids are the most commonly used emulsifying agents in lipid emulsions. Lipid emulsions offer numerous advantages as parenteral drug carriers such as solubilization of highly lipophilic drugs, stabilization of labile drugs against hydrolysis or oxidation, sustained release, and drug targeting. They are biocompatible, biodegradable, and reduce drug side effects by avoiding direct contact of the drug with the body fluid and tissues. Lipid emulsions have been used in parenteral nutrition for more than four decades for delivering fatty acids to patients who cannot eat or metabolize food properly. Examples of marketed formulations are Intralipid, Lipofundin, and Liposyn. Lipid emulsions have been investigated for a number of drugs to treat various disease conditions such as rhizoxin (Stella et al. 1988) and taxol (Tarr et al. 1987) for cancer, physostigmine (Rubinstein et al. 1991) for Alzheimer's disease, and prostaglandine E_1 (Mizushima et al. 1983) for thrombosis therapy. Lipid emulsions of diazepam, propofol, and etomidate are commercially available.

The use of lipid emulsions as ophthalmic vehicles has been explored in the last few years. Lipid emulsions are excellent ocular delivery vehicles as already proved by Restasis®. Restasis® contains cyclosporine A indicated for increased tear production in patients with keratoconjunctivitis sicca. Drugs such as indomethacin (Klang et al. 2000), piroxicam (Klang et al. 1999), and difluprednate (Yamaguchi et al. 2005) have been investigated for ophthalmic lipid emulsions.

Double emulsions are excellent systems for the encapsulation of bioactive compounds. The presence of a reservoir phase inside droplets of another phase can be used to sustain release of active compounds, to protect sensitive molecules from external phase, taste masking, immobilization of enzymes, and for the enhancement of enteral and dermal absorption. The most common double emulsions used are of W/O/W type to entrap water-soluble drugs. Potential applications of double emulsions have been comprehensively reviewed by Khan et al. (2006).

1.3   Colloidal Dispersions

1.3.1   Micelles

Micelles are self-assembling colloidal systems with particle size normally ranging from 5 to 100 nm (Kabanov et al. 1992; Torchilin 2007). They are classified as colloidal dispersion because of their particle size. Micelles are spontaneously formed when amphiphilic molecules are placed in water at a certain concentration and temperature. Property of micellization is generally displayed by molecules that possess two distinct regions with opposite affinities toward a particular solvent (Mittal and Lindman 1991). At a low concentration, the molecules exist separately in a solution. However, when the concentration is increased, the molecules quickly self-assemble to form spherical micelles (Fig. 1.4). The hydrophobic portions of the molecules condense to form the core, whereas the hydrophilic portions constitute the shell or corona of the micelle (Lasic 1992). The concentration at which micellar association ensues is called the critical micelle concentration (CMC) and the temperature below which amphiphilic molecules exist separately is known as critical micellization temperature (CMT). The core of the micelle can solubilize lipophilic substances, whereas the hydrophilic outer portion serves as a stabilizing interface to protect the hydrophobic core from external aqueous environment. The process of micellization leads to free energy minimization of the system as the hydrophobic portions of the molecule are concealed and hydrogen bonds are established between hydrophilic portions in water.

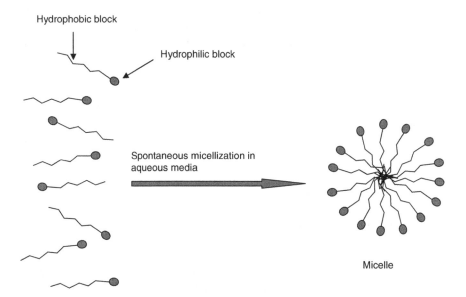

Fig. 1.4 Schematic diagram of spontaneous micellization of amphiphilic molecules in aqueous media

Micelles are attractive candidates as drug carriers for delivering poorly water-soluble drugs. Micelles can solubilize a drug at concentrations much greater than its intrinsic solubility, which results in an increased bioavailability and a reduced toxicity. Incorporation of a drug into a micelle alters release kinetics and enhances the stability of the drug by reducing the access of water and biomolecules. Micelles generally have narrow size distribution and the size can be easily controlled by altering the formulation conditions. Due to their size range, they can be conveniently sterilized by filtration through a membrane with a 0.2 μm cutoff. Specific targeting can be achieved by chemically conjugating a targeting molecule on the surface of a micelle. Passive targeting to tumors can also be achieved due to enhanced permeability and retention effect (EPR effect). Tumors have leaky vasculature and inefficient lymphatic drainage system, which results in a greater accumulation of micelles in tumors compared with normal tissues. Desirable properties of a pharmaceutical micelle include small size, narrow size distribution, low CMC value (low millimolar to micromolar), and high drug loading efficiency. Pharmaceutical micelles can be used through various routes such as parenteral, nasal, oral, otic, and ocular.

1.3.1.1 Surfactant Micelles

Micelles made from surfactants (especially nonionic surfactants) have been commonly used to deliver drugs and biomolecules. They can efficiently entrap hydrophobic drugs and form uniform particles. The water is distributed anisotropically within the micelle, i.e., concentration decreases from the shell to the core (Torchilin 2002). Thus, the position of a drug within the micelle depends on its polarity; more hydrophobic drug tends to stay closer to the core, whereas drugs with slight polarity are located closer to the micelle shell. A limitation of surfactant micelles is that they are not very stable and break apart rapidly upon dilution, which may cause premature release and precipitation of the drug. The stability of surfactant micelles is severely compromised at concentrations lower than their CMC. Surfactant micelles typically have a rather high CMC, which may lead to concerns about surfactant-related toxicity. Micelle destabilization also compromises drug stability as the drug is exposed to blood components immediately upon systemic administration. Thus, surfactant molecules with lower CMC values and greater stability need to be developed to overcome these limitations.

1.3.1.2 Polymeric Micelles

Micelles prepared from block copolymers have attracted much attention lately because of their better stability and biocompatibility (Gaucher et al. 2005). They are made from amphiphilic block copolymers with a large difference in solubility between hydrophobic and hydrophilic portions. Polymeric micelles generally have CMC values that are several orders of magnitude lower than typical CMC values for surfactants. As a result, polymeric micelles show enhanced stability and slower dissociation at lower concentrations compared with surfactant micelles (Jones and

Leroux 1999). Surface functionalization of a polymeric micelle can be easily performed by chemically attaching a targeting moiety. Their low CMC values and lower rate of dissociation also allow for prolonged release of entrapped drug.

Copolymer for micellization can be synthesized by using two or more polymer blocks with contrasting solubility profiles. Poly (ethylene glycol) (PEG) is the most commonly used hydrophilic (shell forming) block (Torchilin 2002; Nishiyama and Kataoka 2003). Various molecular weights of PEG have been used for micelle preparation. PEG is highly biocompatible and forms a highly stable shell to sterically protect the hydrophobic core. It has also been shown to be efficient at escaping recognition by the reticuloendothelial systems (RES), thereby extending the circulation time of micelles in the blood (Kwon et al. 1997). Moreover, PEG copolymers usually have a low polydispersity (Mw/Mn ratio), which enables strict control of micelle size. Surface functionalization of PEG micelles can be easily performed by chemically linking a targeting moiety. Thus, PEG micelles can be used for active targeting to cells and tissues. Copolymers prepared by conjugating PEG with poly lactic acid (PLA) and poly (ethylene oxide) (PEO) have been extensively investigated as micellar vehicles (Yasugi et al. 1999). Other commonly used hydrophilic polymer blocks are poly (N-vinyl-2-pyrrolidone) (PVP), poly (vinyl alcohol) (PVA), and poly (vinylalcohol-co-vinyloleate). Triblock pluronic copolymers with an A-B-A structure (Ethylene oxide)x-(Propylene oxide)y-(Ethylene oxide)x have been extensively characterized (Kabanov et al. 2005). A variety of molecular weights and block lengths of Pluronic copolymers is available commercially. These copolymers have shown promise for delivering drugs and genes in vitro and in vivo.

1.3.1.3 Polymer-Lipid Micelles

A variety of hybrid micelles with lipid core and hydrophilic polymer shell has recently been investigated. Such micelles have shown good stability, longevity, and capability to accumulate into tissues with damaged vasculature (EPR effect). Micelles prepared by conjugation of PEG and phosphatidylethanolamine (PE) have been studied for delivery of anticancer drug Camptothecin (Mu et al. 2005). Such conjugation resulted in formation of very stable micelles having low toxicity and high delivery efficiency. PEG-PE conjugate form micelles with CMC in micromolar range, which is about 100-fold lower than conventional detergent micelles. Polymer lipid micelles can be formed easily by spontaneous micellization in aqueous media similar to surfactant and polymer micelles, and their size can be tailored by varying the molecular weight of the conjugate.

1.3.2 Microemulsions

The term "microemulsion" was first introduced by Hoar and Schulman (1943) to describe a clear solution obtained when normal O/W coarse emulsions were titrated with medium-chain length alcohols. Since then, there has been much dispute about

Table 1.3 Differences between microemulsions and emulsions

Microemulsions	Emulsions
Thermodynamically stable	Thermodynamically unstable
Optically transparent	Cloudy colloidal systems
Interfacial tension 10^{-2}–10^{-4} m Nm^{-1}	Interfacial tension 20–50 mNm^{-1}
May be single or multiple phase	Multiple phase only
Require no energy in their formation	External energy required for formation

the relationship of these systems to solubilized systems (i.e. micellar solutions, surfactant-free solutions) and to emulsions. Danielson and Lindman (1981) define microemulsion as a system of water, oil, and amphiphile which is (an) optically isotropic and thermodynamically stable liquid solution. The main difference between normal coarse emulsions and microemulsions lies in the droplet size of the dispersed phase. Microemulsions have droplets typically in the size range 10–100 nm and because of this small size range, they produce only a weak scattering of visible light and hence, they appear transparent. The features that distinguish microemulsion systems from emulsions are shown in Table 1.3.

Microemulsions are thermodynamically stable systems. The driving force for their thermodynamic stability is the ultralow interfacial tension (10^{-2}–10^{-4}m Nm^{-1}). When the interfacial tension is this low, the interaction energy between droplets has been shown to be negligible and a negative free energy formation is achieved making the dispersion thermodynamically stable. The large interfacial tension between oil in water, which is typically about 50 mNm^{-1}, is reduced by employing surfactants. However, it is generally not possible to achieve the required interfacial tension with the use of a single surfactant. Amphiphiles such as medium-chain length alcohols are added as cosurfactants to achieve the desired interfacial tension. Due to their amphiphilic nature, they partition between the aqueous and oil phase thereby altering the solubility properties of these phases. In addition, by interacting with surfactant monolayers at the interface, they affect their packing, which in turn can influence the curvature of the interface and interfacial free energy.

Depending upon the phase volume ratio and the nature of the surfactant used, a microemulsion can be one of the three types: O/W, bicontinous, and W/O. An O/W microemulsion is formed when the concentration of oil is low and a W/O microemulsion is formed when the concentration of water is low. In conditions where the volumes of oil and water are equal, a bicontinuous microemulsion is formed in which both oil and water exist as a continuous phase. A wide variety of internal structures exists within microemulsion systems. They may be spherical, spheroid, or cylindrical rod-shaped micelles and may exist in cubic, hexagonal, or lamellar phases. The relative amounts of aqueous phase, oil phase, and surfactant required to form a microemulsion can be determined with the aid of triangular/ternary phase diagrams. For example, in Fig. 1.5, each corner of the triangle represents 100% of one of the components. Moving away from that corner reduces the volume fraction of that specific component and any point on one of the axes corresponds to a mixture of two of those components in a defined ratio. Any point inside the triangle represents a mixture of all the three components in a defined ratio. A review written

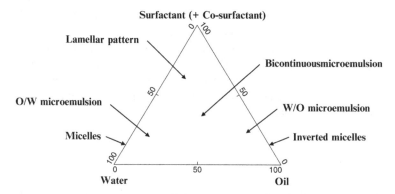

Fig. 1.5 Schematic ternary phase diagram of an oil-surfactant-water system showing microemulsion region and probable internal structures. An O/W microemulsion is formed when the concentration of oil is low, a W/O microemulsion is formed when the concentration of water is low, and a bicontinuous microemulsion is formed if the concentration of oil and water are equal. Depending upon the component concentrations and other characteristics, a variety of internal structures can exist within microemulsion systems

by Forster et al. (1995) describes in detail the physical meaning of the phase behavior of ternary oil/surfactant/water systems, and the concepts related to microemulsion formation and the influence of additives on those microstructures.

1.3.2.1 Theories of Microemulsion Formation

Microemulsions form simultaneously when the interfacial tension between oil and water is reduced to close to zero. The formation and stability of a microemulsion can be affected by various factors, including the nature and molecular weight of surfactant, alcohol chain length and concentration, salinity, and temperature. Three different theories have been proposed to explain the microemulsion formation and stability (Paul and Moulik 1997). They are the interfacial mixed film theory, solubilization theory, and thermodynamic theory. According to the interfacial mixed film theory, the film at the interface is assumed to be a dual film and the type (W/O or O/W) of microemulsion formed depends upon the bending or curvature of the interface (Schulman et al. 1959). Solubilization theory states that oil is solubilized by normal micelles and water is solubilized by reverse micelles (Gillberg et al. 1970). Finally, according to the thermodynamic theory, the free energy of formation must be negative to form a thermodynamically stable microemulsion (Paul and Moulik 1997; Attwood 1994). Surfactants play an important role in reducing the interfacial tension in microemulsions. They can be selected based on the HLB concept or the critical packing parameter (CPP) concept. Surfactants with a low HLB value (3–6) are preferred for the formation of W/O microemulsion, whereas surfactants with high HLB value (8–18) are preferred for O/W microemulsions. The CPP describes a ratio between the hydrophobic and hydrophilic parts of a surfactant molecule and is useful in estimating the nature of formed aggregates. The CPP can be calculated using $CPP = v/a \cdot l$, where v is the partial molecular volume of the surfactant, a is the optimal head group area, and l is the length of the surfactant tail. When the CPP is between 0 and 1, O/W microemulsions are formed, and when the CPP > 1, W/O microemulsions are formed. Bicontinous microemulsions are formed when the CPP = 1. Brij, dioctyl sodium sulphosuccinate, and lecithin are some of the widely used surfactants to stabilize microemulsion formulations.

1.3.2.2 Microemulsion Characterization

Characterization of microemulsions is a difficult task because of the variety of structures and components involved, and the limitations with the available techniques. Therefore, it is preferable to employ a combination of techniques as far as possible. Particle size in microemulsions can be elucidated by small-angle X-ray scattering, small-angle neutron scattering, dynamic and static light scattering, freeze fracture electron microscopy, and neutron scattering techniques. NMR has been used to measure self-diffusion coefficients of the various components. Viscosity measurements can be used to determine the interaction of dispersed droplets and can indicate the pressure of rod-like or worm-like reverse micelles (Yu and Neuman 1995; Angelico et al. 1998). Conductivity experiments can be used to determine the type of microemulsion and to study phase inversion phenomena (Lawrence and Rees 2000).

1.3.2.3 Microemulsion Applications

Microemulsions have attracted large interest in the pharmaceutical industry as ideal delivery systems for a variety of drug molecules because of their thermodynamic stability, simplicity of preparation, and solubilization capacity. Garcia-Celma (1997) has reviewed microemulsions as drug delivery systems for a variety of drug molecules. Because of their unique solubilization properties, microemulsions can improve the solubility and bioavailability of poorly water-soluble drugs. Lipophilic, hydrophilic, or amphiphilic drugs can be effectively solubilized because of the existence of microdomains of different polarity within the system. Microemulsions for oral, dermal, transdermal, parenteral, and pulmonary delivery routes have been developed. When administered via the oral route, microemulsions enhance the bioavailability of poorly soluble drugs by maintaining them in molecular dispersion in the GI tract. In addition, the presence of surfactants increases the membrane permeability of the solubilized drug. The oral efficacy of microemulsions is well demonstrated by the commercially available Cyclosporin A formulation (Neoral®). Recently, Kim et al. (2005) reported that tricaprylin microemulsions improved the oral bioavailability of low molecular weight heparin in mice and monkeys. The microemulsion was composed of tricaprylin (a surfactant mixture of Tween® 80 and Span® 20), heparin, and water. Microemulsions can serve as delivery systems for dermal applications and have been proved to increase the permeation rate of both lipophilic and hydrophilic drugs compared with conventional vehicles such as emulsions or solutions (Tenjarla 1999). Surfactants and cosurfactants play an important role in reducing the diffusional barrier of stratum corneum by acting as permeation enhancers and in addition, the internal mobility of the drug within the vehicle also contributes to the increased permeability. Some of the commonly used surfactants for enhancing permeability are L-α-phosphatidylcholine, Azone®, and Tween® 20. Short-chain alkanols such as ethanol and propylene glycol are used as cosurfactants in microemulsion transdermal formulations. The role of microemulsions in percutaneous penetration of drugs has been summarized in a comprehensive review by Kreilgaard (2002).

1.3.3 Nanosuspensions

Many of the marketed drugs and a large proportion (40%) of potentially bioactive molecules from drug discovery are poorly soluble in aqueous and nonaqueous solvents (Lipinski 2002). Administration of poorly soluble drugs by oral route leads to decreased bioavailability because of the dissolution rate-limiting absorption in the GI tract. A traditional method used for solubility enhancement is the particle size reduction technique based on high shear or impaction such as milling or grinding. The use of this technique can be limited by high polydispersities in particle size, long processing times, and shear-induced particle degradation. Limited success has been achieved by novel techniques such as self-emulsifying systems, liposomes, pH-adjustment, and salting-in processes. However, there is no universal approach applicable to all drugs. Nanosuspensions have emerged as a potential solubility-enhancing technique in the last few years as evidenced by a number of nanosuspension-based formulations in clinical trials and in the market (Rabinow 2004). Nanosuspensions are colloidal dispersions containing drug particles dispersed in an aqueous vehicle in which the diameter of the suspended particle is <1 μm in size (Fig. 1.6). The

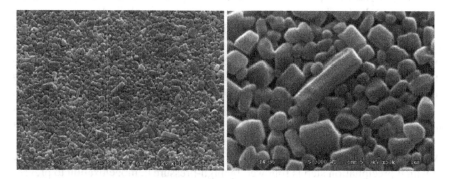

Fig. 1.6 Scanning electron microscope images of megestrol acetate nanoparticles (From Deng et al. (2005), Reprinted with the permission of Cambridge University Press)

basic principle of this technique is to reduce the size of the drug particles to a submicron range. Reducing the particle size to a submicron range increases the surface area to be in contact with the dissolution medium and consequently the dissolution rate. Nanosuspensions have a number of potential benefits compared with conventional methods. Nanosuspensions allow to incorporate a high concentration of drug in a relatively low volume of fluid; provide a chemically and physically stable product; and can be used for controlled and targeted delivery of drugs. In addition, nanosuspensions can be used for drugs that are water insoluble (<0.1 mg/ml) and for drugs insoluble in both water and organic solvents.

1.3.3.1 Preparation of Nanosuspensions

Nanosuspensions can be produced by bottom-up or top-down techniques. In the case of bottom-up technique, the strategy is to build particles from their constituent units, i.e., molecules, atoms. Bottom-up technique is a classical precipitation process in which the drug is dissolved in a solvent, which is subsequently added to a nonsolvent to precipitate the drug crystals. Use of solvents, the difficulty to avoid the formation of microcrystals, and the poor solubility of an increasing number of drugs in all media limit this approach. In the case of top-down technique, the coarse material is subsequently broken down until nanoscopic dimensions are reached. There are two basic techniques widely used: (1) Pearl/ball milling (Nanocrystal technology®, élan) and (2) High-pressure homogenization (Dissocubes®, Skyepharma). In pearl milling, an aqueous suspension of the drug is fed into the mill containing milling pearls made of glass, zirconium oxide, ceramic sintered aluminium oxide, or hard polystyrene with high abrasion resistance. As the pearls are rotated at a very high shear rate, the drug particles are ground into nano-sized particles between the moving milling pearls. Depending upon the hardness, the drug needs to be milled from hours to several days.

In the high-pressure homogenization technique, the drug suspension is forced under pressure through a valve that has a nanoaperture. Typical pressures applied are between 100 and 1,500 bars. As the drug suspension passes through the nano-aperture with high velocity, the static pressure is decreased, leading to the formation of small gas bubbles, which implode as they exit the valve. The cavitation forces created breakdown the drug microparticles into nano-sized particles. It is not possible to obtain the desired particle size for many drugs in a single homogenization cycle. Multiple homogenization cycles are required depending on the hardness of the drug, desired mean particle size, and required homogeneity of the product. To avoid the removal of water after high-pressure homogenization in aqueous media, nanosuspensions were produced using high-pressure homogenization in nonaqueous media or in water with water-miscible liquids (Nanopure®, Pharmasol). A combinative technology (NANOEDGE ™) was introduced by Baxter Healthcare in which precipitation step is followed by high-pressure homogenization to prevent the precipitate from crystal growth. Very recently, Moschwitzer and Muller (2006) reported a new combination method for the production of ultrafine submicron nanosuspensions. This method involves an evaporation step to provide a solvent-free modified starting material followed by high-pressure homogenization. Hydrocortisone acetate nanosuspensions produced using this method with reduced homogenization cycles and the nanosuspensions demonstrated excellent long-term storage stability. Application of supercritical fluid process for production of nano-suspensions has increased in recent years. The most widely used methods are rapid expansion of supercritical solution process, gas antisolvent process, and supercriti-cal antisolvent process. Cyclosporine, budesonide, and griseofulvin nanosuspen-sions have been prepared using these methods (Young et al. 2000; Steckel et al. 1997; Chattopadhyay and Gupta 2001).

1.3.3.2 Characterization of Nanosuspensions

A critical parameter that defines the quality and physicochemical behavior of nanosuspensions is the particle size distribution. The mean particle size and width of distribution can be determined using photon correlation spectroscopy (PCS), laser diffraction (LD), and coulter counter techniques. The measuring range for PCS is 3 nm to 3 μm and LD can measure particles ranging from 0.05 to 80 μm up to 2,000 μm. The coulter counter gives an absolute number of particles per unit volume for the different size classes. The shape and surface morphology of nano-suspensions can be assessed by scanning electron microscopy and atomic force microscopy. The crystalline state of nanosuspensions can be characterized by dif-ferential scanning calorimetry and X-ray diffraction. Determination of zeta potential is important to assess the physical stability of nanosuspensions. For nanosuspension formulations administered intravenously, additional parameters such as surface hydrophilicity/hydrophobicity, sterility, and pyrogenicity need to be determined.

1.3.3.3 Applications of Nanosuspensions

Nanosuspension technology is applicable to all poorly soluble drugs and an out-standing feature of this technology is its simplicity. Nanosuspensions can be applied to various administration routes such as oral, parenteral, pulmonary, otic, ophthalmic, and nasal routes. Administration of poorly water-soluble drug in the form of nanosuspensions has shown increased onset of action, increased bioavail-ability, and dissolution rate. For example, the t_{max} for the nanosuspension formu-lation of naproxen, a nonsteroidal antiinflammatory drug, was 23.7 min versus 33.5 min for unmilled formulation (Liversidge and Conzentino 1995). Nanosuspension formulations of naproxen reduced the time required to achieve C_{max} by approximately 50% compared with marketed suspension (Naprosyn) and tablets (Anaprox) (Merisko-Liversidge et al. 2003). The dissolution rate of naproxen tablet formulation made from naproxen particles ranging from 100 to 600 nm in size was compared with a commercially available product (Aleve) that was prepared from macro-sized naproxen (Jain et al. 2000). Dissolution rate of naproxen from the nanoparticulate formulation was found to be significantly higher compared with the drug release from the marketed product. This study showed that if properly formulated with suitable excipients, nanosuspensions can be processed into conventional dosage forms such as tablets and capsules, using standard equipment. Reducing the mean particle size of a poorly soluble investigational compound from 7 μm to 280 nm showed four times higher bioavailability compared with micrometric size particles of the compound (Jia et al. 2002). Nanoparticle formulation was rapidly absorbed with a t_{max} of 1 h, whereas the t_{max} for microparticle formulation was prolonged for another 3 h. Oleanic acid nanosuspension containing particles with an average particle size of 284.9 nm was reported to have a faster dissolution rate and enhanced therapeutic effect (Chen et al. 2005). Kocbek et al. (2006) developed ibuprofen nanosuspensions prepared by a melt emulsification method traditionally used to prepare solid lipid nanoparticles. Ibuprofen nanosuspensions either in the form of lyophilized powder or granules showed enhanced dissolution rate compared with micronized drug. For lyophilized nanosuspension formulation, more than 65% of the drug dissolved in the first 10 min as opposed to only 15% of the micronized drug (Fig. 1.7).

Drugs that are not absorbed through the GI tract or that undergo extensive first pass metabolism can be administered intravenously as nanosuspensions. Nanosuspensions are ideal for intravenous route and offer many special advantages. A rapid onset of action can be achieved in case of an emergency and high concentrations of the drug can be administered without employing toxic cosolvents and solubility excipients. Nanosuspensions are potential targeted systems as their surface properties can be easily altered. In addition, capillary blockade is avoided, since the particle size in the nanosuspensions is less than 5 μm, which is the inner diameter of the smallest blood capillaries in the body. A chemically stable, intravenously

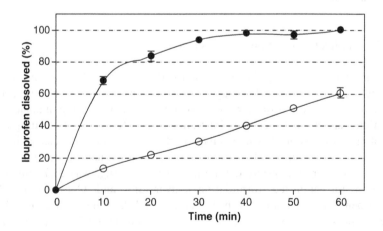

Fig. 1.7 In vitro dissolution profiles of ibuprofen lyophilized nanosuspension (*filled circle*) and micronized ibuprofen (*open circle*). Reproduced from Kocbek et al. (2006) with permission from Elsevier Science

injectable nanosuspension formulation for omeprazole, a poorly soluble, chemically labile drug with a high degradation rate in aqueous media was developed by Moschwitzer et al. (2004). Injectable nanosuspension of the poorly soluble drug tarazepide has been prepared to overcome the limited success achieved using conventional solubilization tecniques, such as the use of mixed micelles or cycodextrins to improve the bioavailability of the drug (Jacobs et al. 2000). Rabinow et al. (2007) evaluated an intravenous itraconazole nanosuspension dosage form, relative to a solution formulation, in the rat. The formulation of itraconazole as a nanosuspension enhanced the efficacy of this antifungal agent, and exhibited altered pharmacokinetics, leading to increased tolerability and high drug levels in target organs. Pulmonary administration of nanosuspensions increases rapid diffusion and dissolution of the drug at the site of action. Budesonide, a poorly water-soluble antiinflammatory corticosteroid, has been successfully formulated as a nanosuspension for pulmonary delivery (Jacobs and Muller 2002). The advances in nanosuspension technology are described in more detail in Chap. 10.

1.3.4 Liposomes

Liposomes are spherical phospholipid vesicles that may range from 20 nm to a few microns in size. Liposomes can be made from natural phospholipids or their synthetic analogs. Liposomal formulations have been developed for a variety of applications such as diagnostic, vaccine adjuvants, and for delivery of small molecules, proteins, and nucleotides (Lasic 1998). The liposomes used for drug or gene delivery are normally unilamellar vesicles with size the range of 50–150 nm. As liposomes have an aqueous core, both hydrophilic and hydrophobic drugs can be delivered using liposomal formulation. The mode of incorporation of the drug inside the liposome depends on the polarity of the drug. Hydrophilic drugs are encapsulated in the aqueous core of the liposome, whereas hydrophobic drugs are entrapped in the phospholipid bilayer (Fig. 1.8).

Surface of liposomes may bear a positive, negative, or neutral charge, depending on the lipid composition and pH. Although neutral liposomes have lower clearance

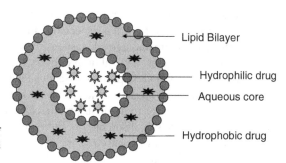

Fig. 1.8 Schematic illustration of localization of hydrophobic and hydrophilic drugs in a liposome

through retuculoendothelial system (RES), they have a high tendency to aggregate. Negatively charged liposomes are highly susceptible to endocytosis by macrophages. Positively charged liposomes, mostly used for gene delivery as polyplexes, interact with serum protein and are subsequently removed by the RES.

Liposomes, unlike micelle, are a thermodynamically unstable system and tend to fuse together and eventually separate out of the aqueous medium on storage. To achieve long-term stability, liposomes with high charge density have been prepared. However, high charge density of liposomes can not provide long-term stabilization in vivo because of the presence of various proteins and enzymes. Another approach to stabilization of liposomes is coating the outer surface of the liposome with an inert hydrophilic polymer (Klibanov et al. 1990; Papahadjopoulos et al. 1991). Such liposomes are called sterically stabilized liposomes. They have the ability to evade the immune system and hence can achieve longer circulation time in vivo. PEG is routinely used to provide steric stability to liposomes for drug and gene delivery.

1.3.4.1 Liposomes Targeting

Liposomal delivery can lead to drastic changes in absorption, distribution, and elimination of the drug. Surface conjugation of PEG results in resistance toward aggregation and prolonged circulation in blood. This longer circulation time results in accumulation of drug at the target site overtime.

Active targeting of liposomes can also be easily achieved by conjugation of a targeting moiety on the surface of liposome. A variety of ligands and receptors has been used for active targeting of liposomes toward specific cells and tissues (Lian and Ho 2001). Antibody-coated liposomes (also referred to as Immunoliposomes) have also been commonly used to achieve specific targeting. Targeting molecule can also be conjugated to PEG through covalent linkage.

1.3.4.2 Applications of Liposomes

Drug Delivery

As the structure of liposomes resembles cell membrane, they were considered attractive delivery system for drugs especially at intracellular level. Liposomes can be formulated as gel, cream, aerosol, suspension, or dry powder and administered through a variety of routes. Several drug delivery systems using liposomes such as Doxil (Doxorubicin in stealth liposomes; Sequus Pharmaceuticals) and Daunoxome (Daunorubicin; Nexstar Pharmaceuticals) are in market. Several others are in an advanced stage of clinical trials or in preclinical studies targeted toward various cancers, autoimmunity, skin disorders etc. Fusogenic, pH-sensitive components can be added to the liposomes to cause release of drug at a certain pH in the tissue or to cause release of drug intracellularly by endosomal destabilization.

Gene Delivery

The use of cationic liposomes to deliver polynucleotides was first reported in 1987 and since they have been commonly used in gene delivery research (Felgner et al. 1987). Cationic liposomes can efficiently condense negatively charged DNA into nanometer-sized particles. Such liposome/polynucleotide complexes are generally referred to as lipoplexes. Neutral lipid dioleoylphosphatidylethanolamine is commonly incorporated along with cationic lipids in a lipoplex. It facilitates endosomal destabilization apart from providing structural stability to the liposome. Formation of lipoplexes also prevents DNA from degradation by extracellular and intracellular nucleases. Cellular entry of lipoplexes primarily occurs through clathrin-mediated endocytosis (Dass and Burton 2003). Liposomes have also been employed for cell-targeting, using a variety of targeting ligands (Talsma et al. 2006; Torchilin 2006). In spite of their high transfection efficiency in vitro, lipoplexes have not been very successful in vivo because of their high toxicity and immunogenicity. Lipoplex formulations have been reported to cause moderate-to-severe toxicities in animal models at both systemic and cellular levels.

Other Applications

Liposomes, depending on their lipid composition and size, can be targeted toward specific immune cells. It is well established that large liposomes are efficiently taken up by macrophages. Thus, liposome formulations provide an attractive tool for antigen delivery for stimulation of both cellular and humoral immune response. Antigens along with immunomodulatory agents can be encapsulated inside liposomes for efficient vaccination.

Liposomes have also been used as a diagnostic tool for transport of tracers such as fluorescent molecules or radio-isotopes for detection of tumors. They also find application in oral delivery of nutrients and in cosmetics.

1.3.5 Nanoparticles

The advent of nanotechnology has opened new avenues for drug and gene delivery in last few years. A variety of nanoscale devices such as nanoparticles, nanotubes, nanogels, and molecular conjugates has been investigated (Ravi Kumar et al. 2004; Murakami and Nakashima 2006; Lemieux et al. 2000). Nanoparticles are the most commonly used nanometer scale delivery systems. They are typically spherical particles in the size range of 1–1,000 nm (Fig. 1.9). The size of these delivery systems confers advantages such as greater and deeper tissue penetration, longer circulation time in blood, enhanced cellular uptake, ability to cross blood–brain barrier, and greater ability to target specific cell types (Kreuter et al. 1995; Davis 1997; Vogt et al. 2006). Drug or gene of interest can be incorporated in a nanoparticle by encapsulation

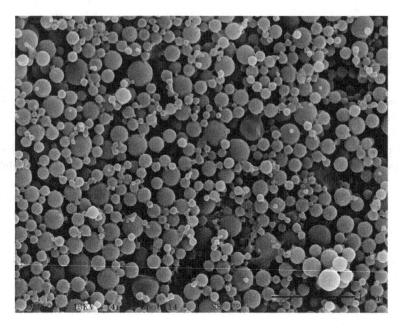

Fig. 1.9 Scanning electron micrograph of polymeric nanoparticles for drug and gene delivery

or surface conjugation. Although majority of nanoparticles used in research have been developed from polymers, certain nonpolymeric materials have also been used.

1.3.5.1 Polymeric Nanoparticles

A variety of polymers such as poly lactide-co-glycolide (PLGA), poly lactic acid (PLA), polyethyleneimine (PEI), polymethacrylates has been investigated for formulation of nanoparticles. Potential applications of polymeric nanoparticles include drug and gene delivery, cell and tissue targeting, pharmacokinetic studies, tissue engineering, vaccine adjuvation, and delivery of diagnostic agents. Majority of polymers used have been biodegradable, which is highly desirable. However, some nonbiodegradable polymers have also been used, especially for gene delivery because of their high transfection efficiency. The type of the polymer used for formulation of nanoparticles is generally determined by the intended application of nanoparticles. For example, if the aim is to deliver the particles systemically and to evade immune recognition, hydrophilic polymers such as PEG may be incorporated. Conversely, if the immune cells are to be targeted, it may be a good idea to use polymers that have high surface charge so as to cause rapid uptake by immune cells. Chemical and physical characteristics of polymers can be tailored to suit the intended application.

Drug delivery remains one of the basic and most common applications of polymeric nanoparticles. Nanoparticles can vehiculate drug of interest by either encapsulation in the matrix or through attachment on the surface. These polymeric nanoparticles possess high drug loading efficiency and high stability to protect and securely deliver the drug in vivo. Polymeric nanoparticles are versatile systems that can be delivered through a variety of routes such as oral, nasal, parenteral, and ocular. Polymeric nanoparticles have also displayed an ability to cross blood–brain barrier (Olivier 2005). Steric stabilization of nanoparticles can be achieved by coating the surface with hydrophilic polymers. It results in increased circulation time and subsequently increased bioavailability. Another important application of polymeric nanoparticles has been delivery of anticancer drugs (van Vlerken and Amiji 2006). Particles with optimum size and surface characteristics can passively travel through the leaky vasculature in tumor tissues and can exhibit passive targeting. Active targeting of tumors can also be achieved by conjugating targeting molecule on the surface of particles.

Gene delivery has been another very extensively investigated area of nanotechnology. Polymeric nanoparticles have been developed that can efficiently deliver genes into cells both in vitro and in vivo, with low toxicity. Nanoparticles for gene delivery can be formulated in a variety of ways. The most common mode is encapsulation of DNA inside the polymeric matrix. Alternatively, DNA can be attached on the surface of nanoparticles by electrostatic interaction or chemical binding. Some cationic polymers such as PEI also condense DNA spontaneously in solution to form polymer/DNA complexes known as polyplexes (De Smedt et al. 2000). PEI has been by far the most commonly used polymer for gene delivery because of its very high efficiency and is widely regarded as the "gold standard" for comparing the efficiency of gene delivery vectors. Cationic charge of polymers facilitates cellular attachment and internalization. Biodegradable polymers such as PLGA and PLA have also been used to achieve sustained gene delivery.

1.3.5.2 Nonpolymeric Nanoparticles

Contrary to polymeric nanoparticles that are generally greater than 100 nm in size, nonpolymeric nanoparticles such as metallic and ceramic nanoparticles are mostly below 100 nm in size. Metal and ceramic nanoparticles have been commonly employed for various biomedical applications.

Magnetic iron oxide particles have been particularly investigated for in vivo imaging apart from other biomedical and bioengineering applications. These particles can be easily synthesized with a narrow size distribution. The surface of iron oxide particles can be coated using a variety of polymeric and nonpolymeric materials to modify surface characteristics (Gupta and Gupta 2005). Surface coating apart from providing stability also facilitates binding of biological ligands and receptors for tissue and cell targeting. Iron oxide nanoparticles have also shown good efficiency in delivering genes in vitro (Dobson 2006). Transfection of cells using magnetic nanoparticles is also known as magnetofection.

Gold nanoparticles have received much attention in recent years in drug delivery as an alternative to lipid and polymer-based delivery systems because of their low toxicity and easy synthesis. The particles size is generally below 50 nm with narrow distribution. Surface functionalization of gold nanoparticles, using a PEG spacer, has been shown to result in rapid cellular uptake and internalization, improved stability, and greater circulation time (Shenoy et al. 2006). Proteins can be delivered using gold nanoparticles by surface attachment using a PEG spacer. Gold nanoparticles surface-functionalized using quarternary ammonium chains have been used for gene delivery and have resulted in very efficient transfection of cells in vitro (Sandhu et al. 2002).

Biocompatible ceramic nanoparticles made from silica, titania, alumina, etc. have also been explored for drug delivery. High porosity of these materials makes them attractive candidates for drug/gene delivery. These nanoparticles have been used for delivery of anticancer drugs and insulin. However, the full potential of these nanosystems is yet to be explored.

1.3.6 Microspheres

Drug delivery using biodegradable polymeric microspheres has gained increased interest in the last two decades. Microspheres are solid, spherical devices containing the drug in a polymer matrix with size ranging from 1 to 1,000 µm. Microspheres are different from microcapsules. In microspheres, the drug is dispersed throughout the polymer matrix, whereas in microcapsules, the drug is the core surrounded by a polymeric membrane. Microspheres are widely used as drug carriers for controlled release and the incorporation of drug molecules into biodegradable polymeric microspheres has many advantages. The polymer matrix can protect drugs, such as proteins, from physiological degradation. Microspheres can control the delivery of drugs from days to months therefore reducing frequent administrations and improving patient compliance and comfort. Different release profiles with desired release rates can be achieved by selecting polymers with different degradation mechanisms. In addition, microspheres can be used to target drugs to a specific site. Microspheres have been proposed as delivery systems for traditional small molecular weight drugs, proteins, enzymes, vaccines, cells, and even delicate molecules such as DNA.

A wide range of natural and synthetic polymers has been used for the microsphere matrix. Some of the natural polymers include chitosan, alginate, albumin, and casein. Synthetic polymers include polyesters, polyanhydrides, polycaprolactones, polyphosphazenes, polyorthoesters, and polycarbonates. Most of the biodegradable polymers are nontoxic; degrade within the body; are easily eliminated without the need for surgical removal. Among the aforementioned polymers, polyesters like PLGA have received more attention because of their excellent biocompatibility and mechanical properties for drug delivery applications. Microsphere products based on PLGA polymer commercially available are Lupron® depot, Risperdal Consta™, Zoladex® depot, Sandostatin LAR® depot, and Decapeptyl® depot. The encapsulated drug release from the microspheres occurs either by diffusion out of the polymer matrix or by degradation of the polymer or both. The release characteristics can be influenced by the molecular weight of the entrapped drug and polymer, size of the microspheres, excipients present in the system, and local environmental conditions such as pH.

1.3.6.1 Microsphere Preparation Techniques

A number of techniques are currently available for preparing microspheres. A well-known and widely used technique is the emulsification-solvent evaporation technique, because it allows encapsulation of the protein as an aqueous solution. In this technique, the polymer is dissolved in a volatile organic solvent such as methylene chloride or acetone. The drug is then dissolved or dispersed in the polymer solution. In the following step, the organic phase is emulsified in an aqueous phase, resulting in an O/W emulsion. The aqueous phase contains appropriate surfactant, which stabilizes the organic solvent droplets formed during the energy input. As the organic solvent is removed either by evaporation or by extraction, solid microspheres are formed. The microspheres can be collected by filtration or centrifugation and lyophilized to obtain free-flowing microsphere product (Fig. 1.10). The O/W technique is appropriate for encapsulating hydrophobic drugs.

Hydrophilic compounds can be encapsulated by W/O/W double emulsion technique. In this technique, an aqueous solution of the drug is added to an organic solution of the polymer to form a W/O emulsion. This emulsion is then added into a large-volume aqueous phase to form a W/O/W emulsion. The resulting emulsion is subjected to solvent removal either by evaporation or by extraction process. Other techniques employed for the preparation of microspheres are polymer phase separation (coacervation), spray drying, milling, and supercritical fluid techniques. For the detailed description of these methods with their merits and limitations, and for the advances in microsphere production technology, the readers are referred to an excellent chapter written by Tewes et al. (2005).

1.3.6.2 Microsphere Characterization

Particle size of microspheres can be determined by light microscopy and coulter counter. Shape and surface morphology can be examined using scanning electron microscopy and atomic force microscopy. The porosity of the microspheres can be measured by air permeability (for pores ranging from 10 to 75 μm) and mercury porosimetry (for pores <10 μm) (Burgess and Hickey 2005). The amount of drug encapsulated can be detected by dissolving the polymer in a suitable solvent and subsequent release of the drug. The release kinetics of the entrapped drug from the microspheres can be determined by in vitro and in vivo methods. In vitro methods include membrane diffusion, sample and separate, and continuous flow methods. In vivo release kinetics is usually determined indirectly from drug plasma levels in suitable animal species. Information regarding in vivo factors that affect release rates and histopathological reactions following injection of microspheres can be obtained from animal studies (Kang and Singh 2005). Injectability and sterility testing are required in addition for microspheres administered by parenteral route.

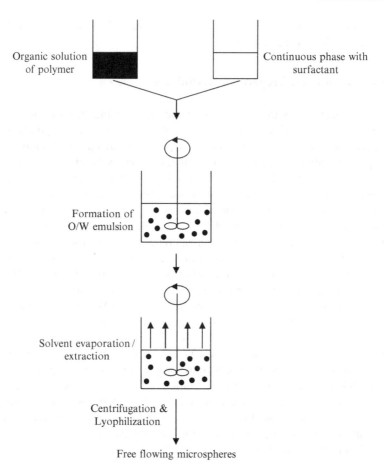

Fig. 1.10 Schematic representation of O/W emulsification-solvent evaporation technique

1.3.6.3 Microsphere Applications

At present, biodegradable microspheres are increasingly investigated as drug delivery carriers for biotechnology-derived compounds like proteins, which are an increasing part of the newly discovered chemical entities. Oral delivery of proteins has always been a significant challenge because of their poor bioavailability. They have very short biological half-lives and are easily degraded by proteolytic enzymes in the gastro-intestinal tract. Therefore, most of the proteins are useful when following a therapeutic regimen of frequent injections throughout a long period. Microspheres have the potential to deliver proteins at a controlled rate for longer duration and enhance their in vivo stability. In addition, they can be used to target specific tissues. Despite intensive research over the last few years, microspheres have some serious limitations such as low protein loadings, high initial burst release, and protein instability during microsphere fabrication. Due to their hydrophilic nature, proteins tend to diffuse to the continuous external aqueous phase during the evaporation step thereby leading to low protein loadings. Burst release is the rapid release of a large part of the encapsulated drug in the first 24 h after administration. Factors affecting burst release include microsphere loading, porosity, addition of excipients, and the rapid diffusion of surface-localized drug molecules. It has also been shown that the size of the microspheres affects the burst release (Ghaderi et al. 1996). Small microspheres exhibit a greater initial burst because of the increased surface area and large microspheres exhibit a comparatively less burst release.

Protein instability during microsphere fabrication remains a major challenge for a formulation scientist, since a number of steps during fabrication imperil the three-dimensional structure of the protein which is crucial for biological activity. In the most widely used W/O/W emulsion method, the formation of water–organic solvent interface in the primary emulsion, use of high shear forces and freeze drying may lead to an altered protein structure. However, alternative fabrication methods such as solid-in-oil-in-water (S/O/W) emulsion, spray drying, gas antisolvent technique, and new cryogenic methods have been developed to encapsulate proteins while preserving protein stability (Jiang and Schwendeman 2001; Bittner et al. 1998; Elvassore et al. 2001; Johnson et al. 1997). To stabilize proteins in microsphere formulations, various additives such as surfactants, sugars, salts, cyclodextrins, and certain amino acids have also been used (Bilati et al. 2005).

Several small molecules and protein-based drugs have been encapsulated in a variety of biodegradable polymers as shown in Table 1.4. In addition, it can be seen

Table 1.4 List of some of the drugs encapsulated in biodegradable microspheres

Drug	Polymer	Administration route	References
Capreomycin sulfate	PLGA	Pulmonary	Giovagnoli et al. (2007)
Naltrexone	PLGA	i.m.	Swainston Harrison et al. (2006)
Vitamin A palmitate and acyclovir	PLGA	Intraocular	Martinez-Sancho et al. (2006)
Parathyroid hormone	PLGA	s.c.	Wei et al. (2004)
Levonorgestrel	PLGA	i.m.	Wang et al. (2005)
Human growth hormone	PLGA	s.c.	Johnson et al. (1997)
Insulin-like growth factor	PLGA	s.c.	Carrascosa et al. (2004)
Naproxen	Polyphosphazene	s.c.	Veronese et al. (1998)
Insulin	Eudragit®	Oral	Morishita et al. (1993)
Insulin	Alginate	Oral	Silva et al. (2006)
Insulin	Chitosan	Oral	Wang et al. (2006)
Insulin	PLGA	s.c.	Kang and Singh (2005)
Insulin	Poly-ε-caprolactone	s.c.	Shenoy et al. (2003)
Insulin	Poly(fumaric-co-sebacic anhydride)	s.c.	Furtado et al. (2006)

that PLGA has been widely used to encapsulate a variety of hormones, enzymes, and vaccines. Chitosan has been shown to be promising for the delivery of proteins by oral route because of its mucoadhesive property and permeation enhancing effect across the biological surfaces. Encapsulating insulin in chitosan microspheres improved protein loading and chemical stability (Wang et al. 2006). Eudragit® microspheres have been used for targeting drugs to the colon, and calcitonin has been successfully delivered (Lamprecht et al. 2004). PLGA has shown to protect the encapsulated vaccines and improve immune response when administered orally. Polymeric microspheres for controlled delivery are discussed in the following section.

1.4 Colloidal Dispersions as Controlled Drug Delivery Systems

Controlled drug delivery systems provide several advantages compared with conventional dosage forms such as reduced dosing frequency, greater bioavailability, better patient compliance, and reduced toxicity. Although the concept of controlled delivery has been around since the 1960s, lack of suitable degradable delivery system is a major deterrent toward its wider acceptability. However, with the development of novel biomaterials with desirable properties such as biocompatibility, biodegradability, and ease of manufacture, a variety of controlled release system has been developed.

1.4.1 Nanosuspensions

While nanosuspension formulations of poorly soluble drugs exhibit prolonged release, nanosuspensions can also be employed for controlled drug delivery. Ibuprofen-loaded polymeric nanoparticle suspensions for intraocular delivery made from Eudragit RS100® showed a controlled release profile in vitro (Pignatello et al. 2002). Ibuprofen nanosuspensions had mean sizes around 100 nm and a positive charge making them suitable for ophthalmic applications. An inhibition of the mitotic response to the surgical trauma was achieved by the nanosuspensions compared with a control aqueous eye-drop formulation.

1.4.2 Nanoparticles

Various types of polymeric nanoparticles have been explored for controlled delivery of therapeutic agents (Kumar 2000). The release kinetics of the drug depends on the nature of both polymer and the drug. More hydrophobic polymers generally tend to release the drug at a lower rate compared with hydrophilic polymers. The release of drug may range from a few hours to several months. These polymers degrade slowly in vivo and the rate of degradation is governed by polymer molecular weight, temperature, pH, and polymer composition. The release rate of drug depends both on diffusion and on polymer erosion. Most commonly employed polymers for controlled delivery are PLGA, PLA, polymethacrylates, chitosan, and PEG. High initial burst release of drug is an area of concern while developing nanoparticulate formulations especially for hydrophilic drugs. High burst release may result in acute elevation of drug concentration in blood followed by little or no release. Efforts have been made to reduce burst release by changing nanoparticle size, surface characteristics, polymer composition, and surfactant type. Nanoparticles are also of value for achieving controlled gene delivery in some cases where sustained expression of a particular protein in small concentrations is desired (Cohen et al. 2000).

1.4.3 Liposomes

Although polymeric drug delivery devices have been the system of choice for achieving controlled delivery of therapeutics, liposomes have also aroused interest as controlled delivery systems. Several modifications have been tried to prolong the release rate of drugs from liposomes. Multivesicular liposomes (MVL) have been used to achieve sustained release of drugs up to several weeks. These liposomes are composed of several layers of phospholipids and therefore have much higher strength and stability compared with conventional unilamellar liposomes. These MLV liposomes range from 5 to 50 µm in size, which is much greater than conventional submicron liposomes (Jain et al. 2005). MVLs have been used for sustained delivery of a wide range of drugs from small molecules to peptides and proteins. PEGylation (PEG conjugation) of liposomes has also resulted in sustained delivery of therapeutics primarily through increasing the circulation time of liposomes in the blood (Cattel et al. 2003). Some other approaches that have been used to achieve controlled delivery from liposomes are encapsulation of liposomes inside polymeric microspheres and entrapment of liposomes in thermosensitive gels.

1.4.4 Microspheres

Controlled drug delivery using polymeric microspheres represents one of the most rapidly advancing areas in the drug delivery field. Polymeric microspheres have been investigated for a wide range of molecules, but a few will be discussed here. Kang and Singh (2005) developed a controlled release insulin microsphere formulation to provide basal insulin level for a prolonged period. A mixture of PLGA microspheres prepared by W/O/W and S/O/W emulsion methods in a ratio of 3:1 w/w improved in vitro release of insulin from microspheres. Insulin-loaded microspheres maintained basal insulin level in diabetic rabbits for 40 days. Alginate-chitosan microspheres were developed to reduce dose/dosing frequency in the management of tuberculosis, which otherwise demands prolonged chemotherapy (Pandey and Khuller 2004). The microspheres contained three frontline antituberculous drugs, rifampicin, isoniazid, and pyrazinamide. Administration of a single oral dose of alginate–chitosan microspheres maintained drug levels in the plasma for 7 days as against 2 days for the nonencapsulated parent drugs. PLGA microspheres containing plasmid DNA both in free form and as a complex with poly (L-lysine) showed controlled release for 20 days (Capan et al. 1999).

References

Angelico, R., Palazzo, G., Colafemmina, G., Cirkel, P.A., Giustini, M., and Ceglie, A. 1998. Water diffusion and head group mobility in polymer-like reverse micelles: evidence of a sphere-to-rod-to-sphere transition. J Phys Chem B. 102:2883–2889.

Ansel, C.H., Popovich, N.G., and Allen, L.V., Jr. 1995. Pharmaceutical Dosage Forms and Drug Delivery Systems. Philadelphia: Williams and Wilkins.

Attwood, D. 1994. Microemulsion. In: Kreuter, J. (ed.) Colloidal Drug Delivery System. New York: Marcel Dekker, pp. 31–65.

Becher, P. 1977. Emulsions: Theory and Practice, New York: Krieger.

Bilati, U., Allemann, E., and Doelker, E. 2005. Strategic approaches for overcoming peptide and protein instability within biodegradable nano- and microparticles. Eur J Pharm Biopharm. 59:375–388.

Bittner, B., Morlock, M., Koll, H., Winter, G., and Kissel, T. 1998. Recombinant human erythropoietin (rhEPO) loaded poly(lactide-co-glycolide) microspheres: influence of the encapsulation technique and polymer purity on microsphere characteristics. Eur J Pharm Biopharm. 45:295–305.

Bjerregaard, S., Wulf-Andersen, L., Stephens, R.W., Lund, L.R., Vermehren, C., Soderberg, I., and Frokjaer, S. 2001a. Sustained elevated plasma aprotinin concentration in mice following intraperitoneal injections of w/o emulsions incorporating aprotinin. J Control Release. 71:87–98.

Bjerregaard, S., Pedersen, H., Vedstesen, H., Vermehren, C., Soderberg, I., and Frokjaer, S. 2001b. Parenteral water/oil emulsions containing hydrophilic compounds with enhanced in vivo retention: formulation, rheological characterisation and study of in vivo fate using whole body gamma-scintigraphy. Int J Pharm. 215:13–27.

Bjerregaard, S., Söderberg, I., Vermehren, C., and Frokjaer, S. 1999a. The effect of controlled osmotic stress on release and swelling properties of a water-in-oil emulsion. Int J Pharm. 183:17–20.

Bjerregaard, S., Söderberg, I., Vermehren, C., and Frokjaer, S. 1999b. Formulation and evaluation of release and swelling mechanism of a water-in-oil emulsion using factorial design. Int J Pharm. 193:1–11.

Burgess, J.D. and Hickey, J.A. 2005. Microspheres: design and manufacturing. In: Burgess, J.D. (ed.) Injectable Disperse Systems. New York: Taylor & Francis Group, pp. 305–353.

Capan, Y., Woo, B.H., Gebrekidan, S., Ahmed, S., and DeLuca, P.P. 1999. Preparation and characterization of poly (D,L-lactide-co-glycolide) microspheres for controlled release of poly(L-lysine) complexed plasmid DNA. Pharm Res. 16:509–513.

Capek, I. 2004. Degradation of kinetically-stable o/w emulsions. Adv Colloid Interface Sci. 107:125–155.

Carrascosa, C., Torres-Aleman, I., Lopez-Lopez, C., Carro, E., Espejo, L., Torrado, S., and Torrado, J.J. 2004. Microspheres containing insulin-like growth factor I for treatment of chronic neurodegeneration. Biomaterials. 25:707–714.

Carrigan, P.J. and Bates, T.R. 1973. Biopharmaceutics of drugs administered in lipid-containing dosage forms. I. GI absorption of griseofulvin from an oil-in-water emulsion in the rat. J Pharm Sci. 62:1476–1479.

Cattel, L., Ceruti, M., and Dosio, F. 2003. From conventional to stealth liposomes: a new frontier in cancer chemotherapy. Tumori. 89:237–249.

Charcosset, C., Limayem, I., and Fessi, H. 2004. The membrane emulsification process – a review. J Chem Technol Biotechnol. 79:209–218.

Chattopadhyay, P. and Gupta, R.B. 2001. Production of griseofulvin nanoparticles using super-critical CO(2) antisolvent with enhanced mass transfer. Int J Pharm. 228:19–31.

Chen, Y., Liu, J., Yang, X., Zhao, X., and Xu, H. 2005. Oleanolic acid nanosuspensions: preparation, in-vitro characterization and enhanced hepatoprotective effect. J Pharm Pharmacol. 57:259–264.

Clausse, D., Gomez, F., Pezron, I., Komunjer, L., and Dalmazzone, C. 2005. Morphology characterization of emulsions by differential scanning calorimetry. Adv Colloid Interface Sci. 117:59–74.

Cohen, H., Levy, R.J., Gao, J., Fishbein, I., Kousaev, V., Sosnowski, S., Slomkowski, S., and Golomb, G. 2000. Sustained delivery and expression of DNA encapsulated in polymeric nanoparticles. Gene Ther. 7:1896–1905.

Danielson, I. and Lindman, B. 1981. The definition of microemulsion. Colloids Surf. 3:391–392.

Dass, C.R. and Burton, M.A. 2003. Modified microplex vector enhances transfection of cells in culture while maintaining tumour-selective gene delivery in-vivo. J Pharm Pharmacol. 55:19–25.

Davis, S.S., Hadgraft, J., and Palin, J.K. 1985. Medical and pharmaceutical applications of emulsions. In: Becher, P. (ed.) Encyclopedia of Emulsion Technology. New York: Marcel Dekker, pp. 159–238.

Davis, S.S. 1997. Biomedical applications of nanotechnology – implications for drug targeting and gene therapy. Trends Biotechnol. 15:217–224.

De Smedt, S.C., Demeester, J., and Hennink, W.E. 2000. Cationic polymer based gene delivery systems. Pharm Res. 17:113–126.

Deng, Z., Zhao, R., Dong, L.-C., and Wong, G. 2005. Characterization of nanoparticles for drug delivery applications. Microsc Microanal. 11(Suppl 2):1934–1935.

Diamond, L. 1970. A comparison of the gastrointestinal absorption profiles of theophylline from various vehicles. Arch Int Pharmacodyn Ther. 185:246–253.

Dobson, J. 2006. Gene therapy progress and prospects: magnetic nanoparticle-based gene delivery. Gene Ther. 13:283–287.

Elvassore, N., Bertucco, A., and Caliceti, P. 2001. Production of insulin-loaded poly(ethylene glycol)/poly(L-lactide) (PEG/PLA) nanoparticles by gas antisolvent techniques. J Pharm Sci. 90:1628–1636.

Felgner, P.L., Gadek, T.R., Holm, M., Roman, R., Chan, H.W., Wenz, M., Northrop, J.P., Ringold, G.M., and Danielsen, M. 1987. Lipofection: a highly efficient, lipid-mediated DNA-transfection procedure. Proc Natl Acad Sci U S A. 84:7413–7417.

Forster, T., Rybinski, W., and Wadle, A. 1995. Influence of microemulsion phases on the preparation of fine disperse emulsions. Adv Colloid Interface. 58:119–149.

Furtado, S., Abramson, D., Simhkay, L., Wobbekind, D., and Mathiowitz, E. 2006. Subcutaneous delivery of insulin loaded poly(fumaric-co-sebacic anhydride) microspheres to type 1 diabetic rats. Eur J Pharm Biopharm. 63:229–236.

Gagnon, M. and Dawson, A.M. 1968. The effect of bile on vitamin A absorption in the rat. Proc Soc Exp Biol Med. 127:99–102.

Garcia-Celma, M.J. 1997. Solubilization of drugs in microemulsions. In: Solans, C. and Kunieda H. (eds.) Industrial Applications of Microemulsions. New York: Marcel Dekker, pp. 123–145.

Gaucher, G., Dufresne, M.H., Sant, V.P., Kang, N., Maysinger, D., and Leroux, J.C. 2005. Block copolymer micelles: preparation, characterization and application in drug delivery. J Control Release. 109:169–188.

Ghaderi, R., Sturesson, C., and Carlfors, J. 1996. Effect of preparative parameters on the characteristics of poly (D,L-lactide-co-glycolide) microspheres made by the double emulsion method. Int J Pharm. 141:205–216.

Gietz, U., Arvinte, T., Haner, M., Aebi, U., and Merkle, H.P. 2000. Formulation of sustained release aqueous Zn-hirudin suspensions. Eur J Pharm Sci. 11:33–41.

Gietz, U., Arvinte, T., Mader, E., Oroszlan, P., and Merkle, H.P. 1998. Sustained release of injectable zinc-recombinant hirudin suspensions: development and validation of in vitro release model. Eur J Pharm Biopharm. 45:259–264.

Gillberg, G., Lehtinen, H., and Friberg, S. 1970. NMR and IR investigation of the conditions determining the stability of microemulsions. J Colloid Interface Sci. 33:40–53.

Giovagnoli, S., Blasi, P., Schoubben, A., Rossi, C., and Ricci, M. 2007. Preparation of large porous biodegradable microspheres by using a simple double-emulsion method for capreomycin sulfate pulmonary delivery. Int J Pharm. 333:103–111.

Griffin, W.C. 1949. Classification of surface-active agents by HLB. J Soc Cosmet Chem. 1:311.

Gupta, A.K. and Gupta, M. 2005. Synthesis and surface engineering of iron oxide nanoparticles for biomedical applications. Biomaterials. 26:3995–4021.

Higashi, S., Shimizu, M., Nakashima, T., Iwata, K., Uchiyama, F., Tateno, S., Tamura, S., and Setoguchi, T. 1995. Arterial-injection chemotherapy for hepatocellular carcinoma using monodispersed poppy-seed oil microdroplets containing fine aqueous vesicles of epirubicin. Initial medical application of a membrane-emulsification technique. Cancer. 75:1245–1254.

Hoar, T.P. and Schulman, J.H. 1943. Transparent water-in-oil dispersions: the oleopathic hydromicelle. Nature. 152:102–105.

Jacobs, C. and Muller, R.H. 2002. Production and characterization of a budesonide nanosuspension for pulmonary administration. Pharm Res. 19:189–194.

Jacobs, C., Kayser, O., and Muller, R.H. 2000. Nanosuspensions as a new approach for the formulation for the poorly soluble drug tarazepide. Int J Pharm. 196:161–164.

Jain, R.A., Wei, L., and Swanson, J. 2000. Solid dose form of nanoparticulate naproxen. US Patent 6165506.

Jain, S.K., Jain, R.K., Chourasia, M.K., Jain, A.K., Chalasani, K.B., Soni, V., and Jain, A. 2005. Design and development of multivesicular liposomal depot delivery system for controlled systemic delivery of acyclovir sodium. AAPS PharmSciTech. 6:E35–E41.

Jia, L., Wong, H., Cerna, C., and Weitman, S.D. 2002. Effect of nanonization on absorption of 301029: ex vivo and in vivo pharmacokinetic correlations determined by liquid chromatography/mass spectrometry. Pharm Res. 19:1091–1096.

Jiang, W. and Schwendeman, S.P. 2001. Stabilization and controlled release of bovine serum albumin encapsulated in poly(D,L-lactide) and poly(ethylene glycol) microsphere blends. Pharm Res. 18:878–885.

Johnson, O.L., Jaworowicz, W., Cleland, J.L., Bailey, L., Charnis, M., Duenas, E., Wu, C., Shepard, D., Magil, S., Last, T., Jones, A.J., and Putney, S.D. 1997. The stabilization and encapsulation of human growth hormone into biodegradable microspheres. Pharm Res. 14:730–735.

Jones, M. and Leroux, J. 1999. Polymeric micelles – a new generation of colloidal drug carriers. Eur J Pharm Biopharm. 48:101–111.

Kabanov, A., Zhu, J., and Alakhov, V. 2005. Pluronic block copolymers for gene delivery. Adv Genet. 53:231–261.

Kabanov, A.V., Batrakova, E.V., Melik-Nubarov, N.S., Fedoseev, N.A., Dorodnich, T.Y., Alakhov, V.Y., Chekhonin, P., Nazarova, I.R., and Kabanov, V.A. 1992. A new class of drug carriers; micells poly(oxyethylene)–poly(oxypropylene) block copolymers as microcontainers for drug targeting from blood to brain. J Control. Release. 22:141–158.

Kang, F. and Singh, J. 2005. Preparation, in vitro release, in vivo absorption and biocompatibility studies of insulin-loaded microspheres in rabbits. AAPS PharmSciTech. 6:E487–E494.

Khan, A.Y., Talegaonkar, S., Iqbal, Z., Ahmed, F.J., and Khar, R.K. 2006. Multiple emulsions: an overview. Curr Drug Deliv. 3:429–443.

Kim, S.K., Lee, E.H., Vaishali, B., Lee, S., Lee, Y.K., Kim, C.Y., Moon, H.T., and Byun, Y. 2005. Tricaprylin microemulsion for oral delivery of low molecular weight heparin conjugates. J Control Release. 105:32–42.

Klang, S.H., Siganos, C.S., and Benita, S. 1999. Evaluation of a positively charged submicron emulsion of piroxicam in the rabbit corneum healing process following alkali burn. J Control Release. 57:19–27.

Klang, S., Abdulrazik, M., and Benita, S. 2000. Influence of emulsion droplet surface charges on indomethacin ocular tissue distribution. Pharm Dev Technol. 5:521–532.

Klibanov, A.M., Maruyama, M., Torchilin, V.P., and Huang, L. 1990. Amphipathic polyethyleneglycols effectively prolong the circulation time of liposomes. FEBS Lett. 268:235–237.

Kocbek, P.S., Baumgartner, S., and Kristl, J. 2006. Preparation and evaluation of nanosuspensions for enhancing the dissolution of poorly soluble drugs. Int J Pharm. 312:179–186.

Kreilgaard, M. 2002. Influence of microemulsions on cutaneous drug delivery. Adv Drug Deliv Rev. 54(Suppl 1):S77–S98

Kreuter, J., Alyautdin, R.N., Kharkevich, D.A., and Ivanov, A.A. 1995. Passage of peptides through the blood–brain barrier with colloidal polymer particles (nanoparticles). Brain Res. 674:171–174.

Kumar, M. 2000. Nano and microparticles as controlled drug delivery devices. J Pharm Pharm Sci. 3:234–258.

Kwon, G.S., Naito, M., Yokoyama, M., Okano, T., Sakurai, Y., and Kataoka, K. 1997. Block copolymer micelles for drug delivery: loading and release of doxorubicin. J Control Release. 48:195–201.

Lamprecht, A., Yamamoto, H., Takeuchi, H., and Kawashima, Y. 2004. pH-sensitive microsphere delivery increases oral bioavailability of calcitonin. J Control Release. 98:1-9.

Lasic, D.D. 1992. Mixed micelles in drug delivery. Nature. 355:279–280.

Lasic, D.D. 1998. Novel applications of liposomes. Trends Biotechnol. 16:307–321.

Lawrence, M.J., and Rees, G.D. 2000. Microemulsion-based media as novel drug delivery systems. Adv Drug Deliv Rev. 45:89–121.

Lemieux, P., Vinogradov, S.V., Gebhart, C.L., Guerin, N., Paradis, G., Nguyen, H.K., Ochietti, B., Suzdaltseva, Y.G., Bartakova, E.V., Bronich, T.K., St-Pierre, Y., Alakhov, V.Y., and Kabanov, A.V. 2000. Block and graft copolymers and NanoGel copolymer networks for DNA delivery into cell. J Drug Target. 8:91–105.

Lian, T. and Ho, R.J. 2001. Trends and developments in liposome drug delivery systems. J Pharm Sci. 90:667–680.

Lipinski, C. 2002. Poor aqueous solubility – an industry wide problem in drug delivery. Am Pharm Rev. 5:82–85.

Liversidge, G.G., and Conzentino, P. 1995. Drug particle reduction for decreasing gastric irritancy and enhancing absorption of naproxen in rats. Int J Pharm. 125:309–313.

Martin, A., Swarbrick, J., and Cammarata, A. (eds.) 1983. Coarse dispersions. In: Physical Pharmacy, 3rd edn. Philadelphia: Lea and Febiger, p. 544.

Martinez-Sancho, C., Herrero-Vanrell, R., and Negro, S. 2006. Vitamin A palmitate and aciclovir biodegradable microspheres for intraocular sustained release. Int J Pharm. 326:100–106.

Masuda, K., Horie, K., Suzuki, R., Yoshikawa, T., and Hirano, K. 2003. Oral-antigen delivery via a water-in-oil emulsion system modulates the balance of the Th1/Th2 type response in oral tolerance. Pharm Res. 20:130–134.

Masuoka, J., Hegenauer, J., Van Dyke, B.R., and Saltman, P. 1993. Intrinsic stoichiometric equilibrium constants for the binding of zinc(II) and copper(II) to the high affinity site of serum albumin. J Biol Chem. 268:21533–21537.

Merisko-Liversidge, E., Liversidge, G.G., and Cooper, E.R. 2003. Nanosizing: a formulation approach for poorly-water-soluble compounds. Eur J Pharm Sci. 18:113–120.

Mittal, K.L., and Lindman, B (eds.) 1991. Surfactants in solution (Vol. 3). New York: Plenum Press.

Mizushima, Y., Hoshi, K., Aihara, H., and Kurachi, M. 1983. Inhibition of bronchoconstriction by aerosol of a lipid emulsion containing prostaglandin E1. J Pharm Pharmacol. 35:397.

Morales, M.E., Gallardo, V., Calpena, A.C., Doménech, J., and Ruiz, M.A. 2004. Comparative study of morphine diffusion from sustained release polymeric suspensions. J Control Rel. 95:75–81.

Morishita, I., Morishita, M., Takayama, K., Machida, Y., and Nagai, T. 1993. Enteral insulin delivery by microspheres in 3 different formulations using Eudragit L100 and S100. Int J Pharm. 91:29–37.

Moschwitzer, J., Achleitner, G., Pomper, H., and Muller, R.H. 2004. Development of an intravenously injectable chemically stable aqueous omeprazole formulation using nanosuspension technology. Eur J Pharm Biopharm. 58:615–619.

Moschwitzer, J. and Muller, R.H. 2006. New method for the effective production of ultrafine drug nanocrystals. J Nanosci Nanotechnol. 6: 3145–3153.

Mu, L., Elbayoumi, T.A., and Torchilin, V.P. 2005. Mixed micelles made of poly(ethylene glycol)-phosphatidylethanolamine conjugate and d-alpha-tocopheryl polyethylene glycol 1000 succinate as pharmaceutical nanocarriers for camptothecin. Int J Pharm. 306:142–149.

Murakami, H. and Nakashima, N. 2006. Soluble carbon nanotubes and their applications. J Nanosci Nanotechnol. 6:16–27.

Myers, D. 1992. Surfactant Science and Technology, 2nd ed. New York: VCH Publishers, Inc.

Nash, R.A. 1988. Pharmaceutical suspensions. In: Liebermann, H.A., Rieger, M.M., and Banker, G.S. (eds.) Pharmaceutical Dosage Forms: Disperse Systems (Vol. 1). New York and Basel: Mercer Dekker, Inc., pp. 151–198.

Nishiyama, N. and Kataoka, K. 2003. Polymeric micelle drug carrier systems: PEG-PAsp(Dox) and second generation of micellar drugs. Adv Exp Med Biol. 519:155–177.

Olivier, J.C. 2005. Drug transport to brain with targeted nanoparticles. NeuroRx. 2:108–119.

Pandey, R. and Khuller, G.K. 2004. Chemotherapeutic potential of alginate–chitosan microspheres as anti-tubercular drug carriers. J Antimicrob Chemother. 53:635–640.

Papahadjopoulos, D., Allen, T.M., Gabizon, A., Mayhew, E., Matthay, K., Huang, S.K., Lee, K.D., Woodle, M.C., Lasic, D.D., Redemann, C., and Martin, F.J. 1991. Sterically stabilized liposomes: improvements in pharmacokinetics and antitumor therapeutic efficacy. Proc Natl Acad Sci U S A. 88:11460–11464.

Paul, B.K. and Moulik, S.P. 1997. Microemulsions: an overview. J Dispers Sci Technol. 18:301–67

Pignatello, R., Bucolo, C., Ferrara, P., Maltese, A., Puleo, A., and Puglisi, G. 2002. Eudragit RS100 nanosuspensions for the ophthalmic controlled delivery of ibuprofen. Eur J Pharm Sci. 16:53–61.

Rabinow, B.E. 2004. Nanosuspensions in drug delivery. Nat Rev Drug Discov. 3:785–796.

Rabinow, B., Kipp, J., Papadopoulos, P., Wong, J., Glosson, J., Gass, J., Sun, C.S., Wielgos, T., White, R., Cook, C., Barker, K., and Wood, K. 2007. Itraconazole IV nanosuspension enhances efficacy through altered pharmacokinetics in the rat. Int J Pharm. 339:251-260

Ravi Kumar, M., Hellermann, G., Lockey, R.F., and Mohapatra, S.S. 2004. Nanoparticle-mediated gene delivery: state of the art. Expert Opin Biol Ther. 4:1213–1224.

Rubinstein, A., Pathak, Y.V., Kleinstern, J., Reches, A., and Benita, S. 1991. In vitro release and intestinal absorption of physostigmine salicylate from submicron emulsions. J Pharm Sci. 80:643–647.

Sandhu, K.K., McIntosh, C.M., Simard, J.M., Smith, S.W., and Rotello, V.M. 2002. Gold nanoparticle-mediated transfection of mammalian cells. Bioconjug Chem. 13:3–6.

Schulman, J.H., Stoeckenius, W., and Prince, L.M. 1959. Mechanism of formation and structure of microemulsions by electron microscopy. J Phys Chem. 63:1677–1680.

Shenoy, D.B., D'Souza, R.J., Tiwari, S.B., and Udupa, N. 2003. Potential applications of polymeric microsphere suspension as subcutaneous depot for insulin. Drug Dev Ind Pharm. 29:555–563.

Shenoy, D., Fu, W., Li, J., Crasto, C., Jones, G., Dimarzio, C., Sridhar, S., and Amiji, M. 2006. Surface functionalization of gold nanoparticles using hetero-bifunctional Poly (ethylene glycol) spacer for intracellular tracking and delivery. Int J Nanomedicine. 1:51–58.

Silva, C.M., Ribeiro, A.J., Figueiredo, I.V., Goncalves, A.R., and Veiga, F. 2006. Alginate microspheres prepared by internal gelation: development and effect on insulin stability. Int J Pharm. 311:1–10.

Steckel, H., Thies, J., and Muller, B.W. 1997. Micronizing of steroids for pulmonary delivery by supercritical carbon dioxide. Int J Pharm. 152:99–110.

Stella, V.J., Umprayn, K., and Waugh, W.N. 1988. Development of parenteral formulations of experimental cytotoxic agents. I. Rhizoxin (NSC-332598). Int J Pharm. 43:191–199.

Swainston Harrison, T., Plosker, G.L., and Keam, S.J. 2006. Extended-release intramuscular naltrexone. Drugs. 66:1741–1751

Talsma, S.S., Babensee, J.E., Murthy, N., and Williams, I.R. 2006. Development and in vitro validation of a targeted delivery vehicle for DNA vaccines. J Control Release. 112:271–279.

Tarr, B.D., Sambandan, T.G., and Yalkowsky, S.H. 1987. A new parenteral emulsion for the administration of taxol. Pharm Res. 4:162–165.

Tenjarla, S. 1999. Microemulsions: an overview and pharmaceutical applications. Crit Rev Ther Drug Carrier Syst. 16:461–521.

Tewes, F., Boury, F., and Benoit, J.P. 2005. Biodegradable microspheres: advances in production technology. In: Benita, S. (ed) Microencapsulation: Methods and Industrial Applications. New York: Taylor & Francis Group, pp. 1–53

Torchilin, V.P. 2002. PEG-based micelles as carriers of contrast agents for different imaging modalities. Adv Drug Deliv Rev. 54:235–252.

Torchilin, V.P. 2006. Recent approaches to intracellular delivery of drugs and DNA and organelle targeting. Annu Rev Biomed Eng. 8:343–375.

Torchilin, V.P. 2007. Micellar nanocarriers: pharmaceutical perspectives. Pharm Res. 24:1–16.

Van Vlerken, L.E. and Amiji, M.M. 2006. Multi-functional polymeric nanoparticles for tumour-targeted drug delivery. Expert Opin Drug Deliv. 3:205–216.

Veronese, F.M., Marsilio, F., Caliceti, P., De Filippis, P., Giunchedi, P., and Lora, S. 1998. Polyorganophosphazene microspheres for drug release: polymer synthesis, microsphere preparation, in vitro and in vivo naproxen release. J Control Release. 52:227–237.

Vladisavljevic, G.T. and Williams, R.A. 2005. Recent developments in manufacturing emulsions and particulate products using membranes. Adv Colloid Interface Sci. 113:1–20.

Vogt, A., Combadiere, B., Hadam, S., Stieler, K.M., Lademann, J., Schaefer, H., Autran, B., Sterry, W., and Blume-Peytavi, U. 2006. 40 nm, but not 750 or 1,500 nm, nanoparticles enter epidermal CD1a+cells after transcutaneous application on human skin. J Invest Dermatol. 126:1316–1322.

Walstra, P. 1983. Formation of emulsions. In: Becher, P. (ed) Encyclopedia of Emulsion Technology. New York: Marcel Dekker, pp. 57–127.

Wang, L.Y., Gu, Y.H., Su, Z.G., and Ma, G.H. 2006. Preparation and improvement of release behavior of chitosan microspheres containing insulin. Int J Pharm. 311:187–195.

Wang, S.H., Zhang, L.C., Lin, F., Sa, X.Y., Zuo, J.B., Shao, Q.X., Chen, G.S., and Zeng, S. 2005. Controlled release of levonorgestrel from biodegradable poly (D,L-lactide-co-glycolide) microspheres: in vitro and in vivo studies. Int J Pharm. 301:217–225.

Wei, G., Pettway, G.J., McCauley, L.K., and Ma, P.X. 2004. The release profiles and bioactivity of parathyroid hormone from poly(lactic-co-glycolic acid) microspheres. Biomaterials. 25:345–352

Yamaguchi, M., Yasueda, S., Isowaki, A., Yamamoto, M., Kimura, M., Inada, K., and Ohtori, A. 2005. Formulation of an ophthalmic lipid emulsion containing an anti-inflammatory steroidal drug, difluprednate. Int J Pharm. 301:121–128.

Yasugi, K., Nagasaki, Y., Kato, M., and Kataoka, K. 1999. Preparation and characterization of polymer micelles from poly(ethylene glycol)-poly(D,L-lactide) block copolymers as potential drug carrier. J Control Release. 62:89–100.

Young, T.J., Mawson, S., Johnston, K.P., Henriksen, I.B., Pace, G.W., and Mishra, A.K. 2000. Rapid expansion from supercritical to aqueous solution to produce submicron suspensions of water-insoluble drugs. Biotechnol Prog. 16:402–407.

Yu, Z.J. and Neuman, R.D. 1995. Reversed micellar solution-to-bicontinuous microemulsion transition in sodium bis (2-ethylhexyl) phosphate/n-heptane /water system. Langmuir. 11:1081–1086.

Chapter 2
General Principles of Suspensions

Mohammad T. H. Nutan and Indra K. Reddy

Abstract Pharmaceutical suspensions are discussed with regard to theoretical considerations (e.g., interfacial properties, electric double layer, wetting, flocculated and deflocculated systems), stability factors (e.g., settling and sedimentation, effect of particle size, crystal growth, and use of structured vehicles, etc.), and rheologic aspects. A thorough understanding of these aspects is important from a stability standpoint, as the stability of suspension systems depends both on the properties of the dispersed solid and on those of the vehicle.

2.1 Introduction

Suspensions are heterogeneous systems containing two phases. The external phase, which is also referred to as the continuous phase or dispersion medium, is generally a liquid (e.g., liquid suspensions) or semisolid (e.g., gels), and the internal or dispersed phase is made up of particulate matter, which is practically insoluble in the external phase. Most pharmaceutical suspensions consist of an aqueous dispersion medium, although organic or oily liquids are also used in some instances. The discussion in this chapter is limited to solid–liquid suspensions in aqueous medium.

A pharmaceutical suspension is a coarse dispersion of insoluble solid particles in a liquid medium. The particle diameter in a suspension is usually greater than $0.5\,\mu m$. However, it is difficult and also impractical to impose a sharp boundary between the suspensions and the dispersions having finer particles. Therefore, in many instances, suspensions may have smaller particles than $0.5\,\mu m$ and may show some characteristics typical to colloidal dispersions, such as Brownian movement (Martin 2006).

M.T.H. Nutan (✉), and I.R. Reddy
Irma Lerma Rangel College of Pharmacy, Texas A&M University Health Science Center,
1010 West Avenue B, MSC 131, Kingsville, TX 78363, USA
e-mail: mnutan@pharmacy.tamhsc.edu

A.K. Kulshreshtha et al. (eds.), *Pharmaceutical Suspensions: From Formulation Development to Manufacturing*, DOI 10.1007/978-1-4419-1087-5_2,
© AAPS 2010

Suspensions are an important class of pharmaceutical dosage forms. The advantages of suspension dosage forms include effective dispensing of hydrophobic drugs; avoidance of the use of cosolvents; masking of unpleasant taste of certain ingredients; offering resistance to degradation of drugs due to hydrolysis, oxidation or microbial activity; easy swallowing for young or elderly patients; and efficient intramuscular depot therapy. In addition, when compared to solution dosage forms, relatively higher concentration of drugs can be incorporated into suspension products. The disadvantages of suspension dosage forms include the possibility of dose variation, requirement of large storage area, and lack of elegance, among others.

To date, numerous theories had been introduced and successfully used to explain the unique behavior of suspension preparations. The objective of this chapter is to discuss the application of these theoretical considerations that can aid in developing stable suspensions.

Pharmaceutical suspensions, depending on the routes of administration, can be classified into three groups: oral suspensions, externally applied lotions, and injectable preparations. Suspensions are also available as aerosols, which can be applied topically on the skin or internally by pulmonary route. Oral and parenteral suspensions, in addition to the regular suspension products may also be prepared as dry powders for reconstitution with water prior to dispensing. This minimizes the rate of drug degradation in aqueous media. Topical suspensions can be liquid preparations (e.g., lotions) or semisolid (e.g., pastes). Parenteral preparations can be formulated in an oily vehicle to sustain drug release.

At present, many drug formulations are available as suspensions. Some therapeutic classes of drug formulations are mentioned below:

- Oral antacid suspensions
- Oral antibacterial suspensions
- Oral analgesic suspensions
- Oral anthelmintic suspensions
- Oral anticonvulsant suspensions
- Oral antifungal suspensions
- Dry antibiotic powders for oral suspensions
- Intramuscular antidiarrheal suspensions
- Intravenous anticancer suspensions
- Intramuscular contraceptive suspensions
- Topical lotions for various skin conditions

In an ideal suspension formulation, insoluble particles should be uniformly dispersed. However, on standing, the solid particles in suspensions get separated from the liquid as sediments. Regardless of the amount of sedimentation, a well-formulated suspension should redisperse uniformly in the continuous phase, upon moderate shaking, for a sufficient period of time. This allows the withdrawal of the correct amount of medication with minimal dose variation. The rate of settling can be decreased by using viscosity improving agents, and the ease of redispersibility can be controlled by using flocculating agents. Products that are too viscous, however, may not be easy to remove from the container and may be too difficult to transfer to the site of application. Furthermore, the drug diffusion process is also expected to be hindered by high viscosity. It, therefore, presents a challenge to the formulator to find a suitable viscosity-imparting agent and a flocculating agent, which, when used at appropriate concentrations, allow the optimum rate of sedimentation and easy redispersion in a quality product. Suspended particles should also be small and uniform in size to provide a smooth and elegant product that is free from a gritty texture.

2.2 Theoretical Considerations

2.2.1 Interfacial Properties

In pharmaceutical suspensions, the solid phase remains as finely divided particles in the dispersion medium. Therefore, a large amount of interface is involved in the formation, which, in turn, affects the stability of suspension preparations. The interfacial properties, therefore, play a vital role in modifying the physical characteristics of a dispersion. The two most important interfacial properties include surface free energy and surface potential.

2.2.1.1 Surface Free Energy

A large surface area offered by finely divided solid materials is typically associated with large amount of free energy on the surface. The relation between the surface free energy and the surface area can be expressed by (2.1):

$$\Delta G = \gamma \Delta A \qquad (2.1)$$

where ΔG is the increase in surface free energy, ΔA is the increase in surface area, and γ is the interfacial tension between the solid particles and the dispersion medium. The smaller the ΔG is, the more thermodynamically stable is the suspension. Therefore, a system with very fine particles is thermodynamically unstable because of high total surface area. Thus, the system tends to agglomerate in order to reduce the surface area and thereby the excess free energy. Surface free energy may also be reduced to avoid the agglomeration of particles, which can be accomplished by reducing interfacial energy. When a wetting agent is added to the suspension formulation, it is adsorbed at the interface. This will result in a reduction of the interfacial tension, making the system more stable.

2.2.1.2 Surface Potential

The stability of lyophobic colloidal systems can generally be explained on the basis of the presence or absence of surface potential. This theory can also be extended to suspension systems (Gallardo et al. 2005). Surface potential exists when dispersed solid particles in a suspension possess charge in relation to their surrounding liquid medium. Solid particles may become charged through different ways. If the suspension contains electrolytes, selective adsorption of a particular ionic species by the solid particles is a possibility. This will lead to the formation of charged particles. For example, in case of a dispersion of rubber particles in water, it may happen that hydroxyl ions (OH^-) will be more adsorbed than hydronium ions (H_3O^+) due to the asymmetric nature of the hydroxyl ion (Tabibi and Rhodes 1996). Occasionally, the surface active agents, which are already adsorbed at the solid–liquid interface, may ionize to give the particles positive or negative charge. Sodium dodecyl sulfate (SDS), for example, is anionic in aqueous medium. Solid particles can also be charged by ionization of functional group of the particles. In this case, the total charge is a function of the pH of the surrounding vehicle. Peptide and protein molecules contain ionizable groups, such as –COOH and $–NH_2$. Dispersed particles of these types of molecules will ionize. The sign and magnitude of the ionization mostly depend on the pH of the surrounding vehicle.

2.2.2 Electric Double Layer

When dispersed particles are in contact with an aqueous solution of an electrolyte, the particles may selectively adsorb one charge species. If the adsorbed species is an anion, the particles will be overall negatively charged. The ions that give the particle its charge, anions in this case, are called potential-determining ions or co-ions. Remaining ionic species in the solution are the rest of the anions and the total number of cations added. This means, there will be excess cations than anions in the dispersion medium. These cations having a charge opposite to that of the potential-determining ions are known as counter-ions or gegenions. They are attracted to the negatively charged surface by electric forces. Gegenions also repel the approach of any further anions to particle surface, once the initial adsorption is complete. These electric forces and thermal motion keeps an equal distribution of all the ions in solution. It results in an equilibrium condition where some of the excess cations approach the surface and the rest of the cations will be distributed in decreasing the amounts as one moves away from the charged surface.

This situation is explained in Fig. 2.1. The part of the solvent immediately surrounding the particles will almost entirely comprise of the counter-ions. This part of the solvent, along with these counter-ions is tightly bound to the particle surface and is known as the Stern layer. When particles move through the dispersion medium, the Stern layer moves along with them and thus the shear plane is the one peripheral to the Stern layer. There are fewer counter-ions in the tightly bound layer than co-ions adsorbed onto the surface of the solid. Therefore, the potential at the shear plane is still negative. Surrounding the Stern layer is the diffuse layer that contains more counter-ions than co-ions. The ions in this layer are relatively mobile and, because of thermal energy, they are in a constant state of motion into and from the main body of the continuous phase. Electric neutrality occurs where the mobile diffuse layer ends. Beyond the diffuse layer, the concentrations of co- and counter-ions

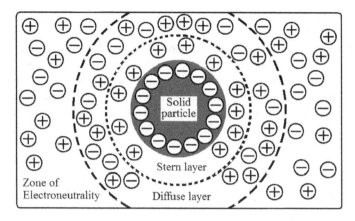

Fig. 2.1 Electric double layer at the solid–liquid medium interface in a disperse system. Adapted from Minko (2006a)

are equal, that is, conditions of electric neutrality prevail throughout the remaining part of the dispersion medium.

Thus, the electric distribution at the solid–liquid interface can be visualized as a double layer of charge. The Stern layer, the first layer is tightly bound to the solid surface and contains mostly the counter-ions. The second layer is more mobile containing more counter-ions than co-ions. These two layers are commonly known as the electric double layer. The thickness of the double layer depends upon the type and concentration of ions in solution. It is important to note that the suspension, as a whole is electrically neutral despite the presence of unequal distribution of charges in the double layer.

Two other situations may arise. Should the concentration of counter-ions in the tightly bound layer be equal to that of the co-ions on the solid surface, then electric neutrality will occur at the shear plane and there will be only one layer of medium and ions, instead of double layer. However, if the total charge of the counter-ions in the Stern layer exceeds the charge due to the co-ions, the net charge at the shear plane will be positive rather than negative. It means electric neutrality will be achieved where the electric double layer ends and the diffuse layer, will contain more co-ions than counter-ions. The charge density at any distance from the surface is determined by taking the difference in concentration between positive and negative ions at that point.

2.2.3 Nernst and Zeta Potentials

The electric double layer is formed in order to neutralize the charged particles in a suspension. As discussed in the previous few paragraphs, the electrical potential at any point in a suspension system depends on its exact location. The potential in the diffuse layer gradually changes as one moves away from a solid particle. This is shown in Fig. 2.2.

The difference in electric potential between the actual or true surface of the particle and the electroneutral region is referred to as the surface or electrothermo-dynamic or Nernst potential (E). Hence, Nernst potential is controlled by the electrical potential at the surface of the particle due to the potential determining ions. The potential difference between the shear plane and the electroneutral region is known as the electrokinetic or zeta (ζ) potential (Fig. 2.3). While Nernst potential has little influence in the formulation of stable suspension, zeta potential has significant effect on it (Brigger et al. 2003). Zeta potential governs the degree of repulsion between adjacent, similarly charged solid dispersed particles. If the zeta potential is reduced below a certain value, which depends on the specific

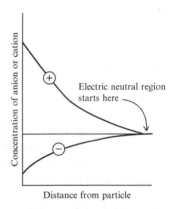

Fig. 2.2 Variation in concentration of cations and anions with distance from a negatively charged suspended particle

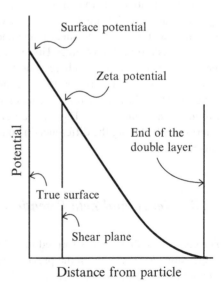

Fig. 2.3 The relationship between Nernst and zeta potentials

system under investigation, the attractive forces between particles due to van der Waals' force, overcome the forces of repulsion and the particles come together to form floccules. This phenomenon is known as flocculation. The magnitude of surface and zeta potentials is related to the surface charge and the thickness of the double layer.

Zeta potential can be measured by microelectrophoresis, in which migration of the particles in a voltage field is observed through a microscope (Black and Smith 1966; Au et al. 1986).

2.2.4 Wetting

Suspensions are prepared by insoluble solids in dispersion medium, mostly water. Some insoluble solids may easily be wet by water and disperse readily throughout the aqueous phase with minimal agitation. Many solid materials, however, are too hydrophobic to be wet and thus, either form large porous clumps within the liquid or remain floated on the surface despite their high density. The low adhesion may also result from a layer of air or greasy materials or other impurities covering the surface of the particles. Fine powders are more susceptible to this problem because of the larger surface area. These particles can only be wet after displacing these adsorbed materials.

Low adhesion of hydrophobic powders materials occurs due to high interfacial tension between the powders and the dispersion medium. Because of this tension, the contact angle between the solid and liquid phases remains very high. By using surface active agents, the interfacial tension can be lowered to decrease the contact angle, resulting in good wetting. The hydrocarbon chain of the surface active agents is adsorbed onto the hydrophobic surface of the particles and the polar end remains with the liquid medium. Thus the interfacial tension can be reduced to get the desired wetting of the solid materials. Wetting of solids can be determined by using Young's equation (Adamson 1990):

$$\cos\theta = \frac{\gamma_{S/V} - \gamma_{S/L}}{\gamma_{L/V}} \tag{2.2}$$

where, θ is the contact angle, γ is the interfacial tension between the different phases, solid (S), vapor (V), and liquid (L). Wetting agents can reduce $\gamma S/L$ and $\gamma L/V$ and thus decrease the contact angle.

Surfactants with HLB value between 7 and 9 can function as wetting agents (Billany 2002). Most surfactants are used at concentrations up to 0.1% as wetting agents. Commonly used wetting agents for oral use are polysorbates and sorbitan esters. SDS is an example of a surfactant that is used externally. Parenteral preparation may contain polysorbates, some of the poloxamers, and lecithin, among others. Too much of the surfactant may produce foam or deflocculated systems, both of which may be undesirable.

Hydrophilic polymers can also be used to improve the wetting of solid materials. Examples are acacia, tragacanth, xanthan gum, bentonite, aluminum–magnesium silicates, colloidal silica, and cellulose derivatives, such as sodium carboxymethyl-cellulose. These polymers form a hydrophilic coating around the solid particles and thus promote wetting. When used at too high concentrations, these materials can cause an undesirable gelling of the suspension.

Wetting can also be improved by using hygroscopic solvents, such as alcohol, glycerol, and glycols, especially, propylene glycols. These agents (also known as levigating agents) lower the liquid–air interfacial tension, helping solvent penetration into the loose agglomerates of powder and thus displace the air from the pores of the individual particles. Mineral oil can be used as the levigating agent when the vehicle is an oil.

When a number of wetting agents are available for a particular preparation, the one, which can efficiently wet the particles at low concentration, is generally selected.

2.2.5 Electrokinetic Phenomena

The presence of interfacial potentials may lead to the existence of four electrokinetic phenomena, which include electrophoresis, electroosmosis, sedimentation potential, and streaming potential. All of these properties are essentially the direct results of the movement of a charged surface with respect to an adjacent liquid phase.

Electrophoresis measures the movement of charged particles through a liquid under the influence of an applied potential difference (Minko 2006b). The zeta potential of the system can be calculated using (2.3):

$$\zeta = \frac{v}{E} \times \frac{4\pi\eta}{\varepsilon} \times (9 \times 10^4) \qquad (2.3)$$

where ζ is the zeta potential in volts, v is the velocity of migration of the particles in cm/s in the electrophoresis tube, η is the viscosity of the medium in poises, ε is the dielectric constant of the medium, and E is the potential gradient in V/cm.

Electroosmosis may be considered the opposite to electrophoresis. In the latter, charged solid particles move relative to the dispersion medium under an applied potential, whereas in electroosmosis, the solid is rendered immobile but the liquid moves relative to the charged surface when a potential is applied.

Sedimentation potential is the potential generated when particles undergo sedimentation. Therefore, basically it is the reverse process of electrophoresis. The streaming potential, which is essentially the reverse process of electroosmosis, is created by forcing a liquid to flow through a stationary solid phase (e.g., plug or bed of particles). Like electrophoresis, the other electrokinetic parameters can be used to determine surface or zeta potential and therefore, the stability of pharmaceutical dispersions (Bhattacharjee et al. 2005).

2.2.6 DLVO Theory

Particle collision in a suspension preparation may occur due to Brownian motion or differential sedimentation rates. The consequence of collision would be either the formation of aggregates or redispersion of the particles. The outcome of collision depends on the attractive or repulsive forces between the particles and determines the quality of the preparation. As mentioned previously, zeta potential plays a very important role in suspension stability. A minimum, known as the critical zeta potential, is required to prepare a stable suspension. A system with low critical zeta potential indicates that only a minute charge is required for stabilization and it will show marked stability against the added electrolytes. The precipitation of suspension can be brought about by adding electrolytes. The precipitating power increases rapidly with the valence of the ions. This is known as the Schulze-Hardy rule.

Derjaguin and Landau and Verwey and Overbeek worked independently and used the knowledge from Schulze-Hardy rule to describe the stability of lyophobic colloids (Derjaguin and Landau 1941, Verwey and Overbeek 1948). This is the classic DLVO theory, which explains the result of particle interaction in lyophobic colloids. Kayes (1977) confirmed that the DLVO theory can also be applied to coarse suspension systems. According to this theory, the potential energy of interaction between particles, V_T is the result of repulsion due to electrical double layer, V_R and attraction due to van der Waals' force, V_A and can be shown by (2.4) (Matthews and Rhodes 1970):

$$V_r = V_R + V_A \tag{2.4}$$

V_R depends on several factors including the zeta potential of the system, the particle radius, the interparticular distance, the dielectric constant of the medium, whereas the factors that affect V_A includes the particle radius, and the interparticular distance.

DLVO theory can be easily understood from Fig. 2.4. Electrical repulsion due to the electric double layer and the attraction due to van der Waals' force are shown in the opposite direction due to their opposite nature of force. At any distance from a particle (h), the net energy (V_T) is calculated by subtracting the smaller value from the larger one. When the net energy curve remains above the baseline, it represents repulsion. On the other hand, attraction can be shown by the curves below the baseline. The maximum repulsive value is known as the energy barrier. In order to agglomerate, two particles on a course of collision must have sufficient momentum to cross this barrier. As the particles overcome the repulsion, they agglomerate due to the attractive force and can be considered trapped due to van der Waals'-London force.

Figure 2.5 shows net energy interaction curves at different situations. Curve A exists when V_R is much larger than V_A ($V_R \gg V_A$). In this situation, the dispersion will be highly stable because of the high net repulsive force. This dispersion is resistant to aggregation (i.e., flocculation or coagulation) as long as the particles do not sediment under gravity. Curve B explains a situation in which a high energy barrier, VM, must be overcome by the particles to form aggregates. If VM greatly exceeds the mean

Fig. 2.4 Net energy of interaction between two
particles as a function of interparticulate distance

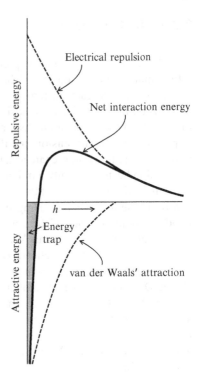

thermal energy of the particles, these particles will not enter P, the primary energy
minimum (Tabibi and Rhodes 1996). The minimum value of VM that can create
this situation corresponds to a zeta potential of more than 50 mV (Flory 1953;
Napper 1967; Raghavan et al. 2000). A very small interparticular distance is found
at P. The high magnitude of energy at P causes the particles to bond tightly together.
Consequently, it is possible that it will compact into a hard cake, which will be very
hard to redisperse. Occasionally, there occurs a secondary minimum of curve B at
S, which is far from the surface of the particle. Loose aggregates can be created at
this point, and this aggregate can usually be broken easily by shaking or dilution.
The aggregation of solid particles in suspension systems can be termed either
as "flocculation" or "coagulation." Some authors like to use these two terms inter-
changeably; others differentiate them by the mechanism of the aggregation process.
However, according to the most generally accepted definition, flocculation occurs
at the secondary energy minimum, S; and coagulation occurs at the primary energy
minimum, P, of the potential energy curve of two interacting particles (Fig. 2.5).
Curve C represents a situation where attractive forces predominate over repulsion
forces all the time ($V_A \gg V_R$); and rapid aggregation will occur.

In addition to electric stabilization, steric stabilization can also be applied
to prepare a stable dispersion. Substances such as nonionic surfactants, when
adsorbed at the particle surface, can stabilize a dispersion, even when there is no
significant zeta potential (Matthews and Rhodes 1970). Therefore, the term for steric

Fig. 2.5 Net energy interaction curves at different situations of attractive and repulsive forces

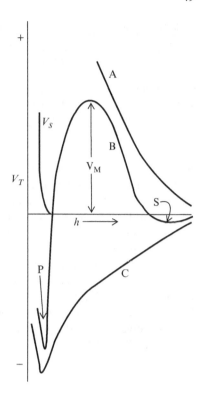

stabilization, V_S, should be added to the equation obtained in DLVO theory, which gives rise to (2.5):

$$V_r = V_R + V_A + V_S \qquad (2.5)$$

2.2.7   Flocculation and Deflocculation

A deflocculated system has a zeta potential higher than the critical value when the repulsive forces supercede the attractive forces. Particles in this type of systems remains suspended for a long period of time, and only a small portion of the solid is found in the sediment due to the force of gravitation. During sedimentation, the smaller particles fill the void between the larger ones; and the particles lowest in the sediment are gradually pressed together by the weight of the particles above. Both situations increase the closeness of the particles; and, thus, they are attracted by a large amount of van der Waals'-London force. A close-packed, hard cake-like residue is formed, which is difficult, if not impossible, to redisperse.

Upon the addition of a small amount of electrolyte, the zeta potential of the system reduces. Once it is below the critical value, the attractive forces supercede

Table 2.1 Comparative properties of flocculated and deflocculated suspensions

Properties	Deflocculated	Flocculated
Particles	Exist in separate entities	Form loose aggregates (flocs)
Sedimentation rate	Slow	Rapid
Sediment structure	Compact	Scaffold-like loose
Redispersion	Difficult	Easy
Supernatant liquid	Cloudy	Generally clear

the repulsive forces, producing flocculation. Flocculation can also be brought about by flocculating agents other than electrolytes (e.g., nonionic surfactants). As shown in Fig. 2.5, flocculation occurs at the secondary energy minimum when the particles are far apart from each other. When the particles collide, loosely packed aggregates of particles or flocs are created. Being larger and heavier than individual particles, these flocs get settled faster than the deflocculated particles. However, since the flocs are loose structures, they will not form hard, cake-like deflocculated suspensions and will resuspend with minute agitation. Generally a clear supernatant liquid is found at the top of the flocculated suspensions. On the other hand, deflocculated suspensions remain cloudy due to the very fine dispersed solids. A comparative study of the properties of flocculated and deflocculated suspensions is shown in Table 2.1.

2.3 Stability Considerations

2.3.1 Settling/Sedimentation

Settling or sedimentation is a very important issue in suspension stability. It is a general trend to reduce the rate of settling, although as mentioned earlier, an inordinately slow rate of settling in a deflocculated suspension may cause the particles to settle as compact residue at the bottom of the container. A commonly accepted theory that explains the settling kinetics in dispersed systems was proposed by George Gabriel Stokes. His theory is known as Stokes' equation or Stokes' law:

$$v = \frac{d^2(\rho_1 - \rho_2)g}{18\eta} = \frac{2r^2(\rho_1 - \rho_2)g}{9\eta} \tag{2.6}$$

where v is the velocity of sedimentation; d and r are the diameter and radius of the particle, respectively; ρ_1 and ρ_2 are the densities of the dispersed phase and dispersion medium, respectively; g is the acceleration due to gravity; and η is the viscosity of the dispersion medium.

The most important parameter is the particle size because it is presented as square form in the equation. Smaller particles yield a low rate of sedimentation. Viscosity of the medium can be increased in order to reduce settling. Common viscosity-increasing agents are cellulose derivatives (e.g., methylcellulose and hydroxypropyl methylcellulose). Other examples would be natural gums (e.g., acacia and tragacanth). The difference in density between the dispersed phase and dispersion medium can affect the rate of settling. A difference of zero means no sedimentation. Since, the density of the dispersed phase cannot be changed, it would be necessary to increase the density of the medium. However, it is rarely possible to increase the vehicle density much above 1.3; therefore, the difference cannot be completely eliminated (Tabibi and Rhodes 1996). However, by using some density modifiers (e.g., sorbitol and mannitol), the difference can be reduced. These agents may also increase the viscosity of the medium. A single parameter or a combination of parameters in Stokes' equation can be altered to reduce settling.

2.3.2   Limitations of Stokes' Law

Stokes' law is a generalized equation that describes how certain factors affect the rate of settling in dispersed systems. However, this law is based upon several assumptions, which may not always hold true for pharmaceutical suspensions.

Stokes' law is valid for diluted pharmaceutical suspensions that are composed of no more than 2% solids (Martin 2006). In a diluted suspension, the solid particles settle without interference from one another in what is termed free settling. In a concentrated suspension, this interference may occur, and may hinder the settling results.

Particle shape and size are also important in Stokes' equation. This equation assumes spherical and monodisperse particles, which may not be encountered in real systems. The following equation was introduced to accommodate particles with different shapes (Alexander et al. 1990):

$$v' = v\varepsilon^n \tag{2.7}$$

where, v' is the corrected rate of settling and ε is the initial porosity of the system, which represents the initial volume fraction of the uniformly mixed suspension. The values of ε range from zero to unity. The exponent n, which is a constant for each system, is a measure of the hindering of the system.

Stokes' equation is invalid if the density difference in the equation is negative (i.e., the particles are lighter than the dispersion medium). In this case, floatation or creaming, which is common in emulsion systems, will occur.

Stokes' equation may not show the real sedimentation rate when the solid content is high. The equation contains only the viscosity of the medium. However, the high solid content imparts additional viscosity to the system, which must be taken into consideration if the correct rate of settling is to be determined. This issue will be discussed later in this chapter.

Dielectric constant is also an important parameter in many situations, but is ignored in Stokes' equation. The electrical potential between two charges is inversely proportional to the dielectric constant of the medium. Therefore, zeta potential is dependent on the dielectric constant of the medium. This factor is important when nonaqueous vehicles, e.g., sesame oil, corn oil, chlorofluoro carbon propellant (in aerosol suspension) are used. In these vehicles with low dielectric constant, the double layer is many times thicker than in an aqueous medium. The zeta potential and, thus, the settling will be different.

Brownian movement. Brownian movement is another factor that can influence the accuracy of the results obtained in Stokes' equation. If the particle in a dispersion is too small, the molecular bombardment can cause a random motion, known as Brownian movement, which is not uniform throughout the dispersion. The displacement due to Brownian movement, D_i can be calculated as:

$$D_i = \frac{RT}{3N\pi\eta r} \times t \qquad (2.8)$$

where R is the molar gas constant, N is Avogadro's number, T is absolute temperature, t is time, η is the viscosity of the medium, and r is the radius of the particle.

Brownian movement counteracts sedimentation to a measurable extent. Therefore, the actual rate of sedimentation may greatly deviate from the one calculated using Stokes' equation. As the particle size is reduced, Brownian movement becomes significant. Below a critical radius, the movement will be sufficient to keep the particles from sedimentation. This radius will, of course, depend on the density and viscosity values in the previous two equations (Tabibi and Rhodes 1996).

Although Stokes' equation does not consider many important parameters, it, nevertheless, provides a very good estimate, based on what further investigation can be carried out to determine the exact rate of sedimentation.

2.3.3 Sedimentation in Flocculated and Deflocculated Systems

As discussed earlier, flocculated suspensions show rapid sedimentation creating loose sediment; whereas, deflocculated suspensions show slow, but compact, sedimentation. The whole process of sedimentation can be viewed in Fig. 2.6. Situations A, B, and C represent a deflocculated system; and D, E, and F show a flocculated system. The deflocculated system shows almost no change in appearance after a few minutes of manufacture (A). After several hours, the whole dispersion is still cloudy; however, a small and compact sediment appears (B). After prolonged storage, a clear supernatant in contact with compact sediment is obtained (C). A flocculated system reacts differently. Within a few minutes of manufacture, a clear supernatant with a distinct boundary from the sediment is evident (D). After several hours, a larger volume of clear supernatant occurs (E). At this point, the size of the sediment is larger than that of the deflocculated system. After prolonged standing, the volume of the sediment shows little change (F).

Figure 2.6 suggests that deflocculated systems have an advantage in that their sedimentation rate is slow, thereby allowing uniform dosing. However, if settling

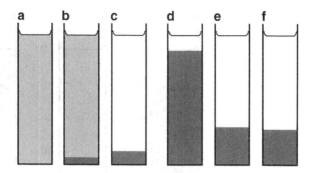

Fig. 2.6 The sedimentation behavior of deflocculated (a,b,c) and flocculated (d,e,f) suspension systems. Adapted from Billany (2002)

occurs, the sediment can be compact and difficult to redisperse. Flocculated suspensions, on the other hand are stilted because the particles separate quickly. In addition, rapid sedimentation may cause inaccurate dosing. However, since the sediment is loose, redispersion can occur, even after long storage. An intermediate condition known as controlled flocculation can be created to obtain the most acceptable product. That concept will be discussed later in this chapter.

Sedimentation parameters. The extent of flocculation of a system can be redefined by the extent of sedimentation. Two important sedimentation parameters are commonly used for that purpose. The sedimentation volume, F, is the ratio of the equilibrium volume of the sediment, V_u to the total volume of the suspension, V_o. Thus,

$$F = \frac{V_u}{V_o} \tag{2.9}$$

The value of F normally ranges from 0 to 1. If the sedimentation volume is 0.8, 80% volume of the suspension is occupied by the loose flocs as the sediment (Fig. 2.7a). The F value of a deflocculated suspension is relatively small, 0.2, as in Fig. 2.7b. Suspension with F value 1 is the ideal system (Fig. 2.7c). The system is flocculated, but there is no sedimentation or caking. This system is also aesthetically elegant since there is no visible clear supernatant. This type of system is said to be in flocculation equilibrium. It is possible to have F values greater than 1, when the ultimate volume of the sediment is greater than the original volume of the suspension. The particles in the suspension create such a loose network of flocs that the floc comes out of the dispersion medium. An extra vehicle should be added to cover the sediment.

The sedimentation volume gives only a qualitative idea about the sedimentation of the suspension because it lacks any meaningful reference point (Hiestand 1964). A more applicable parameter for flocculation is the degree of flocculation, β, which relates the sedimentation volume of the flocculated suspension, F, to the sedimentation volume of the suspension when it is deflocculated, F_∞, by the following expression.

Fig. 2.7 Sedimentation parameters of suspensions at different situations. Adapted from Martin (2006)

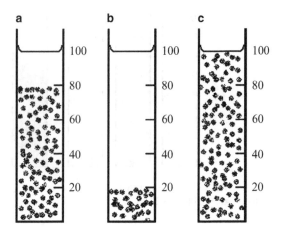

$$\beta = \frac{F}{F_\infty} \qquad (2.10)$$

The sedimentation volume of the deflocculated suspension can be shown by the following equation:

$$F_\infty = \frac{V_\infty}{V_0} \qquad (2.11)$$

where F_∞ is the sedimentation volume of the deflocculated system and V_0 is the ultimate volume of the sediment, which will be relatively small. These two equations can be combined as follows:

$$\beta = \frac{V_u / V_0}{V_\infty / V_0} = \frac{V_u}{V_\infty} \qquad (2.12)$$

The β value for the flocculated suspension in Fig. 2.7 is 80/20 = 4. That is, flocculated suspension creates four times more sediment than the deflocculated system.

The degree of flocculation is a more useful parameter because it compares the suspension under investigation to a standard: the deflocculated state of the system.

2.3.4 Controlled Flocculation

A partially flocculated suspension with sufficient viscosity should yield the product with desirable sedimentation properties. Flocculation should be carefully controlled, and the viscosity should not be too high to make redispersion difficult.

Controlled flocculation can be achieved by a combination of control of particle size and the use of flocculating agents. The importance of particle size was discussed earlier in this chapter. The most common categories of flocculating agents are electrolytes, surfactants and polymers.

The first step in the preparation of suspension is wetting the already reduced particles, followed by the addition of suspension agents. Depending on the charge of the wetting agents and the particles, the resultant dispersion could be flocculated or deflocculated. If it is deflocculated, the next step would be to add flocculating agents.

2.3.4.1 Electrolytes

Electrolytes act by reducing the zeta potential, which brings the particles together to form loosely arranged structures. The flocculating power increases with the valency of the ions. Therefore, calcium ions are more powerful than sodium or potassium ions. However, trivalent ions are less commonly used because of their toxicity. When electrolytes are added to a positively charged deflocculated suspension, zeta potential decreases slowly. At a certain stage, upon persistent addition, it becomes zero. Beyond that limit, zeta potential becomes negative. As zeta potential decreases, the sedimentation volume increases sharply up to a point. The sedimentation volume reaches its maximum value and remains relatively constant within a certain range of zeta potential, where it changes from low positive potential to low negative potential. When the potential becomes too negative, the sedimentation volume decreases again. A microscopic examination of bismuth subnitrate suspension by Haines and Martin (1961a, b, c) showed that flocculation increases with the addition of electrolytes, and the extent of flocculation coincided with the sedimentation volume. Caking was observed at less than the maximum values of sedimentation volume.

Surfactants: Both ionic and nonionic surfactants can be used as flocculating agents. Ionic surfactants cause flocculation by neutralizing the charge on particles. Because of long structure, nonionic surfactants are adsorbed onto more than one particle, thereby, forming a loose flocculated structure.

2.3.4.2 Polymers

Linear and branched chain polymer form a gel-like network that adsorbs onto the surface of dispersed particles, holding them in a flocculated state. Moreover, hydrophilic polymers can also function as protective colloids. In this capacity, flocs are sterically prevented from adhering to one another. Loose sediment is the result. Kellaway and Najib (1981) studied the effect of polyvinylpyrrolidone and acid and alkaline gelatins on the stability of sulfadimidine suspension. At low polymer concentrations, when the drug particles are not completely covered by the polymers, the latter can form bridges between multiple particles. Flocculation then occurs.

At sufficiently high polymer concentration, the particles are fully covered by the polymers, which creates steric stabilization. Some polymers, known as polyelectrolytes, can ionize in aqueous medium. The extent of ionization depends on the pH and the ionic strength of the dispersion medium. These polymers are able to act both electrostatically and sterically. Sodium carboxymethylcellulose was shown to function as a polyelectrolyte. Linear polymers (e.g., sodium carboxymethylcellulose) serve better as flocculating agents; however, coiled polymers (e.g., polyvinylpyrrolidone) are not conducive to flocculation, due to their shape. They do produce steric stability (Kellaway and Najib 1981).

2.3.5   Effect of Particle Size on Suspension Stability

As discussed earlier, controlling the size of the particles is very important for suspension stability. Finely divided particles are necessary to reduce sedimentation. However, improper control of particle size can create several undesirable consequences. Suspensions containing particles greater than 5 µm in diameter will be gritty and unsuitable for parenteral or ophthalmic preparations due to the possibility of irritation. Particles larger than 25 µm in diameter may clog the hypodermic needle, especially when the particles are acicular in shape. As mentioned previously, very fine particles sediment as hard cake if the suspension is not flocculated.

Besides the effects on sedimentation, particle size has other significance. For a concentrated suspension, there is a significant particle–particle interaction, which may lead to more viscous or thixotropic dispersions. Smaller particles can have a greater effect on increasing the viscosity of the system because they have a higher specific surface area than the larger particles.

Particle size can also significantly affect the drug bioavailability. Suspensions are used to deliver poorly soluble drugs. The absorption of these drugs is primarily controlled by the dissolution rate. In such cases, using smaller particles may provide a better rate and extent of drug absorption due to their higher specific surface area.

Smaller particles are also beneficial to achieve better dose uniformity. This difference in size can be significant if small doses are withdrawn from a large container.

Most pharmaceutical suspensions contain polydisperse solids. A narrow particle size distribution is desirable as it yields uniform sedimentation rate, which ultimately will provide better predictability of suspension properties from batch to batch of finished suspensions.

2.3.6   Crystal Growth

Also known as Ostwald ripening, crystal growth sometimes is very important for suspension sedimentation, physical stability, redispersibility, appearance, and bioavailability. Size of all particles in a dispersion may not remain constant throughout its shelf life. One reason for that change would be crystal formation. Particles in dispersion generally are not monodisperse. Rather, there is a range of particle size, not a single value. The surface free energy on smaller particles is comparatively more than that on larger particles. Therefore, smaller particles are more soluble in the dispersion medium. If the temperature rises, more materials are dissolved from the smaller particles, decreasing their size even more. When the temperature goes down, the drug attempts to recrystallize on the surface of existing particles (Tabibi and Rhodes 1996). Thus, gradually the larger particles will increase in size as the smaller particles decrease in size. Thus, a slight fluctuation of temperature may cause the particle size spectrum shift to higher values. This situation is especially true for slightly soluble drugs. This problem can be initially eliminated by using a narrow particle size range. Surface active agents or polymeric colloids can also prevent crystal growth by being adsorbed on the particle surface. SDS and polysorbate 80 were tested in carbamazepine (CBZ) suspension. SDS was found to increase the drug solubility and thus exacerbated the crystal growth problem. Polysorbate 80, however, was able to diminish crystal growth of CBZ when used up to a certain concentration. It was concluded that surfactants can modify the growth rate of CBZ crystals, an effect related to the solubilizing capacity of the surfactants. The less the solubilizing capacity of the surfactant, the more efficient it is in preventing crystal growth (Luhtala 1992). Benzalkonium chloride showed a slightly different relationship to CBZ crystal growth. At low concentration, the chloride accelerated the rate of crystal growth by increasing the drug solubility. However, at higher concentrations, when the peak level of solubility was reached, crystal growth was stopped (Luhtala et al. 1990).

Crystal growth in pharmaceutical suspension may also happen for polymorphic drugs (Kuhnert-Brandstätter 1977). Metastable (i.e., the least stable form of the drug) is the most soluble. As the metastable form changes to more stable form, solubility decreases, and crystallization occurs. This problem can be avoided by excluding the metastable form from the dispersion and by using the most stable polymorph of the drug. Amorphous metronidazole was found to convert to the monohydrate form in an aqueous suspension, thus favoring crystal growth. A combination of suspending agents, microcrystalline cellulose and carboxymethylcellulose was found to prevent the conversion (Zietsman et al. 2007).

2.3.7 Use of Structured Vehicles

Structured vehicles are generally aqueous solutions of natural and synthetic polymers, such as methylcellulose, sodium carboxymethylcellulose, acacia, tragacanth, etc. They are viscosity-imparting agents and basically reduce the rate of sedimentation in dispersed systems. These vehicles are plastic or pseudoplastic in nature. Additionally, some degree of thixotropy is desirable. For a deflocculated suspension, the shear thinning property of the structured vehicle will allow easy redistribution of the particles from the small sediment. However, the ultimate sediment in a deflocculated system in a structured vehicle, as in other vehicles, may form a compact cake, which will not be resuspendable by the shear thinning liquid. Therefore, a structured network, flocculated suspension, in a structured vehicle is preferred (Martin 1961). When the shaking is discontinued, the vehicle goes back to its original higher consistency, and that keeps the particles suspended. A flocculated suspension with no structured vehicle may not look elegant if the sedimentation volume is not close to 1. The concentration of polymer in a structured vehicle depends on the required consistency of the preparation and, therefore, on the particle size and density.

2.4   Rheologic Considerations

2.4.1   Importance

Rheologic consideration is of great importance in the study of the stability of pharmaceutical suspensions because viscosity, as discussed under Stokes' law can modify the sedimentation rate. Maintaining the proper viscosity of suspensions is also important to ensure the accuracy of dosing and ease of application. If the particles in a suspension get settled too fast, it may not be possible to withdraw the needed volume with the correct amount of drug from the dispensing bottle. Externally applied suspension products should spread easily, but they are not expected to be so fluid that they run off the skin surface. Parenteral products have to be fluid enough to pass through hypodermic needles when moderate pressure is applied (high syringeability). Therefore, from a formulator's perspective, a compromise in viscosity of the dispersed preparation should be attained. Rheology can also affect the manufacturing process of suspensions. Highly viscous mixture produces excessive frictional drag on the mixing vessel and other machinery accessories, thereby resulting in wasted energy. Therefore, from the manufacturer's point of view, low viscous products are more productive.

Diluted suspensions behave in the manner of a Newtonian system in that only a small amount of solids when compared to the dispersion medium is present to affect the viscosity of the whole product. Concentrated suspensions, however, typically follow non-Newtonian flow (Vaithiyalingam et al. 2002). Concentrated flocculated suspensions show high viscosity when static, due to high interparticular attraction. A minimum force is required to overcome that attraction. Once that force is applied, viscosity decreases substantially. Beyond that point, viscosity remains constant in the manner of Newtonian systems. These characters are typical to plastic flow. Pseudoplastic flow, which is mainly found in polymeric solutions, shows decreased viscosity as the shearing stress is increased (Nutan et al. 2007). Therefore, both plastic and pseudoplastic systems can be beneficial to formulate stable suspension products because at high stress, such as shaking the bottle, the products become thinner and make withdrawal and application easier.

The principle of thixotropy also can be applied to pharmaceutical suspensions. It is applicable to both plastic and pseudoplastic systems. When the applied force is withdrawn from these systems, the consistency of the products increases and comes back to the original value. However, when they are thixotropic, the rate of increase in viscosity is slower. The rate of movement of fluid at any given shearing stress will vary in thixotropic systems, depending on whether the viscosity is increasing or decreasing. This type of system is desirable for pharmaceutical suspensions. In the bottle, viscosity of the products will be high enough to keep the particles suspended during the shelf-life. When the bottle is shaken, the products become thin and those will remain thin for a sufficient period of time even upon withdrawal of the force to allow an accurate dose withdrawal. This characteristic will also allow the topical products to be applied more easily to the skin.

Concentrated deflocculated suspensions appear to be more structured and viscous with the increase in shearing stress. Therefore, this type of preparation is not desirable when rheology is under consideration.

2.4.2 Factors Affecting Rheology

The desired viscosity of a product depends on the ease of manufacture and the stability of the product. Different approaches can be taken to change the rheological behavior of suspension products. Therefore, a formulator should have exact knowledge of the factors that affect the viscosity.

The most important physical parameters that influence the rheological properties include dispersed phase content; particle shape, size, and size distribution; viscosity and rheological behavior of the dispersion medium; and temperature. Along with the physical factors, there are some chemical parameters that affect the particle interactions and, thereby, the effective dispersed phase fraction, which, in turn, changes the viscosity of the product. Some of these factors will be discussed in the following few paragraphs.

2.4.2.1 Dispersed Phase Content

The content of the dispersed phase φ is the most important physical parameter that affects the viscosity of dispersions (Briceño 2000). It is generally considered the ratio (in fraction or percentage) of volume of the dispersed phase to the total volume of the dispersion. It can also be expressed as weight fraction or weight percentage or weight by weight concentration.

Viscosity increases as φ increases. The contribution of the solid phase content to the overall viscosity can be understood by Einstein equation (Hiemenz and Rajagopalan 1997):

$$\eta_s = \eta_O (1 + 2.5\,\phi) \tag{2.13}$$

where η_s is the viscosity of the suspension and η_o is the viscosity of the dispersion medium. Therefore, the lower the value of φ (i.e., the less the solid content), the closer is the overall viscosity of the suspension to the viscosity of the dispersion medium. The Einstein equation gives good results of the overall viscosity for very dilute solutions ($\varphi < 0.02$) (Briceño 2000).

However, in practice, most pharmaceutical suspensions are more concentrated, and interparticular attractions are more prominent. In such cases, particles remain as aggregates of two or more particles, and the dispersion medium is trapped within the aggregate. This situation reduces the effective volume of the dispersion medium and, thereby, increases the effective volume of the dispersed phase. Effective volume fraction of a solid can be designated as φ_e. Equation (2.4) is a modified version of the Einstein equation that gives more accurate results (Thomas 1965):

$$\eta_s = \eta_O(1 + 2.5\,\phi_e + a\,\phi_e^2) \tag{2.14}$$

where a is the two-body interaction parameter. The third term in the equation accounts for the increase of viscosity due to the formation of doublets.

Particle interactions of a higher order can be expressed by the following semiempirical equation, known as the Thomas equation (Thomas 1965):

$$\eta_s = \eta_O(1 + 2.5\,\phi_e + 10.05\,\phi_e^2 + 0.00273e^{16.6\,\phi_e}) \tag{2.15}$$

This equation gives satisfactory results for φ value of up to 0.6.

With high φ values, the last term in the equation becomes very important since it is in exponential form. Thus, as φ increases, viscosity increases exponentially (Krieger and Dougherty 1959; Thomas 1965; Sherman 1983; Barnes 1997).

A different way to express the viscosity of concentrated dispersions is the Krieger-Dougherty equation (Krieger and Dougherty 1959):

$$\eta_s = \eta_o\left(1 - \frac{\phi_e}{\phi_{e,m}}\right)^{-\phi_{e,m}[\eta]} \tag{2.16}$$

where $[\eta]$ is the intrinsic viscosity and $\varphi_{e,m}$ is the maximum theoretical packing fraction corresponding to a hexagonally packed sphere structure.

In lyophilic colloidal dispersion, in addition to electrostatic and steric stabilization, particles can be stabilized by solvation forces (Kissa 1999). The solvation layer, which surrounds lyophilic surfaces has a special molecular structure which is different from that of the bulk liquid (Churaev et al. 1987). When two particles come sufficiently close to each other, the solvation layers start to overlap, generating a strong nonelectrostatic repulsive force that prevents the particles from coming closer. The thicker the solvation layers, the stronger are the repulsive solvation forces. Consequently, the colloid is more stable. When the solid particles have some degree of solubility in the dispersion medium, the solvation layers appear and cause an apparent swelling of the particles. This swelling leads to an increase in φ and a change in viscosity. It is, therefore, necessary to correct Einstein equation to include this effect. Equation (2.13) was modified to obtain (2.17) as follows (Song and Peng 2005):

$$\eta_s = \eta_o\left\{1+2.5\left(1+\frac{t_i}{R}\right)^3\phi\right\} = \eta_o\left\{1+2.5\left(1+\frac{t_i\times\rho A_{sp}}{3}\right)^3\phi\right\} \quad (2.17)$$

where t_i is the thickness of the solvation layer, ρ is the density of the dispersed phase and A_{sp} is the specific surface area of the solid particles, assuming spherical.

As mentioned earlier, viscosity increases as φ increases, and the rheological properties also appear to be more complicated. As φ increases, the shear thinning and shear thickening behavior become more prominent (Vaithiyalingam et al. 2002; Allahham et al. 2005). If a suspending agent is used in the vehicle, as φ decreases, the shear thinning or shear thickening property of the system is retarded, but the overall viscosity of the product is expected to increase due to an increased volume fraction of the vehicle (Allahham et al. 2005). All of the above-mentioned equations, however, have limitations as they do not take into consideration the non-Newtonian behavior of the system, which may be evident at φ value as low as 0.2. Moreover, these equations do not explain the effect of particle shape, size and size distribution (Briceño 2000). Nevertheless, these equations give fairly good results for complex suspension systems.

2.4.2.2 Particle Shape

As mentioned previously, particle shape can affect the rate of sedimentation. It can also affect the viscosity of suspension. As the particles deviates from spherical shape, the suspension becomes more viscous (Clarke 1967; Illing and Unruh 2004). The maximum volume fraction of the dispersed phase, $\varphi_{e,m}$ decreases as the particles become increasingly irregular in shape. Thus, as evident from (2.16), lower $\varphi_{e,m}$ will yield more viscous system. Nonspherical particles generally show dilatant flow as they have higher contact area among themselves under high shear. However, shear thinning behavior is also possible with the nonspherical particles. A common example is the dispersion of natural or synthetic polymers (Briceño 2000).

2.4.2.3 Particle Size and Size Distribution

In dilute dispersions, viscosity is independent on particle size. However, in concentrated suspension, the effect of particle size on viscosity depends on the counterbalance of hydrodynamic and Brownian forces. An increase in particle size increases the shear thickening property of a levodopa injectable suspension (Allahham et al. 2005). Particle size can also have an effect in thixotropy. In dispersions, smaller particles aggregate at a faster rate than larger particles (Briceño 2000).

Particle size distribution can also play an important role in determining the viscosity of dispersion. Suspension with a wide particle size distribution shows lower viscosity than the one with narrow particle size distribution. The broadening of the particle distribution span was also found to lower shear thickening property

(Allahham et al. 2005). If a suspension has too many smaller particles, they tend to occupy themselves within larger particles and, thus, reduce the interactions between the latter. In this way, smaller particles and the dispersion medium act as a pseudo-continuous phase that carries the larger particles suspended. Therefore, the effective dispersed phase content is reduced, which reduces the viscosity (Briceño 2000). Bimodal particle distributions were also found to reduce the shear thickening behavior (Allahham et al. 2005).

2.4.2.4 Temperature

Temperature can affect the viscosity of the dispersion medium and thus that of the dispersion. Generally, a rise in temperature causes viscosity to go down. Apparent viscosities of suspensions containing suspending agents, sodium starch glycolate and modified corn starch, were found to decrease progressively with increase in temperature over 10–50°C (Sabra and Deasy 1983). The relationship of viscosity to the temperature of a liquid can be shown as:

$$\eta = A\mathrm{e}^{E_V RT} \tag{2.18}$$

where η is the viscosity of the liquid, A is a constant, which depends on the molecular weight and molar volume of the liquid and E_V is the energy of activation required to initiate the flow between molecules (Longer 2006).

Temperature can affect the viscosity of a suspension by also modifying interfacial properties and thus inducing or reducing flocculation. Flocculation increases viscosity. Temperature can also affect the viscosity of a system by increasing the Brownian movement. In addition, temperature causes volume expansion in both the dispersion medium and the solid. However, liquid medium expands more than the solid, leading to a decrease in φ value and a decrease in viscosity (Briceño 2000).

2.5   Conclusion

The therapeutic utility of drugs involves the application of dosage forms/delivery systems, which serve as carrier systems together with several excipients to deliver the active therapeutic agent to the site of action. Suspensions are an important class of pharmaceutical dosage forms that may be given by many routes, including oral, topical, parenteral, and also used in the eye for ophthalmic purposes. Surprisingly large proportions of new drug candidates that are emerging are predominantly water insoluble and, therefore, demonstrate poor bioavailability in the solution dosage form. While suspensions present a viable formulation option for many drugs, particularly for water insoluble, hydrophobic drug substances, there are certain criteria that a well-formulated suspension should meet.

This chapter has dealt in sufficient detail regarding the theoretical considerations of pharmaceutical coarse dispersions (e.g., interfacial properties, electric double layer, wetting, flocculated and deflocculated systems), stability considerations (e.g., settling and sedimentation, effect of particle size, crystal growth, and use of structured vehicles, etc.), and the rheologic considerations. A thorough understanding of these aspects is important from a stability standpoint, as the stability of suspension systems depends both on the properties of the dispersed solid and on those of the vehicle.

Apart from the gravitational sedimentation, the behavior of the solid may be determined by the zeta potential. Concise information on zeta and Nernst potential is provided. Investigations on the influence of temperature on the deterioration of drugs in suspension form are complicated by the change in drug solubility with temperature. The concentration of the drug in solution usually remains constant because, as the reaction proceeds, more of the suspended drug dissolves and the solution remains saturated with respect to undegraded reactant, thus following zero-order kinetics.

Drugs from suspension formulations typically exhibit an improved bioavailability when compared to the same drug formulated as a tablet or capsule. For example, suspensions of antacid preparations offer more rapid onset of action and are, therefore, more effective than an equivalent dose of the same drug in the tablet dosage form. However, course dispersions may not provide optimal bioavailability for certain drug candidates; and, in such cases, formulating them into crystalline nanosuspensions may present a viable option.

In the process of overcoming issues involving solubility, additional pharmacokinetic benefits of the drugs so formulated have come to be appreciated. As such, insolubility issues over the years have provoked a significant change, which now offers novel solutions for innovative drugs of the future. Nanosuspensions have emerged as a promising strategy for the efficient delivery of hydrophobic drugs because of their versatile features and unique advantages. Rapid advances have been made in the delivery of nanosuspensions by parenteral, peroral, ocular and pulmonary routes. Currently, efforts are being directed to extend their applications in site-specific drug delivery.

References

Adamson AW, 1990. Physical Chemistry of Surfaces, Wiley-Interscience, New York, NY.

Alexander KS, Azizi J, Dollimore D, Uppala V, 1990. The interpretation of the hindered settling of calcium carbonate suspensions in terms of permeability. J Pharm Sci, 79: 401–406.

Allahham A, Stewart P, Marriott J, Mainwaring D, 2005. Factors affecting shear thickening behavior of a concentrated injectable suspension of levodopa. J Pharm Sci, 94: 2393–2401.

Au S, Weiner N, Schacht J, 1986. Membrane perturbation by aminoglycosides as a simple screen of their toxicity. Antimicrob Agents Chemother, 30: 395–397.

Barnes HA, 1997. Thixotropy – a review. J Non-Newton Fluid Mech, 70: 1–33.

Bhattacharjee S, Singh BP, Besra L, Sengupta DK, 2005. Performance evaluation of dispersants through streaming potential measurements. J Dispersion Science and Tech, 26: 365–370.

Billany M, 2002. Suspensions and Emulsions, In: Pharmaceutics: The Science of Dosage Form Design, Ed., Aulton ME. 2nd ed., Churchill Livingstone, Philadelphia, PA, pp 334–359.

Black AP and Smith AL, 1966. Suggested method for calibration of Briggs microelectrophoresis cells. J Am Water Works Assoc, 58: 445–454.

Briceño MI, 2000. Rheology of Suspensions and Emulsions, In: Pharmaceutical Emulsions and Suspensions, Eds., Nielloud F and Marti-Mestres. 1st ed., Marcel Dekker Inc., New York, NY, pp 557–607.

Brigger I, Armand-Lefevre L, Chaminade P, Besnard M, Rigaldie Y, Largeteau A, Andremont A, Grislain L, Demazeau G, Couvreur P, 2003. The stenlying effect of high hydrostatic pressure on thermally and hydrolytically labile nanosized carriers. Pharm Res, 20: 674–683.

Churaev NV, Derjaguin BV, Muller VM, 1987. Surface Forces, 1st ed., Springer, New York, NY.

Clarke B, 1967. Rheology of coarse settling suspensions. Trans Inst Chem Eng, 45: T251–T256.

Derjaguin BV and Landau LD,1941. Theory of the stability of strongly charged lyophobic sols and the adhesion of strongly charged particles in solutions of electrolytes. Acta Physicochim USSR, 14: 633–662.

Flory PJ, 1953. Principles of Polymer Chemistry, Cornell University Press, Ithaca, NY.

Gallardo V, Morales ME, Ruiz MA, Delgado AV, 2005. An experimental investigation of the stability of ethylcellulose latex correlation between zeta potential and sedimentation. Euro J Pharm Sci, 26: 170–175.

Haines BA and Martin AN, 1961a. Interfacial properties of powdered material: caking in liquid dispersions I. Caking and flocculation studies. J Pharm Sci, 50: 228–232.

Haines BA and Martin AN, 1961b. Interfacial properties of powdered material: caking in liquid dispersions II. Electrokinetic phenomena. J Pharm Sci, 50: 753–756.

Haines BA and Martin AN, 1961c. Interfacial properties of powdered material: caking in liquid dispersions III. Adsorption studies. J Pharm Sci, 50: 756–759.

Hiemenz PC and Rajagopalan R, 1997. Principles of Colloid and Surface Chemistry, 3rd ed., Marcel Dekker Inc., New York, NY.

Hiestand EN, 1964. Theory of coarse suspension formulation. J Pharm Sci, 53: 1–18.

Illing A and Unruh T, 2004. Investigation on the flow behavior of dispersions of solid triglyceride nanoparticles. Int J Pharm, 284: 123–131.

Kayes JB, 1977. Pharmaceutical suspensions: relation between zeta potential, sedimentation volume and suspension stability. J Pharmacy and Pharmacology, 29: 199–204.

Kellaway IW and Najib NM, 1981. Hydrophilic polymers as stabilizers and flocculants of sulphadimidine suspensions. Int J Pharm, 9: 59–66.

Kissa E, 1999. Dispersion: Characterizations Testing, and Measurement, 1st ed., Marcel Dekker Inc., New York, NY.

Krieger IM, Dougherty TJ, 1959. A mechanism for non-Newtonian flow in suspensions of rigid spheres. Trans Soc Rheol, 3: 137–152.

Kuhnert-Brandstätter M, 1977. Polymorphe and pseudopolymorphe kristallformen von steroidhormonen. Pharm Ind, 39: 377–383.

Longer M. 2006. Rheology, In: Martin's Physical Pharmacy and Pharmaceutical Sciences, Ed., Sinko, PJ, 5th ed., Lippincott Williams & Wilkins, Philadelphia, PA, pp 561–583.

Luhtala S, Kahela P, Kristoffersson E, 1990. Effect of benzalkonium chloride on crystal growth and aqueous solubility of carbamazapine. Acta Pharm Fenn, 99: 59–67.

Luhtala S, 1992. Effect of sodium lauryl sulfate and polysorbate 80 on crystal growth and aqueous solubility of carbamazepine. Acta Pharm Nord, 4: 85–90.

Martin AN, 1961. Physical chemical approach to the formulation of pharmaceutical suspensions. J Pharm Sci, 50: 513–517.

Martin AN, 2006. Coarse Dispersions, In: Martin's Physical Pharmacy and Pharmaceutical Sciences, Ed., Sinko, PJ, 5th Edition, Lippincott Williams & Wilkins, Philadelphia, PA, pp 499–530.

Matthews BA and Rhodes CT, 1970. Use of the Derjaguin, Landau, Verwey, and Overbeek theory to interpret pharmaceutical suspension stability. J Pharm Sci, 59: 521–525.

Minko T, 2006a. Colloids, In: Martin's Physical Pharmacy and Pharmaceutical Sciences, Ed., Sinko, PJ, 5th ed., Lippincott Williams & Wilkins, Philadelphia, PA, pp 469–498.

Minko T, 2006b. Interfacial Phenomena, In: Martin's Physical Pharmacy and Pharmaceutical Sciences, Ed., Sinko, PJ, 5th ed., Lippincott Williams & Wilkins, Philadelphia, PA, pp 437–467.

Napper DH, 1967. Modern theories of colloid stability. Science Progress, 55: 91–109.

Nutan MTH, Vaithiyalingam SR, Khan MA, 2007. Controlled release multiparticulate beads coated with starch acetate: material characterization, and identification of critical formulation and process variables. Pharm Dev Tech, 12: 307–320.

Raghavan SR, Hou J, Baker GL, Khan SA, 2000. Colloidal interactions between particles with tethered nonpolar chains dispersed in polar media: direct correlation between dynamic rheology and interaction parameters. Langmuir, 16: 1066–1077.

Sabra K, Deasy PB, 1983. Rheological and sedimentation studies on instant Clearjel® and Primojel® suspensions. J Pharm Pharmacol, 35: 275–278.

Sherman P, 1983. Rheological properties of emulsions, In: Encyclopedia of Emulsions Technology, vol 1: Basic Theory, Ed., Becher P, Marcel Dekker, New York, NY, pp 215–248.

Song S, Peng C, 2005. Thickness of salvation layers on nano-scale silica dispersed in water and ethanol. J Dispersion Sci Tech, 26: 197–201.

Tabibi SE and Rhodes CT, 1996. Disperse Systems, In: Modern Pharmaceutics, Eds., Banker GS, Rhodes CT, 3rd ed., Marcel Dekker Inc., New York, NY, pp 299–331.

Thomas DG, 1965. Transport properties of suspension: VII. A note on the viscosity of Newtonian suspensions of uniform spherical particles. J Colloid Sci, 20: 267–277.

Vaithiyalingam S, Nutan M, Reddy I, Khan M, 2002. Preparation and characterization of a customized cellulose acetate butyrate dispersion for controlled drug delivery. J Pharm Sci, 91: 1512–1522.

Verwey EJW and Overbeek JTG, 1948. Theory of the Stability of Lyophobic Colloids, Elsevier Inc., Amsterdam, The Netherlands.

Zietsman S, Kilian G, Worthington M, Stubbs C, 2007. Formulation development and stability studies of aqueous metronidazole benzoate suspensions containing carious suspending agents. Drug Dev Ind Pharm, 33: 19–197.

Chapter 3
Commonly Used Excipients
in Pharmaceutical Suspensions

R. Christian Moreton

Abstract The suspension must be physically stable (no appreciable settling) for a sufficient time, chemically stable over the required time (shelf-life), possess a viscosity that allows it to be used for its intended purpose, be easily reconstituted (redispersible) upon shaking, easy to manufacture and be acceptable in use to the patient, care-giver or other user. This chapter deals in-depth, with the role and selection of commonly used excipients in developing stable pharmaceutical suspension dosage forms.

3.1   Introduction

An important consideration in any treatment regime is to ensure that the patient receives the correct dose of medicine. For many patients and many drugs there is an acceptable dose window that allows fixed-dose medicines to be used to treat patients with a wide range of body weights without the need to precisely adjust the dose. However, there are other groups of patients where the "fixed-unit-dose" model may not be appropriate, depending on the drug's therapeutic index and pharmacokinetics, e.g. pediatric patients, geriatric patients, patients with severe renal insufficiency and patients with severe hepatic insufficiency. Oral solid unit dose forms, e.g. tablets and capsules, are not convenient under such circumstances since they are fixed strength unit dose forms. In contrast, oral liquid dose forms do have the in-built flexibility that allows the dose to be tailored to the patients' needs.

Where the drug is sufficiently soluble, a solution dosage form, e.g. a simple mixture, may be used. But not all drugs are sufficiently soluble to allow suitable strength solution medicines to be developed and manufactured with an acceptable shelf-life. In such cases, an alternative approach could be to develop a stable aqueous suspension that will allow consistent dosing of the patient. Pharmaceutical suspensions have several advantages and disadvantages when compared to other dosage forms. Since suspensions are liquids, dose adjustment for patients with renal or hepatic impairment,

R.C. Moreton

FinnBrit Consulting, Waltham, MA 02452-8043, USA

e-mail: TheMoretons@usa.net

A.K. Kulshreshtha et al. (eds.), *Pharmaceutical Suspensions: From Formulation Development to Manufacturing*, DOI 10.1007/978-1-4419-1087-5_3,
© AAPS 2010

or for pediatric or geriatric patients, may be more straightforward. This is an oversimplification of the development of a dosing strategy for a drug candidate. There are many other details that must be considered for a formulation development project to be successful, but it does provide a simple overview of some of the issues.

The presence of water can be detrimental to the stability of the active pharmaceutical ingredient (API). In practice, reconstitutable suspension formulations may be a suitable approach for labile APIs, e.g. β-lactam antibiotic suspensions. Certain APIs may be partially soluble in the aqueous vehicle and this can lead to dissolution and recrystallization due to changes in temperature, which can in turn lead to increases in particle size of the suspended particles ("Ostwald" ripening), physically less stable suspensions, and changes in bioavailability. There are ways to counter such changes and maintain the efficacy of the suspension through a knowledge and understanding of the properties of different excipients.

Aqueous suspensions for the oral delivery of medicines are probably the most obvious application of suspension technology in pharmaceutical development and manufacturing, but there are others. Suspensions have been or are used in the following pharmaceutical products and processes:

- Oral medicines
- Parenteral products
- Topical products, e.g. high protection-factor sunscreens
- Metered-dose aerosol inhalation products
- Suppositories
- Film-coating of tablets
- Sugar-coating of tablets
- Manufacture of hard gelatin capsule shells and fills
- Manufacture of soft gelatin capsule fills and shells
- Granulating suspensions ("slurries") used in the wet granulation of powders for granule, tablet or capsule manufacture

Some of these systems are aqueous, some are non-aqueous. Regardless of the nature of the continuous phase, some of the same principles of formulation and stabilization apply. The suspension must be physically stable (no appreciable settling) for a sufficient time, chemically stable over the required time (shelf-life), possess a viscosity that allows it to be used for its intended purpose, be easily reconstituted by shaking, and be acceptable in use to the patient, care-giver or other user.

Suspensions are also used in the allied fields of cosmetics, veterinary medicine, pest control, including domestic, industrial and agricultural applications, and other industrial uses outside the scope of this book.

3.1.1 Definitions

Suspensions are part of a general class of systems known as "disperse systems"; these are discussed more fully in Chapter 1. However, it is appropriate to remind ourselves of the definition of a suspension and to explain certain other terms.

A disperse system comprises two phases: the disperse phase and the continuous phase (also known as the dispersion phase or vehicle). The disperse phase is said to be suspended in the continuous phase. For a suspension, the disperse phase is a finely divided solid, and the continuous phase is a liquid under the conditions of assessment (temperature and pressure). Suspensions are not the same as, and should not be confused with, colloidal disperse systems. The difference is in the particle size of the disperse phase. Colloidal sols have a particle diameter below about 1 μm (some authorities suggest below 0.1 μm). So-called "coarse suspensions" have particle diameters larger than 1 μm. There is also an effective upper limit for the size of suspension particles which will be influenced by the viscosity of the suspension, the method of stabilization, and formulation of the suspension. If the particles are too large and too heavy, they will not stay suspended for long enough to be useful.

Other disperse systems include: emulsions (liquid in liquid), foams (gas in liquid), liquid aerosols (liquid in gas), and solid aerosols (solid in gas). In all these cases, including suspensions, the disperse phase is finely divided and dispersed through the continuous phase.

3.1.2 Summary of the Theory of Suspensions

In order to understand the excipients and the reasons they are included in pharmaceutical suspension formulations, it is important to understand how suspensions are made and how they are stabilized. The theories of suspensions and their stabilization are covered in more detail in Chapter 2. A brief summary is included here for completeness.

Most suspensions, other than colloidal sols, are inherently unstable. Over time, in the absence of agitation, the solid disperse phase will tend settle out to form a sediment layer at the bottom of the container. Depending on the nature of the particles of the disperse phase, this sediment may be easy to resuspend, or very difficult. There are ways to address the latter issue. But first it is necessary to review some of the theories relating to suspension stability.

3.1.2.1 Stokes Equation

The settling of suspended particles on standing is described by the Stokes equation which can be written as follows:

$$v = \frac{d^2(\rho_s - \rho_0)g}{18\eta_0} \tag{3.1}$$

where v is the terminal velocity of the particle, d is the diameter of the particle, ρ_s is the density of the disperse phase, ρ_0 is the density of the continuous phase (dispersion medium), g is the acceleration due to gravity, and η_o is the dynamic viscosity of the continuous phase.

Since g is a constant, we thus have three opportunities to modify the settling velocity of the particle: by making the particles smaller, reducing the difference in density between the dispersed phase and the continuous phase, and by increasing the viscosity of the continuous phase.

The Stokes equation properly applies to dilute suspensions where the particles do not interfere with each other and are thus free to settle. In most pharmaceutical suspensions, the concentration of particles in suspension is such that we have hindered settling, i.e. the particles interfere with each other during settling. The changed sedimentation velocity is given by

$$v' = v\varepsilon^n \tag{3.2}$$

where v' is the hindered sedimentation velocity, v is the sedimentation velocity from the Stokes equation (3.1), ε is the porosity of the system, n is the measure of hindering.

3.1.2.2 DLVO Theory

The DLVO (Derjaguin-Landau-Verwey-Overbeek) Theory applies to colloidal systems, and is not wholly applicable to "coarse" suspensions. However, it does give some insight as to how certain suspension stabilizers function, and is included for completeness. The theory was developed independently by Derjaguin and Landau, and by Verwey and Overbeek in the 1940s.

The theory is based on the premise that the forces acting on dispersed lyophobic (solvent-hating) colloidal particles are due to electrostatic repulsion and London-type van der Waals attraction forces. As two particles in suspension approach each other, eventually the repulsive forces will become apparent. As the particles are forced closer and closer together, the repulsion will reach the primary maximum and then, if the particles continue to move closer, the attractive forces will predominate and they will move into the primary minimum and coagulate. Provided the particles are maintained in such a manner that they do not approach each other more closely than the primary maximum, the particles should remain suspended. If the primary maximum is large compared to the thermal energy of the particles, the colloidal dispersion should be stable and the particles remain dispersed. The height of the primary maximum, the energy barrier to coagulation, depends on the zeta potential of the particles and the electrolyte concentration in the continuous phase.

Lyophilic colloids are stabilized by a combination of interactions of the electrical double layer and solvation. Both must be weakened for coagulation to occur. Hydrophilic colloids are unaffected by low concentrations of electrolyte, but "salting out" can occur at high concentrations of strongly hydrated ions.

3.1.2.3   Steric Stabilization

Steric stabilization is also known as stabilization by protective colloids. Non-ionic polymeric materials that are adsorbed onto the surface of the particles can stabilize a lyophobic sol. As two particles with adsorbed polymer approach each other, there will be a steric interaction as the adsorbed layers interact, which leads to repulsion. In general the particles do not approach closer than twice the thickness of the adsorbed polymer layer. The effect can be further classified as either enthalpic stabilization or entropic stabilization.

3.1.2.4   Flocculated vs. Deflocculated Systems

As we have already discussed, particle size reduction is a way to improve the physical stability of a suspension, and to reduce the rate of settling. Unfortunately, there is often a disadvantage in having very small (but not colloidal) particles. Although they settle out more slowly, they are typically much more difficult to resuspend. In some cases, resuspension by shaking the bottle is not possible. Large particles in contrast, although they settle out more quickly, are easier to resuspend.

Many disperse phases used in pharmaceutical suspensions are very fine powders which, if allowed to settle out, would be very difficult to resuspend, and awkward for the patient, care-giver or other user. Such disperse phases are said to be "deflocculated." It is possible using formulation techniques to cause the fine particles to form larger agglomerates, also referred to as "flocs" or "flocculates." Such a suspension is said to be "flocculated." These agglomerates comprise particles of disperse phase with continuous phase trapped between the particles. They are therefore less dense than the individual particles, which helps with suspension stability. Truly flocculated suspensions are typically very easy to resuspend, but there may be problems with rapid settling such that it is not possible to obtain an accurate dose, no matter how hard and for how long the suspension is shaken. However, there does seem to be a compromise in that it is possible to obtain a partially deflocculated system, also known as controlled flocculation.

Typically, the changes between flocculated, partially deflocculated and deflocculated suspensions are mediated through changes in pH and the presence of ions, and thus particle charge and particle–particle interactions. An example of such a system is Calamine Lotion of the British Pharmacopoeia where the inclusion of sodium acid citrate creates a partially deflocculated suspension that has adequate suspension stability and yet is easy to resuspend by simply shaking the bottle.

3.1.3   Summary of Factors to be Considered in Formulating Suspensions

The formulation of suspensions is discussed in detail in Chapter 2. In formulating a pharmaceutical suspension several factors must be considered, including:

- Dose of the active drug (API)
- Route of administration
- Patient population
- Physicochemical characteristics of the API (e.g. solubility)
- Biopharmaceutical properties of the API
- The physical and chemical properties of the other components of the formulation

3.1.3.1 Active Pharmaceutical Ingredient

Ideally, the active pharmaceutical ingredient (API) should be insoluble in the continuous phase. If this is the case, and the API is physically and chemically stable in the presence of water, the formulation of a suspension should be straightforward. However, many APIs are sufficiently soluble in the continuous phase for solubility to be a problem. The problem is a consequence of storage temperature variations which can lead to super-saturation and crystal growth (Ostwald ripening). This phenomenon can be counteracted by the use of crystallization inhibitors such as povidone and other polymers (see below). The specific inhibitor for a particular application will depend on the API and the other components of the formulation.

3.1.3.2 Particle Size of Suspended Solids

From the Stokes equation (3.1), it is clear that particle size of the dispersed phase is important. There will be a balance between particle size distribution, viscosity of the continuous phase and the difference in density between the form of the dispersed phase and that of the continuous phase. In general terms, it is desirable to formulate for a partially deflocculated suspension. Under such circumstances, the particle size (d_{90}) of the dispersed phase (API) should be less than 10 μm. With such small particles, the formation of stable floccules is easier.

3.1.3.3 Relative Proportions of Continuous and Dispersed Phases

The appearance of the final product is always a factor. Patients require pharmaceutically elegant products. With medium to high dose drugs, the API is often present in sufficient quantity to form an acceptable, elegant suspension. However, for low dose APIs, the concentration of the API in the suspension is low and may be insufficient to produce a pharmaceutically elegant suspension. In such cases the number of particles in suspension may be increased by adding a suitable excipient in a finely divided form to provide sufficient disperse phase to achieve an acceptable, elegant suspension. Examples of such excipients include finely divided microcrystalline cellulose, added either co-processed with carmellose sodium (so-called "colloidal" grade microcrystalline cellulose) or as a fine milled grade, or very fine crospovidone.

3.1.3.4 Charge Stabilization

Maximum stability of suspensions is observed when there is a balance of negative and positive charges on the particle, i.e. around zero net charge. This corresponds to the maximum sedimentation volume of the disperse phase and coincides with flocculation. Surface charge on the particles will be influenced by system pH.

3.1.3.5 Adsorption of Polymers

Certain water-soluble, hydrophilic colloids can adsorb onto the surface of the suspended particles. The presence of the polymer adsorbed on the surface of the particle, creates a steric effect by preventing the individual particles from getting sufficiently close to each other so that they are prevented from getting to the primary minimum (from the DLVO Theory), and thus coagulate/aggregate and settle out as a defloccu-lated sediment that is difficult to re-disperse.

The particle/continuum boundary is an interface, and is thus possessed of an interfacial energy. Since the driving force for the adsorption of these polymers would be a reduction in interfacial energy, it follows that the polymers that do adsorb onto the surface of the particles, must be able to bridge the energy gap in some way. Thus we require polymers that have both hydrophilic and lipophilic character, i.e. amphiphilic, e.g. poloxamers. Surfactants are also possessed of this amphiphilic character, and adsorb at interfaces.

3.1.3.6 Viscosity of the Continuous Phase

From the Stokes equation (3.1) we can see that an increase in viscosity can mark-edly reduce settling. It is possible to prevent settling for all practical purposes simply by increasing the viscosity. However, for a pharmaceutical suspension there is a further important requirement that the suspension be sufficiently mobile to be useful, e.g. pourable into a measuring cup or spoon. Frequently, simply using increased viscosity alone, while reducing settling of the suspension, does not meet the requirements for its intended use. Nevertheless, the use of suspending agents to increase viscosity is an important part of the approach to the formulation of phar-maceutical suspensions, but probably will be only a part of the overall approach.

3.1.3.7 Viscosity of the Suspension

For many pharmaceutical suspensions, the loading of the disperse phase is such that the viscosity of the formulation is mostly controlled by the addition of viscosity-increasing agents (suspending agents). However, for certain formulations, notably with very high loading of disperse phase, the suspension is more viscous than can be accounted for by the added suspending agent. In simple terms, the relative proportions of the disperse and continuous phases are such that there is insufficient continuous phase to keep the particle–particle interactions of the disperse phase below a threshold level and contribute only minimally to the overall viscosity of the system. This type of viscosity increase can prevent the effective use of the suspension, e.g. it will not pour out of the bottle. Under such circumstances, the addition of a protective colloid with low viscosity in solution may reduce the viscosity of the suspension to within an acceptable range. If this approach is not feasible or does not work, there may be no alternative other than to reduce the concentration of disperse phase in the suspension.

3.1.3.8 Effects of Other Components; Interactions in Aqueous Solution/Suspension

It is important to remember when developing a pharmaceutical suspension formulation that all the components have the potential to interact and thus affect both the chemical and physical stability of the suspension. The presence of the continuous phase ensures that all components are in solution or suspension, and therefore in more intimate contact than is possible in certain other types of dosage form such as powders, granules, tablets and capsules. Many of the additional components of the formulation are polar or ionizable compounds, or contain such compounds (e.g. flavors), and they thus have the potential to interact with the disperse phase, as well as with the other components of the continuous phase.

Thus when formulating suspensions, as with other dosage forms, a "holistic" approach is needed. The whole range of properties of each component must be considered, since each component has the potential to affect the overall stability of the formulation.

3.2 Types of Excipients Used in the Formulation of Pharmaceutical Suspensions

A review of the suspension formulations included in published lists of pharmaceutical formulations (Niazi 2004a, b) provided the list of excipients and frequencies of use shown in Table 3.1. These can be further classified as shown in Table 3.2. Not all these categories will be discussed in detail. The emphasis will be on those categories of excipient linked to the physical stability of the suspension. For example, colors and flavors are used to improve patient acceptance and will not be discussed in such detail; similarly for antioxidants, preservatives and sweeteners. Nevertheless, it is important to remember that even the so-called "minor components," such as colors and flavors, can have an effect on suspension stability. There may also be potential incompatibilities.

Table 3.1 Excipients used in suspension formulations (from Niazi 2004a, b)

Excipient	Total	Excipient	Total	Excipient	Total
Acetic acid (glacial)	2	Ethanol	10	Preservative (unspecified)	12
Acetone	1	Flavor	62	Propylene glycol	16
Aerosil 200	10	Fumaric acid	1	Propyl paraben	28
Ammonia	6	Glycerol	19	Simethicone	11
Antioxidant (unspecified)	1	Hydrochloric acid	3	Sodium acetate	2
Ascorbic acid	1	Hydroxyethyl cellulose	1	Sodium benzoate	9
Aspartame	3	Hydroxypropyl cellulose	1	Sodium bisulfite	1
Bentonite	1	Hydroxypropyl methylcellulose	1	Sodium chloride	5
Benzoic acid	4	Kaolin	2	Sodium citrate	16
Benzyl alcohol	3	Lecithin	1	Sodium gluconate	1
Calcium cyclamate	2	Magnesium hydroxide	1	Sodium hydroxide	6
Calcium hydroxide	1	Maltodextrin	1	Sodium hypochlorite	6
Calcium saccharinate	2	Mannitol	2	Sodium lauryl sulfate	2
Carbopols	4	Methyl cellulose	6	Sodium phosphate	5
Carrageenan	1	Methyl paraben	35	Sodium saccharin	38
Carmellose sodium	13	Microcrystalline cellulose	2	Sorbic acid	2
Castor oil	2	Modified starch	1	Sorbitol	42
Cellulose	1	Pectin	4	Sucrose	34
Citric acid	28	Poloxamers	8	Tetrasodium edetate	1
Cocamide DEA	1	Polysorbate 80	17	Thimerosal	1
Cocamide betaine	1	Polyvinyl alcohol	1	Titanium dioxide	2
Colloidal microcrystalline cellulose	8	Potassium acetate	2	Tragacanth	1
Colors	16	Potassium citrate	2	Veegum	11
Cremophor RH40	9	Potassium metabisulfite	2	Water	86
Crospovidone M	19	Potassium sorbate	3	Xanthan gum	22
Dextrose	1	Povidone	17		
Disodium edetate	7	Pregelatinized starch	1		

Some materials may possess a combination of properties useful in the formulation and manufacture of stable, elegant pharmaceutical suspensions. Under such circumstances, the material may be discussed in more than one of the following sections. As formulation scientists we need to consider the totality of properties possessed by a particular excipient. Even though it is being added for one particular characteristic, the other properties will still be present, and will still influence

Table 3.2 Classification of excipients used in the formulation of pharmaceutical suspensions (adapted from Niazi 2004a, b)

Category	Number of excipients listed	Category	Number of excipients listed
pH adjustment/ buffering agents	15	Soluble hydrocolloid polymers	11
Solvent/co-solvent	3	Non-ionic surfactants	6
Inorganic solids	7	Ionic surfactants	1
Inorganic clays	3	Sequestrants	2
Antioxidants	4	Oils	2
Preservatives	8	(Flavors)	Not categorized
Sweetening agents	6	(Colors)	Not categorized
Polyols	2	(Water)	Not categorized

the formulation. We must also consider each component in relation to all the other components in the formulation, since there is always potential for chemical and/or physical interaction, i.e. the "holistic" approach to pharmaceutical formulation noted above.

3.2.1 Inorganic Clays

The inorganic clays are typically complex hydrated silicates and include smectites (formerly known as montmorillonites; e.g. bentonite, saponite and magnesium aluminum silicate; magnesium aluminum silicate has also been described as a mixture of montmorillonite and saponite; the principal component of bentonite is hectorite), kaolinites (e.g. kaolin) and palygorskite (attapulgite). These are naturally occurring minerals and are mined and purified for use in pharmaceuticals. They can be described as polymeric complexes of hydrated silicates. The organization of the different types of silicate is in the form of small platelets bound together through ionic interactions with cations.

The clays hydrate in water and form gels or sols depending on the concentration. After hydration is complete and the small platelets have separated, they form chains and networks of platelets linked through the difference in charge, between the positive edges of the platelets and their negative faces. This linked platelet structure traps the continuous and disperse phases, and slows sedimentation of the disperse phase through hindered settling. These clays do not form gels in the absence of water.

As a consequence of the charge, particularly on the faces of the platelets, these clays are able to adsorb a range of compounds from solution, including drugs such as warfarin sodium, and cationic preservatives. This property has been used in the formulation of modified release products, but also can lead to loss of potency if suspensions containing such clays are administered concomitantly with these drugs.

Kaolin and attapulgite will both form gels in water. However, traditionally, both kaolin and attapulgite have been used in medicine as antidotes to food poisoning and diarrhoea, using their capacity to adsorb potential irritants and toxins. Kaolin is not much used now and is probably considered old-fashioned. The primary product has been reformulated, and kaolin is replaced with bismuth subgallate. Attapulgite is still used as an anti-diarrhoea medicine, but in this case a heat activated form (more adsorptive capacity) is used.

Both bentonite and magnesium aluminum silicate are used in the formulation of suspensions. They are both complex hydrated silicates and comprise aluminum and magnesium silicates together with other silicates. The differences between the two relate to the relative proportions of aluminum, magnesium and other cations, with bentonite containing comparatively higher levels of aluminum and iron, and magnesium aluminum silicate containing higher levels of magnesium and calcium.

Both materials are used in the range 0.5–10%, but more typically in the range 1–5% in the formulation of aqueous suspensions. They are generally regarded as inert, but their suspension forming properties are influenced by pH and ionic strength. High concentrations of ions can suppress the hydration and the subsequent formation of the platelet network. Magnesium aluminum silicate is unsuitable for use below about pH 3.5 and bentonite below about pH 6.

A key operation in the use of these clay materials in the formulation or preparation of suspensions is their hydration. During hydration, water is adsorbed between the platelets and displaces the cations allowing the platelets to separate and ultimately form the gel network. This process is not instantaneous and due allowance must be made to achieve complete hydration. The hydration time can be reduced by using mechanical energy, e.g. high shear, and heat. As a general rule for the best chance of timely hydration, the clay is added slowly to the water rather than adding water to the clay.

The inorganic clays, being very finely divided with differential charges between the edges and the faces of the hydrated platelets, have the potential to adsorb other components present in the disperse phase, including preservatives. The effective preservative concentration may therefore be reduced and an overage may be required or selection of a preservative with a reduced adsorption propensity. They can also adsorb surfactants and flavor components, either of which may adversely affect the acceptability of the product.

3.2.2 Water-Soluble Hydrocolloids

There are many water soluble hydrocolloids that can act as suspending agents in the formulation of pharmaceutical suspensions. They can be of natural, semi-synthetic or synthetic origin. A list compiled from an examination of Niazi (2004a, b); Rowe et al. (2006), and the USP 30/NF 25, USP and NF Excipients, listed by category is given in Table 3.3. The list is not intended to be all-inclusive. There are other water-soluble hydrocolloids available, but not all are appropriate for the formulation of

Table 3.3 Water-soluble hydrophilic colloids for use as suspending agents (compiled from Niazi 2004a, b; Rowe et al. 2006; and USP 30/NF 25 2007, USP and NF Excipients, listed by category)

Acacia	Hydroxyethyl cellulose	Polyethylene glycol
Agar	Hydroxypropyl cellulose	Poly (vinyl alcohol)
Alginic acid	Hydroxypropyl starch	Potassium alginate
Carbomer	Hypromellose	Povidone
Carmellose sodium	Maltodextrin	Pregelatinized starch
Carrageenan	Methylcellulose	Propylene glycol alginate
Dextrin	Modified starch	Sodium alginate
Gelatin	Pectin	Tragacanth
Gellan gum	Poloxamer	Xanthan gum
Guar gum	Polycarbophil	

suspension products. Water soluble hydrocolloids are used in applications other than suspensions, and other properties they may possess include adhesiveness, bioadhesiveness, and film formation when the solution is allowed to dry under the correct conditions. As a consequence they may also be used as granulating agents, film-forming agents in the coating of tablets, in the formulation of modified release products and in the formulation of bioadhesive drug delivery systems.

In suspensions, water-soluble hydrophilic colloids work by increasing the viscosity of the aqueous continuous phase, and thereby hindering the sedimentation of the disperse phase. Some hydrophilic colloids are capable of forming true gels whereby the network of polymer fibrils is cross-linked through either ionic bonds or hydrogen bonds. These polymers are used at comparatively low levels of incorporation, where the polymer chains are sufficiently far apart that the cross-linking is insufficient to form a viscous gel. In some cases, e.g. gelatin, partial hydrolysis of the polymer chains can also prevent cross-linking because the polymer chains are not long enough to accommodate two cross-links. Hydrolyzed gelatins are available commercially, and they do not gel or "set" on cooling, but they can increase the viscosity of the continuous phase. At higher levels of incorporation, the non-hydrolyzed polymers tend to form rigid, non-pourable gels unsuited to the formulation of suspensions.

Other materials increase the viscosity in aqueous solution via the simple entanglement of adjacent polymer chains, but with-out the cross-linking found in the true gels. Such materials have been said to form a "viscolized" matrix (hydrolyzed gelatins probably fall into this category). They are still capable of forming very viscous systems at high concentration or molecular weight. These materials appear to be more frequently used as suspending agents for pharmaceuticals. This preference may be due to the need to avoid inadvertent formation of a cross-linked gel which would not be pourable.

Polymer chain length is a significant factor in the development of viscosity with all polymers, with the higher molecular weight grades forming more viscous systems which may not be suitable for the formulation of suspensions. These more viscous or more cross-linked systems are used in the formulation of matrix controlled release products for example. In practice, there is an upper limit for the level of incorporation of any polymer that will allow the formulation of a pourable suspension. This limit will depend on the type of polymer, its mode of action, molecular weight, etc.

For large scale manufacture of pharmaceutical suspensions, there are other factors which may be important, such as the hydration of the components, particularly the polymer. Some polymers hydrate more easily and more quickly than others, and this may be an important consideration as time in the equipment must be factored in to production scheduling. For example, hypromellose is easier to hydrate than methylcellulose. In general terms, lower viscosity grades hydrate more easily than high viscosity grades.

The final choice of which polymer to select will probably be based on a combination of chemical compatibility (absence of degradation), physical compatibility (customer or patient acceptability) and eventually personal choice and experience of the formulation scientist, or their colleagues.

Some of the excipients listed in Table 3.3, while they may be expected to increase viscosity in aqueous solution, also work in other ways to help stabilize suspensions. For example, povidone will adsorb onto suspended particles. This in turn may stabilize the suspension through the steric effect, and the povidone is acting as a protective colloid. In addition, povidone can reduce the viscosity of suspensions with high loadings of suspended particles, and this has been used successfully in spray applications. However, the interplay of these different properties may be expected to vary with molecular weight; the higher molecular weight grades may have a different balance of properties compared to lower molecular weight grades.

Poloxamers also increase the viscosity of an aqueous solution, but will act as protective colloids in the presence of suspended particles in an aqueous continuous phase.

There are simply too many water-soluble, hydrophilic colloids to give detailed information on levels of incorporation. In general terms, low molecular weight polymers will require a higher level of incorporation than high molecular weight grades. However, to achieve a pourable suspension, there will be an upper limiting concentration for a particular polymer grade. It may be that a combination of different viscosity grades will provide the optimum balance between suspension stability and pourability.

Many hydrocolloids can potentially interact with other components of the formulation. Polymers such as the cellulosics and povidone can interact with preservatives to reduce preservative efficacy. Certain hydrocolloids are ionic, and they therefore have the potential to interact with components having opposite charge. For example, cationic preservatives will interact with alginates and xanthan gum since they have opposite charge, thereby reducing the effective preservative concentration in the formulation.

3.2.3 Bulking Agents (Auxiliary Suspending Agents)

In some instances, there are insufficient drug particles in a unit dose of suspension to make a pharmaceutically elegant suspension. This is particularly true for the more highly active drugs, where the unit dose is small. Under such circumstances, the formulator will need to add more particles to improve the appearance of the final product, and also to help stabilize the suspension.

There are many potential materials that could be used, but as in any formulation project there are constraints. The excipient must be compatible with the API. In most instances, the final marketed product will require a shelf-life in excess of 2 years from the date of manufacture, and the excipient should not cause the API to degrade. The excipient should not be soluble in the continuous phase, or to be more precise, it should not undergo Ostwald ripening during storage. The excipient should also not reduce the bioavailability of the drug. It is possible for certain drugs to be adsorbed onto excipient particles during storage; however, once in the gastro-intestinal tract, any such adsorption should not interfere with the timely absorption of the drug, unless modified release is specifically required. The other constraints reflect the general requirements for stable suspensions; namely, that the particles remain in suspension for a sufficient length of time to permit accurate dose measurement, and are easy to resuspend after settling, or that they remain in suspension indefinitely.

The different excipients that may be used include inorganic solids, insoluble cellulose derivatives and insoluble synthetic polymers. Examples of auxiliary suspending agents are listed in Table 3.4. The more popular materials are all available as very fine particle size grades, typically prepared by some form of high energy milling, or sometimes by a combination of chemical and physical processing.

The inorganic materials are generally insoluble in water (in reality their solubility is so low that it is referred to as a solubility product, as in the case of dibasic calcium phosphate). Calcium sulfate is an exception in that it does have measurable water solubility. All the inorganic materials listed in Table 3.4 are available as very fine grades, and it is these very fine grades that are typically used in the formulation and manufacture of suspensions. Most often, the very fine particle size materials are obtained commercially by high-energy dry milling.

Microcrystalline cellulose is available as two forms for use in suspensions. It is available as so-called "colloidal" grades co-processed with carmellose sodium (RC581, RC591 and CL611 grades), and it also available as a milled/micronized grade (PH105).

The co-processed grades are manufactured by taking the microcrystalline cellulose slurry immediately after neutralization and mixing it with carmellose sodium in water; the resulting mixture is then spray-dried. The presence of carmellose

Table 3.4 Bulking agents/auxiliary suspending agents (compiled from Niazi 2004a, b; Rowe et al. 2006; and USP 30/NF 25 2007, USP and NF Excipients, listed by category)

Calcium carbonate	Magnesium carbonate
Calcium hydroxide	Magnesium hydroxide
Cellulose	Microcrystalline cellulose
Crospovidone	Silica (silicon dioxide)
Dibasic calcium phosphate	Titanium dioxide

sodium has several advantages. It protects the cellulose microcrystallites during drying, since aggregation during drying can lead to a change in structure of the microcrystalline cellulose (termed "hornification" in the pulp/paper industries) which makes dispersion of the microcrystallites more difficult. Carmellose sodium also aids dispersion of the microcrystallites on subsequent addition to water as part of the manufacturing process. In addition, it is a suspending agent which will help stabilize the suspension during and after manufacture.

The PH105 grade of microcrystalline cellulose is micron-sized and produced by high energy milling in the dry state. However, dispersion of previously dried grades of microcrystalline cellulose requires extended high shear mixing, and the amount of shear required will depend, in part, on the particle size of the dried material, with larger particle size material taking longer to disperse. Re-dispersed, previously dried microcrystalline cellulose usually does not achieve the same degree of dispersion that is available using the colloidal grades, even after extended high shear wet mixing.

Crospovidone is a synthetic, hydrophilic, water-insoluble polymer which is primarily used as a tablet disintegrant. It is cross-linked povidone prepared by a "popcorn" polymerization process. On dispersion in an aqueous medium, water is taken up into the particle structure; the larger the particle, the greater the water uptake on immersion. The grades used in the formulation and manufacture of suspensions are the micronized forms. The very fine particle size appears to minimize the amount of water taken up by the particles.

The level of incorporation required for any auxiliary suspending agent will depend on the nature and composition of the suspension as a whole, the particle size of the components, the required viscosity and the means of administration. The RC/CL grades of microcrystalline cellulose are typically used in the range 0.5–2.5%, the micronized crospovidone can be used up to about 20% of the suspension, similarly for the inorganic materials. These levels of incorporation are guidelines, a place to start on a formulation project. They can be exceeded, and there are products in the market that contain these materials at higher levels of incorporation.

As with other components in the formulation, auxiliary suspending agents have the potential to interact with components from the continuous phase. For example, in water, microcrystalline cellulose can adsorb primary amines. This phenomenon is suppressed in the presence of 0.05 M of a soluble electrolyte. The adsorption is due to the presence of acidic moieties on the microcrystalline cellulose. During manufacture, they are neutralized with ammonia, but in aqueous suspension, and in the absence of electrolyte, the ammonia residues on the microcrystalline cellulose can be displaced by primary amines (Steele et al. 2003).

3.2.4 Polymeric Adsorption Stabilizers

Many different materials are capable of adsorbing onto the suspended particles, e.g. natural gums, cellulosics and non-ionic surfactants. However, not all of them are able to act as protective colloids and provide steric hindrance to caking at a suffi-ciently low concentration. High levels of surfactants, for example, can increase gastro-intestinal motility. Higher molecular weight gums and cellulosics may also cause an unacceptable increase in the viscosity of the system. There are, however, certain polymers, or grades of polymers, that are capable of acting as protective colloids at concentrations that do not markedly increase the viscosity of the system, or increase gut motility, etc. Such materials include poloxamers, lower molecular weight grades of povidone, and low molecular weight grades of some other hydro-philic colloids.

3.2.4.1 Povidone

Povidone is a synthetic, water-soluble, straight-chain polymer of 1-vinyl-2-pyrroli-done. It is available as different grades based on the K-value, which loosely ties to molecular weight, and thus polymer chain length and viscosity. It is also soluble in certain more polar organic solvents. It is generally regarded as non-toxic and has been used for many years, and in many different applications. Low molecular weight, pyrogen-free grades of povidone are also available.

The high molecular weight grades do form viscous solutions in water and in a variety of polar organic solvents, and thus can be used to increase the viscosity of the continuous phase and thereby hinder settling. However, the lower molecular weight grades do not form viscous solutions. They can act as protective colloids without increasing the viscosity of the suspension. Under certain circumstances, they can reduce the viscosity of suspensions containing high levels of suspended solids. This property has been used to good effect in both spray-drying and film coating applications.

Povidone is typically used in pharmaceutical suspensions up to about 5% w/v. As little as 1% povidone may be effective in reducing the viscosity of suspensions containing high levels of suspended solids. It is compatible with a wide range of other materials in solution including inorganic salts and other polymers. Povidone will complex with some small organic molecules, including certain APIs and pre-servatives, and may reduce the efficacy of some preservatives.

3.2.4.2 Poloxamers

The structures of the poloxamers are very different to that of povidone. Poloxamers are a group of closely related linear triple block co-polymers of ethylene oxide and propylene oxide with the general structure:

$$HO(C_2H_4O)_n(C_3H_6O)_m(C_2H_4O)_nH$$

The polyoxyethylene blocks provide the hydrophilic part of the molecule, and the polyoxypropylene blocks the lipophilic part. There are several different grades available which are differentiated by a combination of the lengths of the polyoxy-ethylene and polyoxypropylene blocks and molecular weight. There are five grades generally approved for pharmaceutical use, and included in pharmacopeia mono-graphs: Poloxamers 124, 188, 237, 338 and 407. There are other grades available for use in applications other than pharmaceutical formulations. Poloxamer 124 is a liquid at room temperature, whereas all the other pharmaceutical grades are solids. All the pharmaceutical grades are soluble in water and ethanol, but some are less soluble in other solvents.

Because of their amphiphilic nature, poloxamers are used in a variety of appli-cations in pharmaceutical products, including tablets, liquids and gels. In liquid formulations, besides being used as a stabilizer for suspensions, the poloxamers are also used as emulsifying agents and solubilizers. Grade selection will depend on the nature of the application and the necessary balance between the hydro-philic and lipophilic portions of the molecule (hydrophil-lipophil balance (HLB)). The poloxamers cover a range of HLB values from about 1 to 30. A high HLB value indicates a more hydrophilic balance, a low HLB shows a more lipophilic balance. Poloxamers have some similar properties to non-ionic surface active agents. For example, in aqueous solution, poloxamers possess surface active properties; they will lower the surface tension of an aqueous solution, they will also form micelles.

Typically, in suspension formulations, poloxamers are used at a level of incor-poration of about 5% w/v depending on the intended use.

Poloxamers may be incompatible with phenols and parabens; this has been con-firmed for Poloxamer 188. The effect is presumably due to incorporation of these preservatives into the micelles, and is concentration dependent. Care should be exercised when selecting the preservative for formulations containing poloxamers. The formulations should be evaluated using a preservative efficacy test and micro-bial challenge testing. It will also be important to monitor preservative efficacy during stability testing to confirm the preservative system is effective over the anticipated shelf-life of the product.

The poloxamers are generally regarded as non-irritant, non-toxic materials. They are included in medicinal finished products in both US and European markets. Poloxamers are also used in the formulation of food and cosmetic products.

3.2.5 Surfactants/Wetting Agents

Surfactant is a general name for materials that possess surface activity; in solution they tend to orient at the surface of the liquid. There are several general classes of surfactants: anionic, cationic, amphoteric and non-ionic. Surfactants are amphiphilic molecules, i.e. part of the molecule is hydrophilic, and part is lipophilic. This com-bination of the two opposite affinities in the same molecule causes them to orient to the interface and thereby reduce the interfacial tension between the continuous and disperse phases, such as in emulsions and suspensions. Ionic surfactants work primarily through electrostatic forces, whereas non-ionic surfactants work primarily through steric forces.

3.2.5.1 Cationic Surfactants

With one major exception (see below), cationic surfactants are generally used as preservatives in pharmaceutical formulations because of their antimicrobial properties, rather than surfactants per se, e.g. benzalkonium chloride. They may also be used in medicated throat preparations, e.g. cetylpyridinium chloride. Many cationic surfactants can irritate the mucosa at higher concentrations.

There is one group of cationic surfactants that is well tolerated; the phosphatides, e.g., phosphatidyl choline. These are found in lecithin. Lecithin is a component of cell walls, and is obtained from, e.g., eggs, egg yolks, or soy beans. The lecithins obtained from different sources differ slightly in their composition, and in particular, in the relative amounts of phosphatidyl choline, phosphatidyl ethanolamine, phosphatidyl inositol and phosphatidyl serine. Lecithin has been administered to humans in quantities of several grams per dose and is well tolerated. In suspension formulations for human use, lecithin is typically used at a level of incorporation in the range 0.25–10%. Lecithin and phosphatides are also used in the preparation of liposomes.

Since this type of surfactant is ionic, there is the potential for interaction and incompatibility with other charged species in the suspension.

3.2.5.2 Anionic Surfactants

Anionic surfactants are used in pharmaceutical suspensions, e.g. sodium lauryl sulfate and docusate sodium. However, anionic surfactants can also affect the lower gastro-intestinal tract and act as laxatives above certain concentrations. They are thus restricted to uses requiring low concentrations, e.g. to aid wetting of hydrophobic surfaces, including powders, and as solubilizing agents.

Again, since they are ionic in nature, there is the potential for interaction and incompatibility with other charged species within the suspension.

3.2.5.3 Amphoteric Surfactants

Amphoteric surfactants are zwitter-ionic molecules, i.e. they have both an acidic and a basic group in the same molecule as well as having both hydrophilic and lipophilic parts in the molecule. Examples include cocamidopropyl amino betaine. They are mainly used in the development and manufacture of suspensions used as topical lotions and hair shampoos, and not for internal medicines.

Table 3.5 Types of non-ionic surfactants used in pharmaceutical suspensions (compiled from Niazi 2004a, b; Rowe et al. 2006; and USP 30/NF 25 2007, USP and NF Excipients, listed by category)

Generic name	Synonyms/*sample trade names*
Polyoxyethylene sorbitan fatty acid esters	Polysorbate, *Tween®*
Polyoxyethylene 15 hydroxystearate	Macrogol 15 hydroxystearate, *Solutol HS15®*
Polyoxyethylene castor oil derivatives	*Cremophor® EL, ELP, RH 40*
Polyoxyethylene stearates	*Myrj®*
Sorbitan fatty acid esters	*Span®*
Polyoxyethylene alkyl ethers	*Brij®*
Polyoxyethylene nonylphenol ether	*Nonoxynol®*

3.2.5.4 Non-ionic Surfactants

By far, the largest group of surfactants used in the formulation of pharmaceutical suspensions is the non-ionic surfactants. There are several different types of non-ionic surfactants available. A list of the non-ionic surfactants that have been used in the formulation of pharmaceutical suspensions, compiled from Niazi (2004a, b), Rowe et al. (2006) and the USP 30/NF 25, is given in Table 3.5. The surfactants listed, with the exception of the sorbitan esters, all contain polyoxyethylene moieties of different chain lengths as the hydrophilic component. The lipophilic components of all those listed surfactants are fatty acids or substituted fatty acids.

In addition to the non-ionic surfactants listed, certain block copolymers also have surfactant properties such as poloxamers (see above), and may be used in a similar manner.

The final choice of non-ionic surfactant will depend on a variety of factors, but chief among them will be the HLB value and their chemical compatibility with other components of the formulation. Since many of them are esters, such compounds are susceptible to hydrolysis under conditions of high or very low pH. The polyoxyethylene component brings its own particular problems. There are side reactions that can take place during the manufacture of the polyoxyethylene chains which can give rise to peroxides and/or formaldehyde, and trace levels of these materials can be found in the finished excipients. These traces of peroxides and formaldehyde can have a deleterious effect on APIs and other components that are susceptible to such interaction.

When using non-ionic surfactants in liquid formulations, the choice of preservative is important. Generally, the preservative efficacy of phenolic preservatives is compromised in the presence of non-ionic surfactants due to incorporation of the preservatives in the micelles. Alternative, non-phenolic preservatives should be selected.

3.2.6 pH Modifiers and Buffers

As has been discussed elsewhere, there is often an optimum pH range for both the physical and chemical stability of aqueous pharmaceutical suspensions. This may require modification of the pH during formulation and manufacture, because of the

Table 3.6 Buffering agents, pH modifiers and salts used in pharmaceutical suspensions (compiled from Niazi 2004a, b; Rowe et al. 2006; and USP 30/NF 25 2007, USP and NF Excipients, listed by category)

Acetic acid	Proprionic acid	Sodium citrate
Adipic acid	Potassium acetate	Sodium glycolate
Ammonium carbonate	Potassium bicarbonate	Sodium hydroxide
Ammonium hydroxide	Potassium chloride	Sodium lactate
Ammonium phosphate	Potassium citrate	Sodium phosphate
Boric acid[a]	Potassium metaphosphate	Sodium proprionate
Citric acid	Potassium phosphate	Succinic acid
Diethanolamine	Sodium acetate	Sulfuric acid
Fumaric acid	Sodium bicarbonate	Tartaric acid
Hydrochloric acid	Sodium borate[a]	Trolamine
Malic acid	Sodium carbonate	
Nitric acid	Sodium chloride	

[a]Boric acid and sodium borate are accepted for use as food additives in Europe, and they are included in the US FDA's Inactive Ingredients Database. However, they can cause toxicity in babies and young children; their use is restricted in the UK

components that render the preparation too acidic or too basic. Simply adding an acid or a base may give the required pH, but this may not hold during storage since ions may be adsorbed onto the suspended particles with a consequent drift in pH. Buffer systems are better able to tolerate slight changes in ionic strength and yet maintain the pH within the required range. Typically, the better buffer systems use salts of weaker polyvalent acids, e.g. sodium salts of citric acid, or combinations of sodium salts of citric acid with sodium salts of phosphoric acid. A list of acids, salts and other materials used as buffers, and compiled from Niazi (2004a, b), Rowe et al. (2006) and the USP 30/NF 25, is given in Table 3.6. There are other salts and acids that are not listed but could also be used. The restrictions are that the buffer component should be sufficiently soluble, be approved for use in pharmaceutical products, and be acceptably safe for use in the intended application or route of administration.

Other ionic materials may be included in suspension formulations for various purposes. These other components may influence the pH of the suspension and the buffer requirements.

3.2.7 Preservatives, Antioxidants and Chelating Agents

The term "preservative" in the context of pharmaceutical formulations usually refers to antimicrobial agents used to suppress the growth of micro-organisms in the product. However, in the broadest sense, preservation of pharmaceutical products can require protection against both microbial spoilage, using antimicrobial agents, and chemical degradation, typically using antioxidants and/or chelating agents (sequestrants). Either type of spoilage will give rise to an unacceptable, possibly harmful, product.

3.2.7.1 Antimicrobial Preservatives

Pharmaceutical suspensions with water as the continuous phase are susceptible to microbial spoilage. The source of the microbial growth may be raw materials, e.g. materials of natural origin such a clays (mined) and alginates (extracted from marine algae), process related (airborne contamination or from operators), or during use by the patient. Whilst every effort is made to keep raw material and process related contamination to an acceptable minimum, it is almost impossible to eliminate micro-organisms entirely. In addition, the risk of microbial contamination and subsequent product spoilage is much increased during use by the patient. For these reasons, pharmaceutical suspension formulations contain preservatives. The preservative system has to be effective against both bacteria, and yeasts and molds, and typically, combinations of preservatives are used.

A list of preservatives suitable for use in oral pharmaceutical suspensions and included in Rowe et al. (2006) and the USP 30/NF 25, together with their typical use levels, is given in Table 3.7. This list includes preservatives used for oral and topical medicines. In some instances, different levels may be permitted for different applications and this is also indicated in Table 3.7. There are of course a number of other antimicrobial agents that may be approved for use in other types of product. Information on these other antimicrobial agents is available from, e.g., *The Handbook of Pharmaceutical Excipients*. The mercury-based antimicrobial agents have been omitted. They are still permitted for some very specific applications, but there are concerns regarding long term safety of these agents.

The final choice of preservative will depend very much on the nature of the product and the other components present, and the nature of the micro-organisms likely to be encountered during manufacture and use.

3.2.7.2 Antioxidants

Some components of pharmaceutical suspensions are susceptible to oxidative degradation. Antioxidant molecules which themselves are preferentially oxidized can be included in the formulation to reduce the degradation of these other components. The type of antioxidant will depend on the nature of the formulation. Different antioxidants will be required for aqueous and non-aqueous formulations. A list of antioxidants that may be used in pharmaceutical products, together with indications of usage level, stability and incompatibilities, compiled from Niazi (2004a, b), Rowe et al. (2006), the USP 30/NF 25 and the FDA's Inactive Ingredient Database is given in Table 3.8. The choice of antioxidant will depend on the nature of the formulation, the API and the other excipients.

Table 3.7 Antimicrobial agents used in oral suspension products (compiled from information provided in Rowe et al. 2006, Martindale; the Complete Drug Reference 35th Edition, and the FDA's Inactive Ingredient Database)

Antimicrobial agent	Typical use levels	Comments
Alcohols		
Benzyl alcohol	Up to 5.0% (oral) Up to 3.0% (topical; 50% as gel)	Moderately effective. Most effective at pH <5. Antimicrobial activity may be compromised in the presence of non-ionic surfactants. May be absorbed by polyethylene containers
Chlorobutanol	0.3% (topical) 0.6% (ophthalmic)	Ineffective above pH 5.5. Synergies with other antimicrobial preservatives. May be adsorbed onto suspended solids, e.g. magnesium trisilicate and bentonite. Not used for oral products in the US
Ethanol	Up to 90% (oral and topical)	Below 50% v/v may be bacteriostatic; 50–90% bactericidal. May potentiate other microbial preservatives. Ineffective against bacterial spores
Phenylethyl alcohol	0.5–1% (topical)	Also used as a preservative in vaccines. Effective over a wide pH range. Commonly used in combination with other antimicrobial preservatives, e.g. parabens. Not used for oral products in the US
Phenoxyethanol	0.25–0.5% (nasal, otic and ophthalmic) Up to 1% (topical)	Optimum activity <pH 5; inactive >pH 8. Synergies reported with glycols, chlorhexidine and phenyl mercuric acetate. Less susceptible to non-ionic surfactants than, e.g. paraben preservatives. Not listed under this name in the FDA's Inactive Ingredients Database
Benzoates		Incompatible with quaternary ammonium surfactants, calcium, ferric and heavy metal salts. May be adsorbed by kaolin with resulting decrease in antimicrobial activity
Benzoic acid	0.01–0.75% (oral solution) 0.1% (oral suspension) Up to 0.25% (topical)	Also used in rectal and vaginal preparations. Optimum antimicrobial activity in range pH 2.5–4.5
Potassium benzoate	≤0.1% (food) 0.03–0.08% (carbonated beverages)	Most effective at <pH 4.5. At low pH benzoic acid formation may give rise to a slight taste in food. May be used as an alternative to sodium benzoate in low sodium preparations. Not currently listed in the FDA's Inactive Ingredient Database
Sodium benzoate	0.2–3.75% (oral) Up to 10% (i.m./i.v. injection, more typically around 0.1%) 0.1–0.5% (cosmetics)	Effective range pH 2–5. See also benzoic acid and potassium benzoate

Parabens		In general, the paraben esters are more active against Gram +ve than Gram –ve micro-organisms. More active against yeasts and molds than bacteria. Optimum pH range 4–8, but more active in acidic media. Inactive at higher pH due to formation of phenolate ion. Activity increases with alkyl chain length, but solubility decreases. Activity reduced in presence of non-ionic surfactants. Activity enhanced in presence of propylene glycol, phenylethyl alcohol and edetic acid. Sodium salts are also available with increased solubility, but may need pH buffer. The hydrolysis product, *p*-hydroxybenzoic acid, is inactive. At high concentrations paraben esters can modify taste sensation
Butyl paraben	Up to 0.8% (oral – typically lower) 0.02–0.4% (topical)	Most active and least soluble of the commonly used paraben esters. Commonly used in combination with methyl- and propylparaben
Ethyl paraben		Commonly used in combination with methyl- and propylparaben. Used in both oral and topical formulations, but no concentrations listed in the Inactive Ingredient Database. Widely used in cosmetics
Methyl paraben	0.065–0.75% (i.m./i.v. injection) 0.015–13% (oral) 0.02–18% (topical)	Often used in combination with methylparanem. Also used in inhalation, nasal, ophthalmic, rectal and vaginal preparations
Propyl paraben	0.005–20% (i.m., i.v., s.c. injection) 0.01–20% (oral) 0.01–10% (topical lotion) 30% (topical gel)	Often used in combination with methylparaben. Also used in inhalation, nasal, ophthalmic, rectal and vaginal preparations
Phenolics		Incompatible with non-ionic surfactants
Chlorocresol	Up to 0.75 (topical)	Not recommended for oral products or preparations that come into contact with mucosal tissues. More active at acid pH; inactive above pH 9. Active against Gram –ve and Gram +ve micro-organisms, spores, yeasts and molds. Synergies reported with other antimicrobial preservatives. May be absorbed into rubber closures and some plastics. Incompatible with some APIs, e.g. opiates. Not used for oral products in the US

(continued)

Table 3.7 (continued)

Antimicrobial agent	Typical use levels	Comments
Cresol	0.15–0.3% (parenteral)	Similar to phenol but more active. Active at <pH 9; best in acidic media. Synergies reported with other antimicrobial preservatives. Three isomer o-, m- and p-. m-cresol is the least toxic. Not active against bacterial spores. Not used in oral medicines in the US
Phenol	1% (i.m./i.v. injection)	Most active in acidic media. <1% – bacteriostatic; higher concentrations bactericidal. Inactivated by organic matter
Thymol	0.01% (inhalation liquid)	Approved in at least one oral product on US market. More active than phenol, but less soluble. Irritant to tissues. Has antioxidant properties. Inactivated by organic matter, e.g. proteinaceous matter, and oxidizing agents. Mainly used in topical preparations, mouthwashes and inhalation cold remedies
Quaternary ammonium compounds		Incompatible with anionic surfactants and other organic anions
Benzalkonium chloride	0.01–0.02% (parenteral) Up to 0.2% (topical) 20% (inhalation solution)	Often used in combination with disodium edetate to enhance its activity against *Pseudomonas* bacteria. Better activity against Gram –ve than Gram +ve bacteria. Activity increases with pH; best activity in range pH 4–10. Not active against bacterial spores, but active against some viruses. Also used in ophthalmic, nasal and otic products
Benzethonium chloride	0.01–0.02% (ophthalmic and otic) Up to 0.012% (parenteral)	Active over the range pH 4–10. Activity is enhanced by ethanol. Used mainly in topical products
Cetrimonium bromide	0.005% (ophthalmic)	Used topically up to 1% as an antiseptic for burns and up to 3% in shampoos. More active against Gram +ve than Gram –ve micro-organisms. Activity reduced in acidic media; optimal activity in neutral to slightly alkaline media. Activity enhanced in the presence of ethanol. Inactive against bacterial spores. May be absorbed by suspended solids, e.g. bentonite. Not currently listed in the Inactive Ingredient Database
Cetylpyridinium chloride	0.02–1.5 mg (oral)	Used mainly as an active principle in sore throat remedies. May adsorb onto suspended solids
Sorbates		
Potassium sorbate	0.1–0.65% (oral) Up to 0.47% (topical)	Not inactivated by non-ionic surfactants. Prone to oxidation
Sorbic acid	0.05–0.5% (oral)Up to 2.7% (topical)	Active <pH 6. Synergies reported with other antimicrobial preservatives and glycols. More soluble than sorbic acid. Optimum activity at about pH 4.5; no activity >pH 6

Table 3.8 Antioxidants for use in pharmaceutical preparations (compiled from Niazi 2004a, b; Rowe et al. 2006; and USP 30/NF 25 2007, USP and NF Excipients, listed by category)

Aqueous systems	Comments
Ascorbic acid	Vitamin C; used at a level of 0.01–0.1%. Many incompatibilities including alkalis, heavy metals, oxidizing materials and several APIs. Sodium ascorbate in solution rapidly oxidizes above pH 6
Erythorbic acid	Stereo isomer of L-ascorbic acid. GRAS listed, accepted as a food additive in Europe. Included in Inactive Ingredients Database (oral use – no level specified)
Monothioglycerol	Mainly used in parenteral formulations. Reference in Inactive Ingredient Database is to thioglycerol (*Note*: Merck Index refers to thioglycerol as dimercaptoglycerol; an antidote for heavy metal poisoning)
Sodium formaldehyde sulfoxylate	Stabilized by sodium carbonate. Used in parenteral and topical products in US
Sodium metabisulfite, sodium bisulfite, sodium sulfite	Typically used at levels of 0.001–0.1%, although there are products approved in US with higher levels. Sodium metabisulfite is used at low pH, sodium bisulfite at intermediate pH, and sodium sulfite at high pH. Sodium metabisulfite is incompatible with high pH, *ortho-* and *para*-benzyl alcohol derivatives, e.g. epinephrine, and also chloramphenicol and cisplatin. It may interact with rubber closures. Sodium sulfite is incompatible with low pH, oxidizing agents, vitamin B_1 and many proteins
Potassium metabisulfite	See sodium metabisulfite. Potassium metabisulfite is incompatible with strong acids, water and most metals
Non-aqueous systems	
Ascorbyl palmitate	Generally used at a level of 0.05%. Not soluble in water or oils, but soluble in polar organic solvents. Synergy with alpha tocopherol
Butylated hydroxyanisole	Generally used at a level of 0.01–0.5% (0.0002–0.0005% in i.v. injections). Often used in combination with butylated hydroxytoluene or propyl gallate. Has the incompatibilities of a phenol
Butylated hydroxytoluene	Similar to butylated hydroxyanisole
Propyl gallate	Generally used at levels up to 0.1%. Synergy with butylated hydroxyanisole and butylated hydroxytoluene. Forms complexes with metal ions that may be prevented by the use of a chelating agent
Tocopherol	Also known as alphatocopherol. Generally used at a level of 0.005–0.05%. Synergies with lecithin and ascorbyl palmitate. Incompatible with peroxides and metal ions
Tocopherol excipient	A solution of not less than 50.0% of alphatocopherol in a vegetable oil

3.2.7.3 Chelating Agents

Chelating agents, also known as sequestrants, are molecules that have the ability to form stable complexes with metal ions, particularly di-valent and tri-valent metal ions including trace metals and heavy metals. These metal ions are often implicated in API degradation by acting as catalysts, e.g. Mg^{2+} will catalyze both ester hydrolysis and the Maillard interaction between primary or secondary amines and reducing sugars. Oxidative degradation is also often catalyzed by heavy metals. In addition, certain trace metals are required for microbial growth, and chelation (sequestration) to form complexes can help prevent microbial growth and spoilage, and thus allow lower levels of microbiocidal agents to be used.

The materials used in pharmaceutical applications as chelating agents given in Niazi (2004a, b), Rowe et al. (2006) and the USP 30/NF 25 comprise:

Calcium disodium edetate
Disodium edetate
Edetic acid
Citric acid

Edetic acid is also known as ethylenediaminetetraacetic acid (EDTA). While calcium disodium edetate and disodium edetate are soluble in water, the free acid is only slightly soluble. Citric acid may also be used as part of a buffer system to maintain pH.

These are ionic materials and have the potential to interact with other components of the formulation.

3.2.8 Sweetening Agents

Palatability of oral medicines is an important factor in compliance. There are several components to palatability including flavor, mouth-feel and sweetness. Most patients prefer medicines that are not too bitter but may be slightly "tart" (acidic). Most APIs are bitter. However, for bitterness to develop, the drug must be sufficiently soluble to interact with taste receptors on the tongue. For insoluble APIs in the form of suspensions, this begs the question as to why a sweetener is needed. The answer is straightforward in that many of the minor, but necessary, components of the suspension are also bitter, e.g. preservatives, or very salty, e.g. buffer systems. However, a slight saltiness and a slight bitterness are desirable for palatability.

Traditionally, oral medicines were sweetened using Syrup (concentrated sucrose solution) or honey (contains fructose). However, these materials are inadequate for the formulation of many products because they simply are not able to adequately mask the very bitter taste of many pharmaceutical materials, including APIs and excipients. In addition, the large quantities of both sucrose and fructose required increased the dietary carbohydrate intake, particularly when consumed as soft drinks. There have been campaigns over the years to persuade people to eat and drink more healthily, and to cut down on carbohydrates. Several alternative sweetening agents have been developed over the years to better mask unpleasant tastes in both processed foods and pharmaceuticals. They are also used in low-calorie foods and soft drinks.

Several of the materials classified as sweetening agents are sugar alcohols (also know as polyhydric alcohols, polyols and hydrogenated sugars). These materials in general are noncariogenic, although some may be simply much less cariogenic than sucrose. On administration to patients in large doses, e.g. 10–20 g in one dose, they have laxative properties. Below the laxative threshold there is evidence that they increase gut motility causing the contents of the intestines to pass through more quickly. For dugs which have a restricted window of absorption in the gastro-intestinal tract (GIT) this can cause a reduction in bioavailability.

Several of the commonly used sweetening agents are ionic and have the potential to interact with other components of the suspension. Some sweetening agents are more stable than others in aqueous solution. These will be important factors in the final selection of the sweetening agent. A list of sweetening agents for use in oral pharmaceutical suspensions, compiled from Niazi (2004a, b), Rowe et al. (2006) and the USP 30/NF 25, is included in Table 3.9. The regulatory position for the different sweetening agents varies with country and/or region. This is also addressed in Table 3.9.

The use of artificial sweetening agents in pediatric oral medicines is under active debate. However, the use of sucrose ("sugar") is also under a cloud because of its cariogenic potential.

3.2.9 Flavoring Agents

Flavors, as mentioned above, are used to improve the palatability of oral medicines. One problem that can arise with oral suspensions is that the suspension may produce a "cloying" sensation in the mouth. While this is not the same as a bitter taste, it can nevertheless cause problems for the patient and affect compliance. This can be a particular problem with high levels of inorganic components. Flavors can help reduce this "cloying" taste and thereby improve palatability, and ultimately patient compliance.

There are many different flavors, and most flavors are complex mixtures of many components. Today most flavors are developed by specialist flavor houses, and typically the flavor is formulated for each individual application. Since flavor will be part of the suspension continuous phase, it has the maximum potential for interaction, and some flavor components may cause stability issues (physical or chemical) for the suspension. Flavor development and compounding is a specialist discipline. When deciding on which particular flavor is appropriate, the flavor specialist would benefit from knowledge of the other likely components in the suspension, just as the formulation scientist would benefit from knowledge of the components of the flavor. Unfortunately, information on the flavor components is not easily shared because it is considered proprietary. In some cases, even a confidential disclosure agreement is not sufficient to encourage the sharing of such information. Nevertheless, as much information as possible should be exchanged to reduce the chances of a stability failure for the suspension formulation. Flavors can

Table 3.9 Sweetening agents used in oral pharmaceutical suspensions (compiled from Niazi 2004a, b; Rowe et al. 2006; USP 30/NF 25 2007, USP and NF Excipients, listed by category; and Belitz et al. 2004)

Sweetening agent	Sweetness (sucrose=1)	Comments
Acesulfame K	180–200×	Stable in acid pH. Stable in solution at pH 3–3.5. Included in the Inactive Ingredient Database, and permitted in certain foods in the US
Alitame	ca. 2,000×	Not approved for use in pharmaceuticals in US or Europe. Unstable at low pH, more stable at higher pH. Almost non-nutritive
Aspartame	180–200×	Unstable in the presence of moisture. Optimal stability at pH 4.5; unstable to either acid or base. GRAS listed and recognized as a food additive in Europe. Included in the Inactive Ingredient Database
Dextrose	0.5×	Dextrose includes a range of very similar materials such as dextrates and high DE (dextrose equivalents) corn syrup. Reducing sugar. Cariogenic. Included in the Inactive Ingredient Database
Fructose	ca. 1.7×	The sweetest tasting of the natural sugars. Most stable at pH 3–4. Included in the Inactive Ingredient Database. Cariogenic
Galactose	0.3–0.5×	Obtained from lactose. Monographs in NF 25 and Ph.Eur 5.5
Inulin	Slightly sweet	GRAS listed. Non-caloric. Suitable for use by diabetic patients. Natural material of plant origin. Oligofructan
Isomalt	0.5×	Sugar alcohol (polyol). GRAS listed and recognized as a food additive in Europe. Noncariogenic
Lactitol	0.3×	GRAS listed and recognized as a food additive in Europe
Maltitol	ca. 1×	Noncariogenic sugar alcohol. GRAS listed and recognized as a food additive in Europe. Heat stable
Maltose	ca. 0.3×	Recognized as a food in the US, and not subject to GRAS or Food Additive regulations
Mannitol	0.5×	GRAS listed and recognized as a food additive in Europe. Included in the Inactive Ingredient Database. –ve heat of solution
Neohesperidin dihydrochalcone	1,500–1,800×	GRAS listed and recognized as a food additive in Europe. Limited solution stability; optimum pH 2–6
Saccharin (inc. salts)	ca. 500×	Recognized as a food additive in Europe. Included in the Inactive Ingredient Database. Can leave a metallic aftertaste in about 20% of users

Sodium cyclamate	30×	Recognized as a food additive in Europe. Included in the Inactive Ingredient Database. Formerly under suspicion as a carcinogen, but study design questionable and results not confirmed
Sorbitol	ca. 0.5–0.6×	Hydrogenated sugar; non-cariogenic. GRAS listed and recognized as a food additive in Europe. Included in the Inactive Ingredient Database. Reacts with iron oxide leading to discoloration
Sucralose	300–1,000×	GRAS listed and approved for food use in UK. Included in the Inactive Ingredient Database. Stable above pH 3; optimally pH 5–6. Non-cariogenic
Sucrose	1	Cariogenic. Will show the reactions of a reducing sugar due to traces of "invert sugar"
Tagatose	ca. 0.9×	Epimer of D-fructose. Prepared from D-galactose by isomerization under alkaline conditions
Thaumatin	2,000–3,000×	A structure of five related proteins stabilized by disulfide cross-link. Disulfide link cleavage destroys sweetness. Delayed onset and licorice-like aftertaste. Synergies with other sweetening agents. Used in foods and pharmaceutical products. GRAS listed and accepted as a food additive in Europe
Trehalose	ca. 0.45×	GRAS listed. May be used in some food applications in the UK. Mainly used as a lyoprotectant in the formulation of therapeutic proteins. Also used in topical products and cosmetics
Xylitol	ca. 1	Hydrogenated sugar; non-cariogenic. GRAS listed and recognized as a food additive in Europe. Included in the Inactive Ingredient Database. –ve heat of solution

adsorb onto finely divided solids, thus reducing their effectiveness. They can also be absorbed by packaging.

Flavor preferences vary with age, but the citrus flavors appear generally acceptable to most age groups.

3.2.10   Coloring Agents

Pharmaceutical colors come in two types; soluble dyes and insoluble pigments. For pharmaceutical suspensions intended for oral use, soluble dyes are often used; however, pigments may also be used and would be part of the disperse phase. Soluble dyes have the potential to interact with other components of the formulation.

The list of colors acceptable for use in pharmaceuticals differs in different countries and/or regions. For further information see Mroz (2006) and Galichet (2006).

3.3   Non-aqueous Suspensions

Non-aqueous suspensions, as the name implies, do not have an aqueous continuous phase. Nevertheless, many of the guidelines for forming a stable suspension still apply. The rules that may not apply are related to pH and ionic charge, since these phenomena would not be expected to be manifest in the absence of water. Non-aqueous suspensions may be stabilized by the use of viscosity-increasing agents, adsorbed polymers (steric stabilization) and the incorporation of auxiliary suspending agents. Obviously, suspending agents that rely on the interaction with water will not be effective, e.g. the inorganic clays and the hydrophilic colloids. There are other materials that can be used in non-aqueous systems. Surfactants, because of their amphiphilic nature, may still be used, although with rather different HLB values compared to aqueous suspensions.

Several examples of more commonly used non-aqueous suspensions are discussed in more detail in the following sections.

3.3.1   Oily Suspensions

For oily suspensions, as the name implies, the continuous phase is an oil, typically, a vegetable oil. Since the continuous phase is not water, we are restricted to increasing the viscosity of the system and steric factors. Vegetable oils are more viscous than water (typically in the range 35–40 mPa.s compared to about 1 mPa.s for water) and this aids in the formulation of stable oily suspensions. Surfactants and certain polymers can be used to improve suspension stability. However, the surfactants will be more lipophilic (lower HLB) than for aqueous suspensions.

Oily suspensions are not commonly used as pharmaceutical finished products. However, there is an increasing use of such formulations in the oral dosing of animals during safety testing. Many of the new APIs are increasingly poorly water-soluble, and administration of a simple aqueous suspension does not give adequate bioavailability (often around 10% or less). Using a suspension of the API in a vegetable oil (e.g. corn oil) can improve the bioavailability of certain APIs to around 30–40%. This may still not be sufficient for a commercial product, but it will at least allow the safety evaluation to start with a reasonable expectation that sufficient absorption will occur in the animals. However, for longer term safety studies, e.g. greater than 3 months duration, even vegetable oil-based suspensions may not be suitable due to the high volume of oil administered on a daily basis. Similar oily suspension formulations may be used for the early phase 1 human studies; however, to progress beyond phase 1, and certainly beyond phase 2, a bioavailability greater than 50% is desirable. (Self-emulsifying, self-microemulsifying or nanoparticulate drug delivery systems may provide improved bioavailability; they may also be required for longer term safety studies.)

Another area in which oily suspensions are now being considered is in the formulation of pediatric medicines; particularly for APIs having poor solubility and/ or bioavailability from aqueous formulations.

3.3.2   Suppositories

Suppositories are typically prepared using semi-solid bases. Frequently, the API is present as a disperse phase. During manufacture of the suppositories, the base is melted and the API incorporated to form a suspension. This suspension must be stable over a sufficient period to allow the proper filling of the molten suspension into the suppository molds. The suspension should also be stable through the setting of the base in the mold, so that the API remains evenly dispersed throughout the suppository. Suppository bases may be lipophilic (hydrophobic) or hydrophilic semi-solids prepared from synthetic or semi-synthetic triglycerides, and other materials.

Microcrystalline wax may be used as an auxiliary suspending agent in the formulation of suppository melts using lipophilic suppository bases. Fine grades of silica may be appropriate for use as auxiliary suspending agents in hydrophilic suppository bases based on polyethylene glycols or modified triglycerides.

3.3.3   Metered Dose Suspension Inhalation Devices

Pressurized metered dose inhalers (sometimes known as aerosol inhalers) for pulmonary delivery come in two forms depending on the nature of the contents of the pressurized container, solutions or suspensions. Both types have been marketed. In general, the inhaled steroids are formulated as suspensions due to poor solubility, the inhaled β-agonists may be formulated as either solutions or suspensions. The formulation continuous phase is the propellant mix. These are mobile liquids in the pressurized container. Again, since water is not present, suspension stabilization must rely on methods that do not require water. There is also a further restriction in that we do not want to increase the viscosity of the system as this would interfere with the ability of the device to generate droplets of the required size for inhalation. On the plus side, the formulation does not have to remain suspended indefinitely. Immediately prior to administration, the device is shaken to re-suspend the particles prior to use. The suspension is only required to be sufficiently stable to allow an accurate dose to be delivered to the holding chamber. In practice, the suspensions are formulated and stabilized using non-ionic surfactants, i.e. steric stabilization.

3.4 Novel Excipients

Excipients, under the terms of the US Food Drug and Cosmetics Act are considered to be drugs, since the definition of a drug in the Act includes components of drug products. In common with all drugs, before they can be offered for sale, their safety must be assessed. The US FDA has issued a Guidance document (May 2005) that details the non-clinical safety studies necessary for new chemical excipients to be approved for use in human (and veterinary) medicines. Similar considerations have applied in the European Union since the mid 1970s. Japan also has similar restrictions on the entry of new pharmaceutical excipients into the market place.

Before we consider novel excipients in more detail, it is important to understand that "new" can mean several different things in the context of excipients:

1. New chemical excipients never before used in man or animals
2. Use of recombinant and other technologies to produce new versions of existing excipients
3. New excipients formerly used in animals and now being considered for use in man
4. New chemical excipients where the human metabolism and elimination of the component parts of the molecule are predictable and well understood
5. A new route of administration for an existing excipient already used in man
6. New co-processed combinations of existing excipients that are already approved for human use and the intended route of administration
7. New grades of existing excipients already approved for use in man

The amount of safety testing varies significantly depending on both the route of administration and the category of a new excipient (see above). Excipients intended for parenteral use will require greater scrutiny than say a new excipient intended for topical administration. For those categories listed above, category 1 would require the most scrutiny; category 7 may require little to none. However, for all new excipients proper safety and risk assessments are required.

In recent years, there have been some new excipients introduced. However, at the time of writing this chapter, the only major excipient that has been launched in the US market, and that classifies as a category 1 excipient, is probably Captisol® (sulfobutyle-ther β-cyclodextrin from Cydex, Inc. for solubilization of poorly water soluble compounds). There are other excipient technologies that are being tested, but so far they remain in development (e.g. Eligen™ from Emisphere Technologies, Inc. for oral protein delivery). The reasons for the paucity of new chemical excipients have been discussed elsewhere (Moreton 1996) and relate to the high cost of safety testing, relative to the development and commercialization timelines, and likely return on investment.

Recent examples of most of the other categories are also available, including recombinant gelatin for use as an adjuvant (category 2), semi-synthetic glyceride esters (category 4), Captisol® from i.v. use to oral use (category 5), silicified microcrystalline cellulose (category 6), and low-density microcrystalline cellulose (category 7).

In recent years, with the exception of Captisol®, most of the new excipients introduced have been co-processed materials. There have also been several new semi-synthetic glyceride esters.

There are three interrelated major barriers to the introduction of new excipients:

1. Safety
2. Regulatory
3. Technical

The pharmaceutical industry is notoriously conservative. Nobody wishes to be first in the introduction of a new excipient. There must be an overwhelming unmet technical need that cannot be solved in any other way to be able to justify the inclusion of a new excipient in a formulation to senior management. There are risks with any new material. There will be the fear that the new excipient ultimately will not be acceptable to the regulatory authorities for some reason. This would mean costly delays to the launch of a new medicinal product.

In particular, the safety of the new chemical excipient is less well understood, and there is the risk that the excipient will fail some aspect of the overall safety assessment. The time required to properly assess the safety of a new chemical excipient combined with the time taken to develop a new drug, is such that the new excipient would probably be off-patent before the new medicine is launched if the safety testing of the excipient and API are carried out sequentially. This would obviously be of least risk to the pharmaceutical company, but makes it an uneconomic proposition for the excipient manufacturer if patent protection is no longer available when the excipient is included in its first approved medicinal product. The alternative to this is to "piggy-back" the safety testing of the excipient onto that of the API; but then the risk increases dramatically for the pharmaceutical company. Nevertheless, this is what was done for Captisol®.

From a regulatory standpoint, because it is a new excipient, it will not have a pharmacopeia monograph. This automatically creates extra work because, under the structure of the ICH Common Technical Document, excipients not having a pharmacopeia monograph are highlighted in a separate section of the submission, and a lot more information must necessarily be included in the submission. In the United States where a Drug Master File (DMF) system is in place, it is possible for the excipient manufacturer to file much of the confidential information in a Type 4 DMF, and for the IND (Investigational New Drug) or NDA (New Drug Application) to include a letter of authorization from the excipient manufacturer to allow the FDA reviewer to access the excipient DMF during review of the IND or NDA. This is not possible in countries where no DMF system is available, e.g. Europe.

With any new excipient, there are always technical challenges. Continuity of supply, analytical methods, scale-up of manufacture, understanding the link between composition and functionality; these are all issues that must be addressed during the development of any new excipient. The paradigm has changed. The advent of the US FDA's Quality-by-Design (QbD) initiative now requires that we understand far more about the influence of excipient variability on the performance of the manufacturing process and of the finished product (see below).

3.5   Excipients for Nanosuspensions

Nanosuspensions are suspensions where the particle size of the disperse phase is less than 1 μm (some authorities suggest 0.1 μm), i.e. measured in nm. They are not new; they were previously referred to as colloidal systems or colloidal suspensions. Only the introduction of the term "nano" is new. Because of their very small size, in the absence of other factors, nanoparticles should remain in suspension for extended periods. However, like other disperse phases they will aggregate and coalesce to form larger particles, and Ostwald ripening may also occur. Once they settle out, the particles will be very difficult to re-suspend. Nanosuspensions therefore need to be stabilized. Many of the same materials, surfactants and polymers used to stabilize conventional suspensions may also be used to stabilize nanosuspensions. However, since interfacial phenomena are involved and the surface area of the disperse phase is very much increased in nanosystems, the level of incorporation of the polymers and surfactants may need to be increased accordingly. In addition, the DLVO theory will apply to lyotropic nanoparticles (see Sect. 3.1.2.2).

Nanosuspensions will be discussed in more detail in Chap. 10. As can be seen from Chap. 10, nanosuspensions may be classified as either systems containing solid nanoparticles, or those based on lipids. The methods of manufacture are very different for the two types of system, but some of the factors affecting nanosuspension stability are similar for both.

3.6   Quality by Design

Quality by design (QbD) is an initiative under the US FDA's umbrella initiative, Quality in the twenty-first century. The umbrella initiative was promulgated to help the pharmaceutical industry adopt more modern and efficient manufacturing techniques, while simultaneously reducing the incidence of product manufacturing failures and recalls, and reducing the number of pre-approval change and changes being effected submissions.

The basic premise of QbD is that quality cannot be tested into a product, project or service; it must be built in, i.e. by design. In this respect, formulation development projects are no different from any other product, project or service. A fundamental part of QbD is the "design space" concept. From ICH Q8, "design space" may be defined as follows:

> **Design Space:** The multidimensional combination and interaction of input variables (e.g., material attributes) and process parameters that have been demonstrated to provide assurance of quality. Working within the design space is not considered as a change. Movement out of the design space is considered to be a change and would normally initiate a regulatory post-approval change process. Design space is proposed by the applicant and is subject to regulatory assessment and approval.

In simple terms, QbD is the application of sound scientific principles to the formulation and process design, development and scale-up of medicinal products. If we are able to define the limits of a formulation and processing that will ensure an acceptable product that meets specification throughout its regulatory shelf-life, and we are able to submit acceptable study reports as part of the NDA, or perhaps supplemental NDA (sNDA), or equivalents in other regions, and these reports are accepted by the FDA or appropriate regulatory body, then working within the defined design space would not be considered a change, and could be reported via the Annual Report. This represents a significant reduction in regulatory work load for both the pharmaceutical industry and the regulatory agencies.

This is a new initiative, and it will take some time before industry becomes familiar with its practical working. One question that does arise is, "Where does QbD start?" It is this author's opinion that QbD starts with the preformulation screen. Much of the QbD work will be carried out at laboratory scale, and at intermediate scale (i.e. greater than one tenth production scale). However, the preformulation screen is where we start a product formulation project and it is the foundation upon which we build our formulation and process design, and thus development and scale up. It therefore follows that, if the preformulation is flawed, formulation and process design development and scale up may also be flawed. Good preformulation underpins any QbD project.

3.7 Summary and Conclusions

The formulation of pharmaceutical suspensions is challenging in many ways. Suspensions are complex systems that must be stabilized in some way. The key to any successful, robust, stable formulation is the understanding of the API, excipients and unit processes, and the interactions between all these components; suspensions are no exception.

There are many excipients that may be used in the formulation of pharmaceutical suspensions. There may be more than one way to achieve a successful, robust, stable formulation. However, without understanding the excipients, some of the tools the formulation scientist must work with, it is unlikely the suspension formulation project will be successful.

There is room for new excipients. There are, and will continue to be, unmet needs of formulation science. It is not realistic to expect many new chemical excipients because of the costs and time for safety evaluation. However, there are opportunities for other types of new excipient such as new semi-synthetic glycerides and co-processed materials that will address some of the unmet needs.

The FDA's QbD initiative and other parts of their umbrella "Quality in the twenty-first century" initiative have placed the emphasis on good science and proper identification of the design space. Excipients will be a very important part of any design space.

References

Belitz, H.-D., Grosch, W., and Schieberle, P. 2004. Food Chemistry, 3rd Edition, Springer-Verlag, Berlin, Heidelberg.

Galichet, L.Y. 2006. Iron Oxides, In Handbook of Pharmaceutical Excipients, 5th Edition, Rowe, R.C., Sheskey, P.J., and Owen, S.C. (eds.), Pharmaceutical Press, London, and American Pharmaceutical Association, Washington, DC, 364–365.

Guidance for Industry: Nonclinical Studies for the Safety Evaluation of Pharmaceutical Excipients, Food and Drug Administration, Rockville, MD, May 2005.

ICH Q8: Pharmaceutical Development, Step 4 Document, International Conference on Harmonization of Technical Requirements for Registration of Pharmaceuticals for Human Use, Geneva, Switzerland, November 2005.

Inactive Ingredient Database, US Food and Drug Administration – Center for Drug Evaluation and Research, http://www.accessdata.fda.gov/scripts/cder/iig/index.cfm.

Moreton, R.C. 1996. Tablet excipients to the year 2001: a look into the crystal ball, Drug Dev. Ind. Pharm. 22(1), 11–23.

Mroz, C. 2006. Coloring Agents, In Handbook of Pharmaceutical Excipients, 5th Edition, Rowe, R.C., Sheskey, P.J., and Owen, S.C. (eds.), Pharmaceutical Press, London, and American Pharmaceutical Association, Washington, DC, 192–200.

Niazi, S.K. 2004a. Handbook of Pharmaceutical Manufacturing Formulations: Volume 3 Liquid Products, CRC Press, Boca Raton, FL.

Niazi, S.K. 2004b. Handbook of Pharmaceutical Manufacturing Formulations: Volume 5 Over-the-Counter Products, CRC Press, Boca Raton, FL.

Rowe, R.C., Sheskey, P.J., and Owen, S.C. 2006. Handbook of Pharmaceutical Excipients, 5th Edition, Pharmaceutical Press, London, and American Pharmaceutical Association, Washington, DC.

Steele, D.F., Edge, S., Tobyn, M.J., Moreton, R.C., and Staniforth, J.N. 2003. Adsorption of an amine drug onto microcrystalline cellulose and silicified microcrystalline cellulose samples. Drug Dev. Ind. Pharm. 29(4), 475–487.

United States Pharmacopeia 30 – National Formulary 25, United States Pharmacopeia Convention, Inc., Rockville, MD, 2007.

Chapter 4
Pharmaceutical Development of Suspension Dosage Form

Yusuf Ali, Akio Kimura, Martin J. Coffey, and Praveen Tyle

Abstract This chapter begins with a definition of the Target Product Profile which defines the goal for the drug product and guides formulation research and development activities. Formulation research begins with the generation of the preformulation information (physicochemical properties) and leads to the selection of a prototype suspension formulation. This chapter also outlines the formulation optimization and development activities required to iteratively refine the formulation composition and process as the project proceeds to the point of filing an NDA.

4.1 Introduction

Pharmaceutical suspensions are used for topical, oral, and injectable administration routes. They consist of a drug delivery system in which solid particles are dispersed in a continuous fluid medium either as discrete particles or as a network of particles. Depending on the route of administration the particle size in the suspension may vary from less than 1 μm for ophthalmic use to about 10–100 μm for dermatological use, and up to 200 μm for oral use.

Many of the recently discovered active pharmaceutical ingredients are quite hydrophobic with limited solubility. They may also be quite distasteful. Other drugs may also have quite a high chemical degradation precluding them to be administered as aqueous solutions, and in this case, it may be possible to synthe-

Y. Ali (✉)
Santen Incorporated and Santen Pharmaceutical Co., Ltd, 555 Gateway Drive, Napa, California, 94558, USA
e-mail: yali@santeninc.com

A. Kimura
Santen Incorporated, Osaka, Japan

M.J. Coffey
Bausch & Lomb, 1400 North Goodman Street, Rochester, New York, NY 14609, U.S.A.

P. Tyle
Novartis Consumer Health, 200 Kimball Drive, Parsippany, NJ 07054, U.S.A.

A.K. Kulshreshtha et al. (eds.), *Pharmaceutical Suspensions: From Formulation Development to Manufacturing*, DOI 10.1007/978-1-4419-1087-5_4,
© AAPS 2010

size an insoluble derivative. In other cases, some drugs are required to be present in the gastrointestinal tract or in the precorneal pocket with long residence time. For such drugs, a suspension is an ideal delivery system as it provides better chemical stability and larger surface area and is often more bioavailable than aqueous solutions, tablets, and capsules.

Formulation of an elegant, stable, preserved, safe, and effective suspension is a technically challenging task compared aqueous solutions, tablets, and capsules. Pharmaceutical suspensions are thermodynamically unstable systems. Thus, preparation of such systems is often associated with problems of physical stability, content uniformity, sedimentation, caking, resuspendibility, and crystal growth. Furthermore, issues related to the masking of bitter taste and undesirable odor of the pharmaceutical ingredient must be taken into consideration.

Some desirable attributes of a suspension are described by Bhargava et al. (1996).

1. It should be safe, effective, stable, and pharmaceutically elegant during the shelf life of the product.
2. The drug should not have a quick sedimentation rate. Furthermore, it should resuspend easily upon shaking and it must not cake.
3. Physical attributes such as particle size, particle size distribution, viscosity should remain fairly uniform throughout the shelf life of the product.
4. Its viscosity must promote free and uniform flow from the container. The product must have appropriate substantivity that it spreads freely over the affected area.
5. Resuspension should produce a homogeneous mix of drug particles such that there is a content uniformity with each dose.

In order to design a suspension product that addresses the desirable attributed described earlier, a formulator must have a systematic approach in the design of experiments so that he or she can rapidly identify a prototype formulation during the product development phase.

4.2   Target Product Profile Definition

The most important part of any product development project is accurately and clearly defining the Target Product Profile (TPP). A good TPP will not only help to guide formulation research and development activities, but will also speed up development activities because clearly stated priorities enable decisions to be quickly made based on available information. The elements of a good TPP for a pharmaceutical suspension dosage formulation include the following:

- Administration route (topical vs. injectable)
- Dosing frequency (QID, BID, QD)
- Dosing protocol (shake immediately before use, dose within 4 h of shaking)
- Preservation (single-use, unpreserved vs. multidose, preserved)
- Define safety/efficacy requirements for market success (better than product X, gentle preservative)

- Market-image package type (opaque, clear, bottle size, fill volume)
- Shelf-life and storage conditions
- Any esthetic requirements (color, texture, comfort, taste)
- Target Market (US, Europe, Japan, Global)
- Target Customer (physician-administered vs. patient-administered)

In addition, it is helpful to further define which product features are desired and which features are required for product development. As a result, there may be a Minimal Acceptable Product Profile (MAPP) which is less restrictive than the TPP.

4.3   Formulation Research

The goal during the formulation research stage is to generate the information about the stability, compatibility, and possibly the drug delivery characteristics of different formulation technologies. These research experiments will involve characterization of the active pharmaceutical ingredient (API) in the solid form and in a suspension formulation, evaluation of prototype suspension vehicles (with particular emphasis on the preservation of the vehicles), and evaluation of the compatibility between the API and the suspension vehicle.

4.3.1   Characterization of API

One of the first formulation research activities is to evaluate the physicochemical properties of the selected active pharmaceutical ingredient (API). The physico-chemical properties on interest include pKa, octanol-water partition coefficient, solubility in various liquids (solvents, surfactant systems, pH), dissolution rate, chemical stability (including photodegradation) of the solid API, chemical stability (including photodegradation) of the dissolved API as a function of pH, and solid properties of the API such as crystallinity, polymorphism, melting point, absolute and bulk density, particle size and surface area, hygroscopicity, and flow characteristics. In addition, if a sterile formulation will be developed, the effect of typical sterilization methods on the API should be evaluated (dry heat sterilization, autoclaving, ethylene oxide treatment, gamma irradiation).

The API characterization information will be helpful for guiding the formulation research work as well as guiding what process will be used for producing and controlling the API as an ingredient in the formulation. For a successful suspension formulation development project, the physicochemical properties of the API should indicate what pH range, particle size for the API, and solubilization technology (if any) are most likely to provide a chemically stable suspension formulation with a reasonable probability of delivering the drug to the target tissue.

4.3.2 Controlling the Particle Size of the API

The distribution of primary particles and aggregates of API in a suspension formulation should be controlled in order to (a) provide uniformity in the dosing of the product, (b) provide reproducible drug delivery characteristics, and (c) to bring about pleasing or comfortable formulation characteristics. A smaller particle size would be required when a suspension is administered through a narrow-gage needle (e.g., 30 Ga = 140 micron ID) or injected in a small volume (e.g., intravitreal injection ≤50 mcL). For low-solubility drugs, the particle size distribution will control the dissolution and hence, the drug delivery characteristics of the formulation, unless the particle size is kept below the threshold where the drug is able to dissolve faster than the dissolved drug is absorbed from the formulation. However, it may be desirable to keep the particles larger if a prolonged delivery formulation is desired. In addition, depending on the route of administration, there may be regulatory requirements for API particle size in the appropriate Pharmacopeia. For example, the following table summarizes the requirements for an ophthalmic suspension according to the Japanese and European Pharmacopeias (Table 4.1)

Generally, in order to achieve the target particle size and its distribution, one of two approaches will be employed (Fig. 4.1).

One method for controlling the particle size is mechanical comminution of previously formed larger, crystalline particles (e.g., by grinding with a mortar-pestle, air-jet micronization or wet-milling with ceramic beads). Another method is the production of small particles, using a controlled association process (e.g., spray drying, precipitation from supercritical fluid, and controlled crystallization). The process used to obtain the desired particle size distribution can have dramatic effects on the properties of the drug product. Communition methods may generate heat that can generate

Table 4.1 Compendial particle size requirements for ophthalmic suspensions

Acceptance criteria
JP: Particles with diameter of 75 μm or more should not be observed
EP: Particles with diameter of 20–50 μm should be 20 or less per 10 μg active ingredient
Particles with diameter of 50–90 μm should be 2 or less per 10 μg active ingredient
Particles with diameter of 90 μm or more should not be observed per 10 μg active ingredient

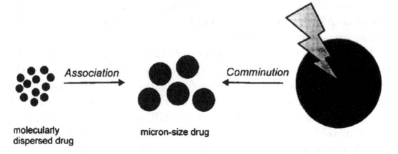

Fig. 4.1 Illustration of the two principle methods for the production of a micron-size drug powder

amorphous regions or polymorphic changes in the API particles that can, in turn, affect dissolution and drug delivery characteristics. When a change is made from one comminution method to another, the API behavior may change significantly. For example, jet-air micronization can frequently result in triboelectrification (charging) of the API particles. This charging of the particles may cause a formulation that was designed to be flocculated to become deflocculated or vice versa.

It is not practical to fully optimize the method for controlling the particle size independent of the formulation, nor is it practical to select the final particle size method prior to beginning formulation research and development. Therefore, a formulator must keep in mind how things will change during the planned development, and how scale-up activities prepare for those necessary changes. The ultimate goal is to develop the formulation and particle size control method that will be used for manufacturing the marketed drug product and therefore, the earlier this compatibility can be tested and verified, the better.

When formulation research is started, the formulation scientist typically has very little API available for the evaluation of particle size methods. Some simple, small-scale experiments may help to indicate what particle size control methods may be viable options. For example, grinding a small amount of drug substance with a mortar-pestle to evaluate how easily a material can be ground (brittleness) and evaluation of the crystallinity of the drug substance before and after grinding may indicate whether comminution methods may be viable. Likewise, small-scale experiments with dissolving and precipitating the drug may indicate whether a controlled precipitation process may produce a suitable crystalline particle.

4.3.3 Evaluation of Suspension Platforms

The evaluation of the various suspension platforms and vehicles is the stage of formulation research where the formulation components that will most significantly affect physical stability and drug delivery are evaluated. At this stage of research, formulations with the API dispersed in lipids may be compared to formulations with API dispersed with high-viscosity polymers and simple, low-viscosity suspensions. In some cases, these platform vehicles may be evaluated for drug delivery characteristics and in other cases, they may be evaluated solely for physical stability (e.g., settling and resuspendability). Regardless of the method used for the selection of the suspension platform, the goal of this research is to select the principle formulation components (e.g., selection of specific surfactant and polymer stabilizers) that will provide the desired formulation characteristics.

In case of suspensions, the understanding of interfacial properties, wetting, particle interaction, zeta potential, aggregation, sedimentation, and rheological concepts are required for formulating an effective and elegant suspension. Accordingly, these factors affecting physical stability should be considered in addition to chemical stability of drug substance, in the selection of each additive. And special attention should be paid on buffer agent and dispersant in terms of securing the physical stability. Index in evaluating physical stability is cohesive property and redispersibility. These items, which affect effectiveness and stability, are of great importance, requiring sufficient verification. About compatibility with preservatives, it is necessary to study the effect on preservative effectiveness by preservatives adsorbing to API as well as the afore mentioned physical stability.

4.3.4 Evaluation of Prototype Vehicles and Preservatives

Many of the commonly used preservatives are cationic (BAK, chlorhexidine, cetrimide, polyquaternium-1); are surface active (BAK, chlorhexidine, cetrimide); are ionizable (benzoic acid, sorbic acid, dehydroacetic acid); or are low in solubility (parabens) and as a result their performance can be dramatically affected by the pH, ionic strength, use of other highlycharged formulation components (e.g., sodium citrate, sodium carboxymethyl cellulose), presence of other surfactants, and the adsorption of the preservative to the surface of the suspended solids. In addition, many widely used, effective preservatives are losing favor with physicians and consumers (e.g., BAK, thimerosal, and parabens). Therefore, the preservation of a suspension vehicle can be a very difficult and time-consuming problem to solve. If a preserved, multidose product is to be developed, it is recommended that a preserved, prototype vehicle be previously identified with a suitable choice of pH, ionic strength, surfactant, and polymer. In addition, the development of the prototype vehicle in a suitable package system can also reduce the risk of discovering later an incompatibility requiring reformulation. Use of a preserved, prototype vehicle as a starting point for further formulation development can significantly reduce development time.

The container/closure and package system characteristics should be evaluated with a prototype formulation in order to select an appropriate package for use in product development. These characteristics are a subset of the characteristics that will need to be eventually tested to demonstrate suitability of the final container/closure system.

The following tests should be used to evaluate the protection for the formulation provided by the container/closure, the safety and compatibility of the container/closure, and the performance of the container/closure:

- USP<661> Containers – light transmission
- USP <671> Containers – permeation (water vapor permeation)
- Seal Integrity and leakage should be evaluated. In the case of a screw-cap closure, the correlation between application and removal torques and leakage (or weight loss) from the container should be evaluated. Inverted storage of the container should be considered to exaggerate the contact between the formulation and the closure seal.
- Extractable/Leachables should be monitored in formulations on stability. Accelerated studies including inverted storage (to exaggerate the contact between the formulation and the closure) should be considered for this purpose.
- Evaluation of loss of functional excipients to semipermeable containers should be monitored in formulations on stability. For example, loss of antioxidant or

preservative resulting from sorption to or diffusion into a plastic bottle or rubber seal could adversely affect the suitability of the container/closure system.
- Dosing uniformity from container/closure should be evaluated. Many container/ closure systems may affect the uniformity of the suspension dose because of the shape of the container and its effect on resuspendability or because of the way the suspension is delivered from the container. Simulated use testing is recommended to evaluate the ability of the suspension to be resuspended and delivered after settling in the container for an appropriate period of time. Freeze/thaw cycling, high-temperature storage, and inverted storage of the container may also be helpful in evaluating potential failures for the container/closure to dependably provide uniform dosing of the suspension.

For pharmaceutical suspension, it is sufficient to study the small-scale (lab-scale) manufacturability with its reproducibility taken into consideration before starting the preclinical studies. In this stage, the reproducibility on the API's particle control in the suspension and its stability should be confirmed.

4.4   Formulation Optimization and Development

4.4.1   Optimization and Development Strategy

In formulation optimization and development, many factors need to be adjusted simultaneously to develop a drug product that (1) is stable, (2) is compatible with the manufacturing capabilities and selected container/closure, and (3) meets the design goals as defined by the Minimal Acceptable Product Profile or the Target Product Profile. This is shown schematically in Fig. 4.2.

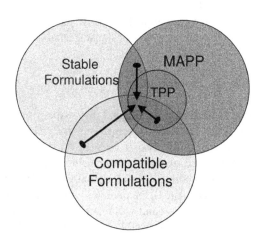

Fig. 4.2 Optimization scheme of the drug product to achieve a stable, compatible, and desirable product

4.4.2 Optimizing and Developing a Stable Drug Product

There are three important attributes for the stability of the drug product: chemical stability, physical stability, and preservative efficacy (i.e., microbiological stability). In order to achieve simultaneous stability for the drug product in all the three categories, it is generally best to start by optimizing the variable with the most restrictive constraint first. For example, if the formulation research information indicates that chemical stability necessitates the use of a narrow pH range, then optimizing the chemical stability first is likely the best approach.

4.4.2.1 Chemical Stability

There are primarily three chemical stability issues that are frequently encountered: hydrolysis, oxidation, and photodegradation. The formulation parameters that can be varied to address these chemical stability issues are briefly:

- **Hydrolysis:** (1) reduce solubility of the drug in the vehicle, (2) adjust the pH to avoid acid or base catalysis, or (3) reduce the storage temperature. A fourth option is to remove the water, at least temporarily, by free-drying (lyophilizing) the formulation, but this changes the nature of the suspension product from a "ready-to-use" to a reconstitutable product and is only considered as a last resort.
- **Oxidation:** (1) add an antioxidant to the formulation, (2) remove oxygen from the manufacturing process and package, (3) use a more protective package, or (4) reduce the storage temperature.
- **Photodegradation:** (1) reduce the solubility of the drug in the vehicle (if photodegradation occurs to drug in solution), or (2) use a more protective package and/or storage condition.

We often select drug candidates to develop as suspensions, rather than solutions, in order to avoid the stability issues associated with the dissolved drug. The selection of a suitable pH and solubility of the drug in the vehicle will often be based on balancing the chemical stability, the comfort of the drug product, and the drug delivery characteristics of the suspension.

4.4.2.2 Physical Stability

It is important to understand that suspensions are kinetically stable, but thermodynamically unstable, systems. The physical stability of suspensions is also discussed in Chap. 6 of this book (see Sect. 6.11).

When left undisturbed for a long period of time the suspension particles will aggregate, sediment, and eventually cake. When a suspension is very well dispersed (i.e., deflocculated), the particles will settle as small individual particles. This settling will be very slow and will result in a low-volume, high-density sediment that may be difficult or impossible to redisperse. When the particles are held together in a loose open structure, the system is said to be in the state of flocculation. The flocculated particles will settle rapidly and form a large-volume, low-density sediment that is readily dispersible. Relative properties of flocculated and deflocculated particles in suspension are provided in Table 4.2 (Bhargava et al. (1996)).

The flocculation state of a suspension product is primarily controlled by the nature of the surface of the suspended particles. The surface charge (i.e., zeta potential) of the particle may be adjusted to move between a flocculated and deflocculated state. Also, adsorption of surface active polymers or surfactants can stabilize suspensions by preventing the removal of water from between the particles. An example of how to modify the zeta potential of a suspension to between a deflocculated and a flocculated state is given by Haines and Martin (1961) and Martin (1961). Briefly, the adsorption of a charged surfactant (e.g., BAK) to the surface of a suspended particle provides charge–charge repulsion resulting in a deflocculated suspension. Then an oppositely charged flocculating agent (e.g., phosphate) is added at increasing levels to shield these surface charges and to reduce the zeta potential to zero, at which point flocculation is observed. The formulation factors that can be adjusted to affect the physical stability of the formulation include the following:

- **Flocculation/Deflocculation:** (1) add charged surface-active polymer or surfactant, (2) add oppositely charged flocculation agent, (3) add nonionic surface active polymer or surfactant, (4) adjust ionic strength of vehicle, (5) if drug has a pKa, adjust pH to modify surface charge.
- **Sedimentation rate:** (1) increase the viscosity of the vehicle, (2) decrease the particle size of the drug, (3) develop a structured vehicle which does not settle.

Particle size of the active agent plays a key role in the physical stability and bioavailability of the drug product. The rate of sedimentation, agglomeration,

Table 4.2 Relative properties of flocculated and deflocculated particles in suspension

Deflocculated	Flocculated
1. Particles exist in suspension as separate entities	Particles form loose aggregates
2. The rate of sedimentation is slow, because each particle settles separately and particle size is minimal	The rate of sedimentation is high, because particles settle as a floc, which is a collection of particles
3. Sediment is formed slowly	Sediment is formed rapidly
4. The sediment eventually becomes very closely packed as a result of the weight of the upper layers of the sedimenting material. Repulsive forces between particles are overcome and a hard cake is formed that is difficult, if not impossible, to redisperse	The sediment is loosely packed and possesses a scaffold-like structure. Particles do not bond tightly to each other and a hard, dense cake does not form. The sediment is easy to redisperse, so as to reform the original suspension
5. The suspension has a pleasing appearance, because the suspended material remains suspended for a relatively long time. The supernatant also remains cloudy, even when settling is apparent	The suspension is somewhat unsightly because of rapid sedimentation and the presence of an obvious, clear supernatant region. This can be minimized if the Volume of sediment is made large. Ideally, the volume of sediment should encompass the volume of suspension

and resuspendability are affected by particle size. The most efficient method of producing small particle size is dry milling. However, wet milling may be desirable for potentially explosive ingredients. Other methods of particle size reduction include micro pulverization, grinding, and controlled precipitation.

In general, a suspension contains many inactive ingredients such as dispersing and wetting agents, suspending agents, buffers, and preservatives. Wetting agents are surfactants that lower the contact angle between the solid surface and the wetting liquid. Wetting agents generally have HLB values of 7–10. Solubilizing surfactants generally have HLB values >13. Generally used wetting and solubilizing agents are provided in Table 4.3.

Suspending agents are used to prevent sedimentation by affecting the rheological behavior of a suspension. An ideal suspending agent should have certain attributes: (1) it should produce a structured vehicle, (2) it should be compatible with other formulation ingredients, and (3) it should be nontoxic. Generally used suspending agents in suspension include cellulosic derivatives (methylcellulose, carboxymethyl cellulose, hydroxyethyl cellulose, and hydroxypropyl methylcellulose), synthetic polymers (carbomers, polyvinylpyrrolidone poloxamers, and polyvinyl alcohol), and polysaccharides and gums (alginates, xanthan, guar gum, etc.) (Table 4.4).

The manufacture of the suspension includes sterilization of the micronized active drug by dry heat, exposure to gamma radiation or ethylene oxide, or in some cases steam sterilization of the concentrated slurry followed by ball-milling (Ali et al. 2000). In general, key steps of manufacturing suspensions are (a) preparation of a dispersion

Table 4.3 Wetting and solubilizing agents (from *USP* 31-*NF* 26). (HLB of surfactant indicated in parentheses)

Nonionic surfactants	Cationic surfactants
Nonoxynol 9 (13)	Benzalkonium chloride
Octoxynol 9 (13.5)	Benzethonium chloride
Poloxamer 188 (29)	Cetylpyridinium chloride (26)
Poloxamer 407 (22)	
Polyoxyl 10 oleyl ether (12.4)	*Anionic surfactants*
Polyoxyl 20 cetostearyl ether (15.7)	Docusate sodium (10)
Polyoxyl 35 castor oil (13)	Sodium lauryl sulfate (40)
Polyoxyl 40 hydrogenated castor oil (15)	
Polyoxyl 40 stearate (16.9)	
Polysorbate 20 (16.7)	
Polysorbate 40 (15.6)	
Polysorbate 60 (14.9)	
Polysorbate 80 (15)	
Sorbitan monolaurate (8.6)	
Sorbitan monooleate (4.3)	
Sorbitan monopalmitate (6.7)	
Sorbitan monostearate (4.7)	
Sorbitan sesquioleate (3.7)	
Sorbitan trioleate (1.8)	
Tyloxapol (12.9)	

Table 4.4 Suspending and/or viscosity-increasing agents (from *USP* 31-*NF* 26). (Charge of agent is indicated in parentheses)

Natural polymers	Cellulose derivatives
Acacia (–)	Carboxymethylcellulose (Calcium, Sodium) (0)
Agar (–)	Hydroxyethyl cellulose (0)
Alamic acid (–)	Hydroxypropyl cellulose (0)
Alginic acid (–)	Hypromellose (0)
Carrageenan (–)	Methylcellulose (0)
Corn syrup solids (0)	Microcrystalline cellulose (0)
Dextrin (0)	
Gellan gum (–)	*Synthetic polymers*
Guar gum (0)	Carbomer (–)
Maltodextrin (0)	Polyethylene oxide (0)
Pectin (–)	Polyvinyl alcohol (0)
Sodium alginate (–)	Povidone (0)
Starch, (Corn, Potato, Tapioca, Wheat) (0)	
Propylene glycol alginate (–)	*Inorganics* (*Clays*)
Tragacanth (–)	Attapulgite (–)
Xanthan gum (–)	Bentonite (–)
Magnesium aluminum silicate (–)	
Miscellaneous	Silicon dioxide (0)
Aluminum monostearate	
Gelatin (+/–)	

of the drug; (b) Preparation of the structured vehicle, followed by addition of the drug dispersion; (c) addition of the other adjuncts; and (d) homogenization. Continuous mixing is required during the filling process to assure homogeneity of the dosage form. Some of these steps may include aseptic handling of sterile materials. The manufacture of ophthalmic suspension is more complicated than conventional solutions.

It is also valuable to consider the transfer of technology to Pilot plant and the support for manufacture of clinical trial materials as the next development stage.

Points to be considered at this stage are as follows:

1. Be consistent with the property of the suspension used for preclinical studies.
2. Design/select the manufacturing equipment in such a manner as to avoid contamination and to make it easy for cleaning.
3. Conform the manufacturing method, equipment, and quality control to GMP requirements.

4.4.2.3   Preservative Efficacy

The preservation of pharmaceutical products became a major concern about 50 years ago, when several reports concerning the microbial contamination of such products were published (Morse and Schonbeck 1968; Morse et al. 1967). At that time, it became necessary to establish standards and methods for acceptable amounts and types of bioburden in nonsterile pharmaceuticals. Cosmetic manufacturers carried out important work in this regard, as cosmetics are generally more prone to contamination than pharmaceutical products.

Sterile and nonsterile multidose suspension products can become contaminated with environmental microorganisms such as bacteria, virus, and fungi through consumer use and also during manufacturing. Therefore, it is necessary to preserve such products. The addition of antimicrobial agent(s) to the pharmaceutical dosage form is required to destroy or inhibit the growth of the microorganisms.

Some formulation factors that can be adjusted to affect the preservative efficacy of a formulation are as follows:

- **Water activity:** Reducing the amount of unbound water in the formulation can inhibit microbial growth, but may only be a useful modification in formulations that are significantly nonaqueous (e.g., an ointment or suspension in oil).
- **Nutrition:** If possible, remove any components that may be nutritive to the bacteria and fungi (carbohydrates, proteins, lipids, organic acids, inorganic salts, and vitamins).
- **Adjusting pH:** Bacteria thrive in the pH range 5.5–8.5. Yeasts and filamentous fungi prefer the pH range 4–6. Use of a good antifungal preservative at a low pH or a good antibacterial preservative at a high pH may provide a better broad spectrum preservation.
- **Storage Temperature:** Molds and Yeasts prefer temperatures of 18–25°C, whereas bacteria prefer 30–37°C. Many molds and bacteria are inhibited or destroyed when stored at temperatures of 42–50°C. Storage at 23°C will encourage molds and yeasts to grow and will keep bacteria static, but alive. Temperatures of 4–15°C will hold microbial contaminants in dormancy. Most preservatives are effective when the microorganisms are active. Hence, preservative efficacy is severely hampered by lower temperatures and will increase at higher temperatures.
- **Ionic strength:** In addition to affecting microbial nutrition, the salts in a formulation may affect ionizable preservatives (quaternary ammonium compounds and organic acids). Quaternary ammonium compounds will generally perform better at lower ionic strengths, where the shielding of the charges will not inhibit the electrostatic interaction with the organisms.
- **Preservative Concentration:** The concentration of the preservative should be optimized to provide adequate preservation, but minimizing the amount of preservative required is desirable to minimize potential cellular toxicity that could result from the preservative.
- **Packaging:** Preservatives may partition into plastics or rubbers, reducing the concentration in the formulation. In addition, volatile preservatives may be lost through permeable packaging and light sensitive compounds may be degraded in transparent or translucent packaging. Brown and amber glass has higher metal content, which can deactivate some preservatives (e.g., sorbic acid)
- **Processing:** Loss of preservative to sorption in processing, adequate control of pH and temperature during processing, and the order of component addition can affect the preservative efficacy of the drug product.

Table 4.5 Antimicrobial preservatives (from *USP* 31-*NF* 26)

Quaternary ammonium compounds	Phenolics
Benzalkonium chloride	Chlorocresol
Cetrimonium bromide	Cresol
Cetylpyridinium chloride	Phenol
	Thymol
Organic acids	
Benzoic acid	*Alcohols*
Potassium benzoate	Chlorobutanol
Potassium sorbate	Benzyl alcohol
Sodium benzoate	Phenoxyethanol
Sodium dehydroacetate	Phenylethyl alcohol
Sodium propionate	
Sorbic acid	*Organic mercurial compounds*
	Phenylmercuric acetate
Parabens	Phenylmercuric nitrate
Butylparaben	Thimerosal
Ethylparaben	
Methylparaben	
Methylparaben sodium	
Propylparaben	
Propylparaben sodium	

Although a formulator has many options for selecting a preservative for use in a pharmaceutical formulation, there is generally a tendency to try to use preservatives (1) for which there is a lot of safety information, (2) which is available from many suppliers at a relatively low price, and (3) for which there are no significant regulatory issues. As a result, the first choice is to try to select something that is a compendial ingredient. Table 4.5 lists and classifies the excipients identified in the USP/NF as antimicrobial preservatives.

Several researchers have prepared lists of properties for the ideal preservative (Croshaw 1977; Wodderburn 1964). This list includes (1) the preservative should be effective at low concentrations against a wide spectrum of microorganisms, (2) should be soluble in the formulation at the desired concentration, (3) should be nontoxic and nonsensitizing at the required concentration, (4) should be compatible with formulation and packaging components, (5) should not have any effect on the color, odor, viscosity, or settling characteristics of the suspension, (6) should be stable over wide ranges of pH and temperature during the shelf life of the product, and (7) should be relatively inexpensive. It is also desirable that dispersed parenteral formulations be free of pyrogens. Pyrogens are substances that produce fever in humans and animals. The vast majority of pyrogens are bacterial endotoxins (Cherian and Portnoff 1998). The endotoxins are composed of lipopolysaccharides and part of the membrane layer of gram-negative bacteria. The dispersed parenteral should be prepared pyrogen-free by dry-heat depyrogenation of glass containers and by filtration of the suspending medium through alumina and other adsorbents.

4.4.2.3.1   Preservative Testing and Compendial Requirements

Criteria for the effectiveness of a preservative system are expressed as the percentage of reduction in viable cells in a specific amount of time. At this time, there is not one harmonized criterion that is accepted globally for product preservation. Table 4.6 summarizes the criteria that are most widely used for antimicrobial preservative testing – from the USP, EP, and JP. The EP criteria are the most stringent among the three. Prior to the adoption of the EP criteria, several European countries had their known requirements.

4.4.3   Optimizing and Developing a Compatible Drug Product

A pharmaceutical disperse system rarely contains drug and a vehicle alone. Usually, it contains several adjuvants to enhance several attributes such as physical and chemical stability, microbiological integrity, and esthetic characteristics. The main considerations for the selection of such adjuvants or excipients are the compatibility with the drug molecule, stability at formulation pH, ease of processing, compatibility with packaging components, and cost. Furthermore, these excipients should be preferably selected from various compendial lists.

In order to prepare an optimal dispersion, the formulator should keep in mind the compatibility of (1) wetting agents, (2) deflocculent or dispersing agents, (3) protective colloids, and (4) inorganic electrolytes. In addition to this list, compatibility with tonicity agents, buffers, and packaging components should be kept in mind.

In addition to the experiments performed during the selection of a container/closure system for development, additional testing will be required to fully demonstrate the suitability of a container/closure and package system. The suitability of the container/closure should be evaluated in terms of protection, safety, compatibility, and performance.

The following are the items that should be taken into consideration when finalizing the container for suspension products.

1. Conformity to applicable tests, regulations, and guidelines established in each country for containers, including JP, USP, and EP. It is important to collect relevant information about the additives and pigments used for the container, including safety information and usage experience in other products.
2. Compatibility between the contents and the container/label.

   – Evaluation items: heat stability, light stability, adsorption, and aggregation.

3. Leachables from container/label into the contents
4. Sterilization method of containers.

   – Sterility
   – Deterioration of resin
   – Effect on the content quality
   – Shelf life

5. Specification of container/label.

Table 4.6 Summary of preservative efficacy testing requirements from the USP, EP, and JP

	USP	EP	JP
Bacteria			
Escherichia coli ATCC8739, NBRC3972 *Pseudomonas aeruginosa* ATCC9027, NBRC13275 *Staphylococcus aureus* ATCC6538, NBRC13276	Sterile parenterals (Cat.1): −1log/7d, −3log/14d, NI/28d Non-sterile parenterals (Cat.2): −2log/14d, NI/28d Orals (Cat.3): −1log/14d, NI/28d	No testing requirement, but recommended for orals. Sterile parenterals: A: −2log/6 h, −3log/24 h, NR/28d B: −1log/24 h, −3log/7d, NI/28d Non-sterile parenterals: A: −2log/2d, −3log/7d, NI/28d B: −3log/14d, NI/28d Orals: −3log/14d, NI/28d	Sterile parenterals (IA): −3log/14d, NI/28d Non-sterile parenterals (IB): −2log/14d, NI/28d Oral aqueous (IC): −1log/14d, NI/28d Non-aqueous formulations (II): NI/14d, NI/28d
Yeasts/Molds			
Candida albicans ATCC10231, NBRC1594, JCM2085 *Aspergillus brasiliensis* ATCC16404, NBRC9455	Sterile parenterals (Cat.1): NI/7d, NI/14d, NI/28d Cat.2 and Cat.3 products: NI/14d, NI/28d	Parenteral and opthalamic preperations: A: −2log/7d, NI/28d B: −1log/14d, NI/28d Topical Preperations: A: −2log/14d, NI/28d B: −1log/14d, NI/28d Orals: −1log/14d, NI/28d	All products: NI/14d, NI/28d

- Appearance
- Dimension
- Weight and others
- Content uniformity of a dose (a drop when administered)
- Integrity
- Weight loss (in case of half-permeable container used)

6. Compatibility with the manufacturing equipment for filling, packaging, and labeling processes.
7. Usability of the container by user.

- Ease of handling
- Ease of identification

8. Manufacturing cost.

The following tests should be done to demonstrate suitable protection for the formulation provided by the container/closure:

- USP<661> Containers — light transmission
- USP <671> Containers — permeation (water vapor permeation)
- Seal Integrity and leakage should be evaluated. In the case of a screw-cap closure, the correlation between application and removal torques and leakage (or weight loss)

from container should be evaluated. Inverted storage of the container should be considered to exaggerate the contact between the formulation and the closure seal.
- The automated filling equipment used for filling and sealing the container/closure should be validated appropriately. The seal integrity for the suspension filled on automated equipment should be confirmed by evaluating application and removal torques or by confirming the appropriate stability with regard to leakage or weight loss.
- Container closure integrity should also be confirmed from the standpoint of microbial stability. For example, the microbial limits or sterility of the suspension placed on stability should be confirmed.

The following tests should be done to demonstrate that the container/closure exhibits suitable safety and compatibility:

- USP<661> Containers – physicochemical testing for plastics

  ○ ROI
  ○ Heavy metals

- USP <87> Biological Reactivity Tests, in vitro

  ○ Cytotoxicity

- Extractable/Leachables should be monitored in formulations on stability. Accelerated studies including inverted storage (to exaggerate the contact between the formulation and the closure) should be considered for this purpose.

  ○ Headspace gas chromatography (HS-GC) with flame ionization detection (FID) is a useful method for the detection of volatile leachables in the drug product.
  ○ Gas chromatography-mass spectrometry (GC-MS) using both selected ion monitoring (SIM) and full scan mode is useful for the analysis of specific targeted model compounds and monitoring unknown peaks, respectively.

- Evaluation of particulate contamination coming from the container/closure should be evaluated for injectable formulations.
- Evaluation of loss of functional excipients to semipermeable containers should be monitored in formulations on stability. For example, loss of antioxidant or preservative, resulting from sorption to or diffusion into a plastic bottle or rubber seal could adversely affect the suitability of the container/closure system.

The following tests should be used to demonstrate that the container/closure exhibits suitable performance:

- The bottle holdup/deliverable volume should be evaluated for the container.

  ○ USP <698> Deliverable volume
  ○ In addition, it may be informative to evaluate the number of doses that may be delivered under simulated use in the case that shaking of the suspension (and the effect of entrapped air on the suspension density) affects the number of deliverable doses.

- Dosing uniformity from container/closure should be evaluated under simulated use. Many container/closure systems may affect the uniformity of the suspension dose because of the shape of the container and its effect on resuspendability or because of the way the suspension is delivered from the container. Simulated use testing is recommended to evaluate the ability of the suspension to be resuspended and delivered after settling in the container for an appropriate period of time. Freeze/thaw cycling, high-temperature storage, and inverted storage of the container may also be helpful in evaluating potential failures for the container/closure to dependably provide uniform dosing of the suspension.

a. Documentation Considerations

Throughout formulation and process development, the progress and key experiments should be periodically documented appropriately. Documentation of the information generated and the decisions made is particularly useful at key project milestones like the production of GLP or GMP supplies. A formulation/process development document should be produced at these points in development. Regulatory considerations and documents are discussed in more detail in Chap. 9 of this book.

4.5 Appendix I: Commonly Used Antimicrobial Preservatives

A brief description of the most commonly used preservatives and their method of use is described here. A detailed information on the most commonly used preservatives can be found in various texts (Martindale 1982). These commonly used preservatives are for pharmaceutical use. One should keep in mind that therapeutic antibiotics are rarely employed as preservative agents in pharmaceutical dosage forms because of concerns about the development of microbial resistance to antibiotics and adverse reactions to specific classes of antibiotics. The exception to this rule is in the preservation of some vaccines where the use of neomycin or streptomycin has been approved.

4.5.1 Quaternary Amines

4.5.1.1 Benzalkonium Chloride

Benzalkonium Chloride (BAK or BAC) is most widely used in ophthalmic, otic, and nasal products. It is generally used in the concentrations of 0.004–0.02%. Benzalkonium Chloride contains a mixture of alkyl chain lengths (mostly C_{12}, C_{14}, and C_{16}) of dimethyl benzylammonium chloride. The antimicrobial activity of BAK is significantly dependent upon the alkyl composition of the homolog mixture. Pure homologs of BAK, particularly the C_{12} chain length, have also been used in some ophthalmic preparations.

BAK is active from pH 4–10, very stable at room temperature and readily soluble in water at typical use concentrations, and may be sterilized by autoclaving with loss of efficacy. BAK is active against a wide range of bacteria, yeasts, and fungi, but it is ineffective against some *Pseudomonas aeruginosa* strains. However, combined with disodium edetate (0.01–0.1% w/v), benzyl alcohol, phenylethanol, or phenylpropanol, the activity against *Pseudomonas aeruginosa* is increased. Benzalkonium chloride is relatively inactive against spores and molds, but is active against some viruses, including HIV.

Because it is a cationic surfactant, BAK may lose part of its activity in the presence of citrates or phosphates or other anionic species, above its critical micelle concentration, in the presence of high concentrations of other surfactants, and in the presence of suspended solids (to which it may adsorb, reducing the solution concentration). BAK has been shown to be adsorbed to various filtering membranes, especially those that are hydrophobic or anionic, and it may partition into polyvinyl chloride or polyurethane containers.

BAK is usually nonirritating, nonsensitizing, and well tolerated at the concentrations normally employed on the skin and mucous membranes, but it has been associated with adverse effects when used in some pharmaceutical formulations. Ototoxicity can occur when BAK is applied to the ear and prolonged contact with the skin can occasionally cause irritation and hypersensitivity.

4.5.1.2   Cetrimonium Bromide

Cetrimonium Bromide, a.k.a. Cetrimide, is a water-soluble (up to 10%) preservative used in the concentration range of 0.005–0.01%. Cetrimide contains a mixture of alkyl chain lengths (C_{12}, C_{14} and C_{16}) of trimethylammonium bromide. Cetrimide has characteristics similar to those of BAK; however, it has been found to support the growth of resistant strains of *P. aeruginosa*. It is most effective at neutral or slightly alkaline pH values, and its activity is enhanced in the presence of alcohols. Cetrimide has variable antifungal activity, is effective against some viruses, and is inactive against bacterial spores.

4.5.1.3   Cetylpyridinium Chloride

Cetylpyridinium Chloride (CPC) is a water-soluble quaternary ammonium preservative used in the concentration range of 0.001–0.05%. It is used in mouthwashes, nasal sprays, and formulations for inhalation, oral, and transdermal drug delivery. As an active ingredient of antiseptic oral mouthwashes, it has a broad antimicrobial spectrum with a rapid bactericidal effect on gram-positive pathogens and a fungicidal effect on yeasts, but there are gaps in its effectiveness against gram-negative pathogens and mycobacteria. CPC, like BAK and Cetrimide, is both cationic and surface active and may strongly adsorb to surfaces, interact with other surfactants, or bind with anionic species resulting in a lower effective concentration in solution.

4.5.2 Organic Acids

4.5.2.1 Benzoic Acid/Potassium Benzoate/Sodium Benzoate

Benzoic acid is used as a preservative at concentration range of 0.1–0.5%. It demonstrates antimicrobial activity only in unionized form. Furthermore, it is only effective in acidic formulation with optimal activity at or below pH 4.5. It has moderate activity against gram-positive bacteria, molds, and yeasts, but is less effective against gram-negative bacteria. Benzoic acid is incompatible with quaternary compounds and nonionic surfactants.

4.5.2.2 Sorbic Acid/Potassium Sorbate

Sorbic acid is used as a preservative in the concentration range of 0.05–0.2%. It has been used extensively in the food industry especially as an antifungal agent. Sorbic acid is one of the least toxic preservatives, as it is readily metabolized by mammalian cells via beta and omega oxidation (Deuel et al. 1954). The undissociated sorbic acid exhibits a greater preservative activity than the dissociated molecule (Bandelin 1958; Wickliffe and Entrekin 1964; Ecklund 1983; Bell et al. 1959). Sorbic acid is effective at acidic pH, preferably in the range of 5.5–6.5. However, sorbic acid is unstable at this pH. Therefore, formulators should be careful when employing sorbic acid as a preservative. It is sensitive to oxidation, resulting in discoloration (McCarthy and Eagles 1976). It is also unstable at temperatures above 38°C. The compatibility of sorbic acid with polyvinyl and polypropylene containers should be carefully investigated.

4.5.2.3 Sodium Dehydroacetate

Sodium Dehydroacetate is used as a preservative typically in the concentration range of 0.02–0.2%. It is similar to sorbic acid and benzoic acid; however it is a better anti-microbial at higher pH because of having a pKa of about 5.3. It is effective in acidic formulations. However, it is incompatible with nonionic surfactants, and cationic excipients and may be adversely affected by the metals present in brown glass.

4.5.2.4 Parabens

Parabens are esters of p-hydoxy benzoic acid. They have been used in pharmaceutical and cosmetic products over the last 50 years. Most commonly used parabens are methyl- and propylparaben. The concentration range for parabens preservatives mostly depends on their water solubility but is typically in the range of 0.005% (for butyl) to 0.20% (for methyl). Parabens are safe and effective against molds, yeasts and gram-positive bacteria. The usefulness of parabens is limited by their water solubility. They can also be lost due to absorption to packaging components such as rubber stoppers. Parabens are incompatible with nonionic surfactants. They are effective in the pH range of 4–8 (Gucklhorn 1969).

4.5.3 Alcohols

4.5.3.1 Benzyl Alcohol

Benzyl alcohol has moderate antimicrobial activity at neutral- to -acidic pH. It has also been used as a cosolvent. Its use in ophthalmic formulations is limited, as it causes ocular discomfort and it has rather slow antimicrobial efficacy. Benzyl alcohol is discontinued in pediatric parenterals because of safety concerns.

4.5.3.2 Chlorobutanol

Chlorobutanol is mostly used in ophthalmic ointments as it has good solubility in petrolatum. Chlorobutanol is quite unstable at pH greater than 6 and thus is less desirable to be used in ophthalmic solutions and suspensions. Furthermore, it is volatile and can be lost through the headspace of plastic containers.

4.5.3.3 Phenylethyl Alcohol

Phenylethyl alcohol is rather a mild preservative. Its activity against fungi and certain *Pseudomonas* species is almost bacteriostatic. It is commonly used in the concentration range of 0.25–0.5%. Additionally, it is used in combination with other preservatives such as benzalkonium chloride.

4.5.4 Phenols

4.5.4.1 Chlorocresol

Chlorocresol is used as a preservative at a concentration range of 0.1–0.2%. It has good antimicrobial efficacy against both gram-positive and gram-negative bacteria as well as molds and yeasts. The antimicrobial activity of chlorocresol decreases with increasing pH and it has no activity above pH 9. Chrlorocresol is generally used in injectables. Chlorocresol is absorbed in rubber closures and is incompatible with polyethylene and polypropylene containers. Chrlorocresol solutions turn yellow with exposure to light.

4.5.4.2   Cresol

Cresol consists of a mixture of cresol isomers, predominantly *m*-cresol and other phenols obtained from coal tar or petroleum. It is a colorless, yellowish to pale brownish-yellow, or pink-liquid, with a characteristic odor similar to that of phenol, but more tarlike. An aqueous solution has a pungent taste. Cresol is used at 0.15–0.3% concentration as an antimicrobial preservative for intramuscular, intradermal, and subcutaneous injectable pharmaceutical formulations. It is also used as a preservative in some topical formulations and as a disinfectant. Cresol is not suitable as a preservative for preparations that are to be freeze-dried.

4.5.4.3   Phenol

Phenol is one of the oldest preservative. However, it is now mostly used as a disinfectant. Phenol is toxic, and irritating and has unpleasant smell. It is also incompatible with nonionic surfactants.

phenol exhibits antimicrobial activity against a wide range of microorganisms such as Gram-negative and Gram-positive bacteria, mycobacteria and some fungi, and viruses; it is only very slowly effective against spores. Aqueous solutions of 1% w/v concentration are bacteriostatic, while stronger solutions are bactericidal. Phenol shows most activity in acidic solutions; increasing temperature also increases the antimicrobial activity. Phenol is inactivated by the presence of organic matter.

Phenol is a reducing agent and is capable of reacting with ferric salts in neutral- to-acidic solutions to form a greenish complex. Phenol decolorizes dilute iodine solutions, forming hydrogen iodide and iodophenol; stronger solutions of iodine react with phenol to form the insoluble 2,4,6-triiodophenol.

Phenol is incompatible with albumin and gelatin, as they are precipitated. It forms a liquid or soft mass when triturated with compounds such as camphor, menthol, thymol, acetaminophen, phenacetin, chloral hydrate, phenazone, ethyl aminobenzoate, methenamine, phenyl salicylate, resorcinol, terpin hydrate, sodium phosphate, or other eutectic formers. Phenol also softens cocoa butter in suppository mixtures.

4.5.4.4   Thymol

Thymol is also known as isopropyl-m-cresol and is similar to m-cresol in its properties and use. Thymol is found in oil of thyme and, like the plant, has a very strong aroma. It is widely used in mouthwashes and has also been used as a preservative in the inhalation anesthetic Halothane at 0.01%. In its safety assessment of thymol and other cresols, the Cosmetic Ingredient Review Expert Panel concluded that thymol and other cresols should be used at 0.5% or less (Alan 2006). Thymol, as well as others cresols, may increase the penetartion of other ingredients (Alan 2006).

4.5.5 Mercurial Compounds

4.5.5.1 Phenylmercuric Acetate

Phenylmercuric acetate occurs as a white- to- creamy white, odorless or almost odorless, crystalline powder, or as small white prisms or leaflets. Phenylmercuric acetate is used as an alternative antimicrobial preservative to phenylmercuric borate or phenylmercuric nitrate in cosmetics (in concentrations not exceeeding 0.0065% of mercury calculated as the metal) and pharmaceuticals. It may be used in preference to phenylmercuric nitrate owing to its greater solubility.

Phenylmercuric acetate is also used as a spermicide.

phenylmercuric acetate is a broad-spectrum antimicrobial preservative with slow bactericidal and fungicidal activity similar to that of phenylmercuric nitrate

Incompatible with: halides; anionic emulsifying agents and suspending agents; tragacanth; starch; talc; sodium metabisulfite; sodium thiosulfate; disodium edetate; silicates; aluminum and other metals; amino acids; ammonia and ammonium salts; sulfur compounds; rubber; and some plastics.

Phenylmercuric acetate is reported to be incompatible with cefuroxime and ceftazidime.

4.5.5.2 Phenylmercuric Nitrate

Phenylmercuric salts are used as antimicrobial preservatives mainly in ophthalmic preparations, but are also used in cosmetics, parenteral, and topical pharmaceutical formulations.

Phenylmercuric salts are active over a wide pH range against bacteria and fungi and are usually used in neutral- to -alkaline solutions, although they have also been used effectively at a slightly acid pH. In acidic formulations, phenylmercuric nitrate may be preferred to phenylmercuric acetate or phenylmercuric borate, as it does not precipitate.

Phenylmercuric nitrate is also an effective spermicide, although its use in vaginal contraceptives is no longer recommended.

A number of adverse reactions to phenylmercuric salts have been reported and concerns over the toxicity of mercury compounds may preclude the use of phenylmercuric salts under certain circumstances.

phenylmercuric salts are broad-spectrum, growth-inhibiting agents at the concentrations normally used for the preservation of pharmaceuticals. They possess slow bactericidal and fungicidal activity. Antimicrobial activity tends to increase with increasing pH, although in solutions of pH 6 and below, activity against Pseudomonas aeruginosa has been demonstrated. Phenylmercuric salts are included in several compendial eye drop formulations of acid pH.

Activity is also increased in the presence of phenylethyl alcohol, and in the presence of sodium metabisulfite at acid pH. Activity is decreased in the presence of sodium metabisulfite at alkaline pH (1–3). When used as preservatives in topical creams, phenylmercuric salts are active at pH 5–8 (4).

Bacteria (Gram-positive): good inhibition, more moderate cidal activity. Minimum inhibitory concentration (MIC) against Staphylococcus aureus is 0.5 µg/mL.

Bacteria (Gram-negative): inhibitory activity for most Gram-negative bacteria is similar to that for Gram-positive bacteria (MIC is approximately 0.3–0.5 µg/mL). Phenylmercuric salts are less active against some Pseudomonas species, and particularly Pseudomonas aeruginosa (MIC is approximately 12 µg/mL).

Fungi: most fungi are inhibited by 0.3–1 µg/mL; phenylmercuric salts exhibit both inhibitory and fungicidal activity; for example, for phenylmercuric acetate against *Candida albicans*, MIC is 0.8 µg/mL; for phenylmercuric acetate against *Aspergillus niger*, MIC is approximately 10 µg/mL.

Spores: phenylmercuric salts may be active in conjunction with heat. The BP 1980 included heating at 100°C for 30 min in the presence of 0.002% w/v phenylmercuric acetate or phenylmercuric nitrate as a sterilization method. However, in practice this may not be sufficient to kill spores and heating with a bactericide no longer appears as a sterilization method in the BP 2001.

4.5.5.3 Thimerosal

Thimerosal is a mercurial preservative. It is used in the concentration range of 0.002–0.02%. Several years ago, it was extensively used in ophthalmic products because of its rapid broad-spectrum antimicrobial activity, particularly against *P. aeruginosa*, a virulent pathogen of the eye. Thimerosal is quite stable in the pH range of 6–8. It is incompatible with other mercurial preservatives, nonionic surfactants, and rubber closure.

References

Alan, A.F. Final report on the safety assessment of sodium p-chloro-m-cresol, p-chloro-m-cresol, chlorothymol, mixed cresols, m-cresol, o-cresol, p-cresol, isopropyl cresols, Thymol, o-cymen-5-ol, and carvacrol. Int. Jour. Of Tox. 25 (suppl.1): 29–127 (2006).

Ali, Y., Beck, R., Sport, R. Process for manufacturing ophthalmic suspension, U.S. Patent 6,071,904, 2000.

Bandelin, F.J. *J. Am. Pharm. Assoc., (Sci. Ed.), 46:*691–694 (1958).

Bell, T.A., Etchells, J.L. and Borg, A.F. *J. Bacteriol.,* 77:573 (1959).

Bhargava, H.N., Nicolai, D.W., Oza, B.J. Topical suspensions, in: H.A. Lieberman, R.M. Rieger, G.S. Banker (Eds.), Pharmaceutical dosage forms: Dispense systems, vol. 2, Marcel Dekker, Inc., 1996, pp. 183–241.

Cherian, M. and Portnoff, J. Scale-up of dispensed parenteral doseage forms, in: H.A. Lieberman, M.M. Rieger and G.S. Banker (Eds.), Pharmaceutical dosage forms: Dispense systems, vol. 3, Marcel Dekker, Inc., 1998, pp. 395–422.

Croshaw, B. *J. Soc. Cosmet. Chem.,* 28:3–16 (1977).

Deuel, H.J. et al., *Food Res.,* 19:1–12 (1954).

Ecklund, T. *J. Appl. Bacteriol.,* 54:383–389 (1983).

Gucklhorn, I.R. Antimicrobials in cosmetics. Manufacturing Chemist & Aerosol News. 40:71–75 (1969).

Haines, B.A. and Martin, A.N. *J. Pharm. Sci.,* 50:228 (1961).

Martin, A.N. *J. Pharm. Sci.,* 50:513 (1961).

Martindale, *The Extra Pharmacopoeia,* 28th Ed. (J.E.F. Reynolds, ed.), The Pharmaceutical Press, London, 1982.

McCarthy, T.J. and Eagles, P.F.K. *Cosmet. Toiletries,* 91 (June):33–35 (1976).

Morse, L.J. and Schonbeck, L.E. *N. Engl. J. Med.,* 278:376–378 (1968).

Morse, L.J., Williams, H.L., Grenn, F.P., Eldridge, E.E., and Rotta, J.R. *N. Engl. J. Med.,* 277: 472–473 (1967).

Wickliffe, B. and Entrekin, D.N. *J. Pharm. Sci.,* 53:769–773 (1964).

Wodderburn, D.L. *Adv. Pharm. Sci.,* 1:195–268 (1964).

Chapter 5
Preclinical Development for Suspensions

**Sudhakar Garad, Jianling Wang, Yatindra Joshi,
and Riccardo Panicucci**

Abstract This chapter summarizes the significance of suspension in preclinical development. Majority of the preclinical studies are carried out using suspension. Therefore, it is important to know the physical form change, particle size distribution, ease of manufacturability and physico-chemical stability for the molecules used in preclinical studies. Here, the impact of physicochemical properties and formulation on the oral exposure *in vivo* and toxicity of drug candidates were reviewed in line with other ADME parameters (absorption, distribution, metabolism and elimination). From drug discovery perspective, the latest development of *in vitro* and *in vivo* approaches and the opportunity/limitation to assess the potential risks of drug candidates are summarized. Strategy to apply multiple ADME and formulation tools in lead optimization and candidate selection in drug discovery were also demonstrated. Authors focused more on oral suspension, however, there are a number of other dosage forms where suspension can be applied such as topical, parenteral, and inhalation.

5.1 Introduction

The goal of drug discovery groups is to identify molecules with good solubility, permeability, bioavailability, efficacy, and safety. With the advent of high throughput screening technologies and hydrophobic molecular targets, a large percentage of new chemical entities (NCEs) discovered by scientists have limited water solubility. Such NCEs often present significant hurdles in developing formulations with a good exposure, acceptable absorption variability, and limited food effect. While the solubility can be increased by a number of formulation approaches which are viable

S. Garad (✉), J. Wang, and R. Panicucci
Novartis Institute for Biomedical Research, Cambridge, MA, USA
e-mail: sudhakar.garad@novartis.com and garadsd@yahoo.com

Y. Joshi
Teva Pharmaceuticals, North Wales, PA, U.S.A. (formerly at Novartis Pharmaceuticals, NJ, U.S.A.)

A.K. Kulshreshtha et al. (eds.), *Pharmaceutical Suspensions: From Formulation Development to Manufacturing*, DOI 10.1007/978-1-4419-1087-5_5,
© AAPS 2010

for preclinical or clinical use, depending on the dose, route of administration, and the target patient population (i.e., children, infants, geriatrics, etc.), the preferred formulation may be a suspension or a powder for suspension. The focus of this chapter is on the tools, technologies, and approaches utilized in transition of compounds from drug discovery through preclinical development of molecules and on critical considerations for the development of suspension formulations.

5.2 *In Vitro* Absorption, Distribution, Metabolism, Excretions and Toxicity Profiling

5.2.1 *Physicochemical Characterization of New Chemical Entities in Suspensions*

In today's era, with the large percentage of molecules in preclinical development showing solubility and/or permeability issues, the formulation scientists are faced with the challenge of developing formulation approaches to maximize exposure. During lead optimization and lead selection, the amount of new chemical entities (NCEs) available is rather small (a few milligrams to few hundred milligrams), the time lines are generally very aggressive, and the resources available are rather limited. Therefore, the formulation scientists have to resort to high throughput technologies to limit the use of NCEs and maximize the value of the information. The first attempt often is to improve solubility by a number of formulation approaches which are viable for preclinical or clinical use. Depending on dose, several solvents, surfactant, and other excipients can be employed as vehicles for animal studies. The goal for solution-based approaches is to ensure that the amount of solvents and surfactants used are within the allowable limits and the stability of the compound in the formulation is acceptable. However, often the solubility of molecules is too low to develop a solution formulation. In such cases, a suspension may be the most viable approach at this stage of discovery. Only in very exceptional cases, there are time and resources available to develop special drug delivery systems to optimize exposure.

A good understanding of physicochemical properties of drug candidates is essential for developing suspension formulations leading to suitable pharmacokinetic (PK) and pharmaco-dynamic (PD) performance. Without a good knowledge of physicochemical properties and thorough characterization of the formulation, the profiling and PK/PD results are difficult to interpret. The reliable data serve as surrogate filters for the downstream, more extensive *in vitro* (e.g., cellular or functional assays) and *in vivo* assays (Di and Kerns 2006; DeWitte 2006; Wang and Faller 2007). Most of the assays that are considered critical for quality guard in early drug discovery are described elsewhere (Wang et al. 2007).

Suspensions are generally developed in preclinical assessment of NCEs for the ease of development, handling, and probability of maximizing exposure. Suspension is a heterogeneous system consisting of two phases. The continuous or external phase is generally a liquid or semisolid, and the dispersed or internal phase is made of particulate matter that is mostly insoluble but dispersed throughout the continuous phase. In medicinal suspension products, the drug which is generally contained in the dispersed or internal phase, is intended for absorption when given orally or applied topically. The dispersed phase may consist of discrete particles or it may be an aggregate of particles resulting from an interaction between the particles. Most of the suspensions separate on standing. The formulator's goal is to make stable suspension to reduce the rate of settling and/or permit easy re-suspendability (Lachman et al. 1970).

Bioavailability of many molecules can also be greatly increased by particle size reduction. In general, for poorly soluble molecules, the rate of absorption is known to decrease in the following order: solution > suspension > capsules > compressed tablets > coated tablets. This is true especially for the molecule which belongs to poorly soluble to sparingly soluble category (Nash 1988). There are a number of advantages and disadvantages of suspensions used in preclinical studies.

5.2.1.1 Advantages

(a) Easy to formulate with standard techniques
(b) Higher dose formulations feasible
(c) Control over particle size range (less than 1 μm would be difficult for other dosage form)
(d) Most excipients used in suspension have GRAS (Generally Regarded as Safe Excipients) status
(e) Chemical stability of NCE can be improved, over a solution formulation: those NCEs are unstable in solution, long term stability can be improved by formulating as a suspension

5.2.1.2 Disadvantages

(a) Change in physical form and particle size during storage
(b) Sedimentation/caking may cause the difficulty to redisperse
(c) Content uniformity/dosing variability
(d) Variable exposure in animal models
(e) Salts of weak bases difficult to be stabilized physically in suspension, because of low pKa. A salt with a pKa less than 3.2 tends to convert back to free base

While it is very common to dose suspensions of NCEs in animal models during preclinical evaluation, often in the physical form of molecules is not well characterized. As a result, any changes in the physical form of the molecules may yield variable results from one study to another. Besides the physical form, bioavailability of an NCE from suspension formulation may also be impacted by poor dissolution rate, gut metabolism, permeability or first pass metabolism. Therefore, the authors suggest that to have clarity of information on the cause of varying bioavailability from one study to another, physical form of the molecule must be characterized and a pharmacokinetic assessment of the solubilized NCE given as IV and/oral must be performed relative to a suspension formulation. For long term preclinical studies, suspensions are usually prepared once in a week to avoid processing errors. While it is desired to eliminate processing errors, storage of suspensions for a long period may lead to changes in physical form of the materials; it has been observed that a lot of amorphous compounds crystallize over time during storage, or one crystalline form converts to another form (polymorphism) or multiple forms are being formed. Therefore, to ensure reliable results from one preclinical study to another, physical stability of suspension must be evaluated, leading to qualification of storage time. Physical stability of a suspension is another important attribute. While most suspension would settle with time, an expert formulator would develop suspension that can be easily suspended upon shaking. If this were not to be the case, the physical stability of the suspension must also be evaluated to ensure that the storage time qualified is adequate.

The most commonly used excipients in suspension formulations are celluloses with or without surfactants. The main function of the excipients is to keep particles suspended and to avoid formation of aggregates or clumps. Celluloses used in suspension usually have little or no impact on solubility or bioavailability enhancement (Ecanow et al. 1969).

5.2.2 Particle Size Distribution

Particle size characterization of preclinical candidates is another attribute that is often not evaluated. For compounds that show dissolution-dependent exposure, particle size may have a significant impact on absorption. It is generally preferred to have a narrow particle size distribution for compounds with poor solubility, preferably below 10 μm and rarely a broad distribution from 1 to 100 μm as show in Fig. 5.1. The effect of particle size on exposure is generally evaluated in animals

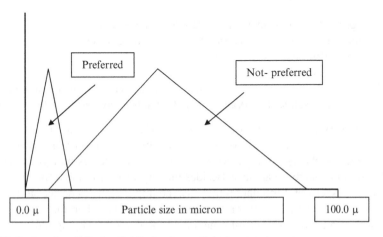

Fig. 5.1 Preferred particle size distribution in preclinical suspensions

during preclinical evaluation. In this study, the molecule is dosed as a suspension with a different particle size. If there is a significant improvement in exposure with reduction in particle size, the micronized molecule can be used for further development. The greatest effect of particle size is often observed for low solubility low dose new chemical entities (van de Waterbeemd et al. 2006). However, for molecules with solubility >1 mg/mL in water and/or aqueous medium in the physiological pH range, particle size is little or no effect on dissolution and/or bioavailability.

Particle–particle interactions in suspension are defined by a number of terminologies such as flocculation, coagulation, and aggregations. Flocculates can be characterized by a fibrous, fluppy, open network of aggregated particles. Their structure is quite rigid; hence these aggregates settle quickly. However, particles with rigid structure are re-dispersed easily because particles are sufficiently far from one another to preclude caking (Lieberman et al. 1998). Coagule is characterized by a tight packing produced by surface film bonding, which settles slowly and is difficult to re-disperse easily; vigorous shaking or stirring is required to redisperse coagule (Lieberman et al. 1998). Dispersed or deaggregated suspension contain individual particles that are dispersed as discrete entities. This type of suspension sediment slowly, depending on the density and zeta potential.

5.2.3 Wetting

Suspension is essentially an immiscible system; it requires some compatibility between the dispersed phase and the medium. A new chemical entity in suspension with poor wetability could result in variable exposure in animal models. Poor wetting may cause particles to float on the surface or to settle quickly in the stored/dispensed containers, which may lead to non-uniform dosing and hence variable exposures. Also, poor wetting may slow dissolution of the drug and thus limit absorption. Therefore, achieving adequate wetting of solid particles by the medium is critical in the development of suspension formulations. The physical stability of a suspension demands that the net energy of interaction between the dispersed particles be repulsive, or in thermodynamic terms, the total energy of interaction between the particles be positive. This means that the particles in suspensions require very good wetting, which is related to their hydrophilicity. Wetting of hydrophobic material is also prevented by the air entrapment on the surface of the particles. The new chemical entities which are hydrophilic wet easily by a polar medium, however a majority of new chemical entities in preclinical studies are hydrophobic. Hydrophobic materials are extremely difficult to wet and often float on the surface of the liquid medium. Wetting can be improved by using a number of surface modifier, which decrease the interfacial tension between solid/solid particle and solid/liquid interface. *The mechanism of surfactant involves* the preferential adsorption of the hydrocarbon chain by hydrophobic surface, with the polar moiety being directed towards aqueous phase. There are number of mechanism involved in adsorption of surfactants at the solid/liquid interface such as (a) ion exchange, which is substitution of previously adsorbed ions on the solid by surfactant ions of identical charge; (b) ion pairing which is adsorption of surfactant ions on the surface of the opposite charge not occupied by counter ions; (c) acid base interactions, mainly hydrogen bonds; (d) adsorption by polarization of pi electrons when the adsorbate has aromatic rings in its molecule and the adsorbent possess positive sites; (e) adsorption by dispersion forces occurs via London-van der Waals dispersion forces between adsorbent and adsorbate molecules; (f) hydrophobic bonding, which is the bonding between molecules that are adsorbed onto or adjacent to other molecules already adsorbed on the solid adsorbent (Fernando et al. 2000). Celluloses such as sodium carboxymethyl cellulose, methyl cellulose, hydroxypropyl cellulose (Klucel), hydroxypropyl methylcellulose are commonly used as a hydrophilic medium. These polymers exert viscosity modifying effect. It is very important to select the right percentage of cellulose and grade to avoid gelling and thixotropy effect. Thixotropy is referred to formulation of a gel like structure that is easily broken and becomes fluid upon agitation.

5.2.4 Crystal Habit

Shape and size give rise to descriptive terms applied to the typical appearance, or agglomeration of crystals called crystal habit. Crystal habit is classified on the basis of geometry of the agglomerates. There are six major crystal habits such as cubic, hexagonal, tetragonal, orthorhombic, monoclinic and triclinic. All these different crystal habits can be distinguished by difference in observed extinction such as isotropic, parallel and symmetrical (Brittain 1999a, b). While it is not very critical to understand the crystal habit for preclinical new chemical entities, it's important to understand the crystal habit, size, density and flow properties of an NCE selected for development. During preclinical evaluations, NCEs are synthesized at milligram scales and end product is recovered by either rotary evaporation or lyophilization. The resultant NCE in the flask or vial often ends up as an amorphous form or semi-crystalline form with lots of solvent adsorbed on the surface. If an amorphous or an unstable semi-crystalline NCE is used in a suspension, conversion into the stable form will inevitably occur over time.

5.2.5 Ionization Constant (pKa)

Ionization constant, pKa, is a useful thermodynamic parameter to monitor the charge state of drug candidates. pKa data are utilized not only for better understanding of binding mechanisms, but can also help predict ADME properties of NCEs due to the pH gradient of 1.0–7.0 in the human gastro-intestinal (GI) tract (Avdeef 2003). For instance, lipophilicity, solubility pH profile, permeability, and human ether-a-go-go related gene (hERG) binding affinity are modulated greatly by pKa.

While pKa can be successfully predicted by *in silico* tools owing to the high dependence on the molecular structure of NCEs, the prediction is less accurate for early discovery compounds in comparison to commercial drugs, thereby justifying the needs for *in vitro* determination. A range of experimental approaches (Wan and Ulander 2006; Wang and Faller 2007; Wang and Urban 2004; Wang et al. 2007) such as potentiometric titration (Avdeef 2003), capillary electrophoresis (CE) (Cleveland et al. 1993; Ishihama et al. 2002) and Spectral Gradient Analyzer (Box et al. 2003) are widely applied to pKa determination at varying throughput, cycle time, sample requirement, and cost at different stages of drug discovery.

5.2.5.1 Potentiometric Titration

It generally records the pH changes, with a glass electrode, caused by introducing a known volume of titrants to the well-mixed solution of a drug candidate. This methodology is tedious as it requires a lengthy process due to long and repetitive equilibrium steps after titrant additions. However, the pKa values obtained are reliable, and therefore, this methodology is considered to be the gold standard despite a number of limitations. This methodology is only suitable for compounds with good solubility, as potentiometry requires concentrations in the 0.1–1 mM range. Although co-solvents have been used to circumvent solubility, to determine pKa, it still requires several titrations to extrapolate the data from different water and cosolvent mixtures to aqueous solution (zero cosolvent concentration). An additional problem with potentiometric titration is that the increasing number of NCEs in early discovery are delivered as salts with protogenic counter-ions, like acetate or fumarate. Finally, this methodology requires materials in mg scale, and therefore, it is more useful in the late discovery than in preclinical or early development.

5.2.5.2 Spectral Gradient Analyzer

This technique is based on a continuously flowing pH-gradient and a UV-DAD (diode array detector). It was developed recently and allows for a much higher throughput (Box et al. 2003). The method establishes a stable time-dependent pH gradient by rapidly mixing acidic and basic buffers, during which drug candidates pre-dissolved in organic solvent are introduced at different pH conditions. The assay works effectively with poorly soluble NCEs by using co-solvents in the media. The pKa data measured using this "rapid-mixing" approach correlate well with those from the potentiometric titration method. In addition, this method is useful for measuring pKa for early discovery compounds as it requires small amount of material, and allows for high throughput. The limitation is that it is only suitable for ionizable compounds, which induce a change in the UV spectra. In other words, not only will a UV chromophore be required, but also the chromophore may have to be located close enough to the ionization center within an NCE. As per our experience, this fast method allows full characterization of ionizable groups for ca. 70% of the discovery output and usually gives high quality data, that are consistent with those obtained by potentiometric titration. Neither this "rapid mixing" nor the potentiometric titration approach is compound specific, so compounds with purity or stability issues or counter ions containing similar UV chromophore may be problematic for this approach.

5.2.6 Lipophilicity (Log P/D)

Lipophilicity, as expressed by the logarithms of partition coefficient or distribution coefficient (Log P or Log D) of NCEs in a lipophilic phase (e.g., octanol) and aqueous phase, is valuable in explaining behavior of some NCEs in the absorption, distribution, metabolism, excretions and toxicity (ADMET) assays (Avdeef 2003). While shake-flask is the conventional method for Log P (or Log D) determination, the dual-phase potentiometric titration approach is also widely accepted, particularly during late drug discovery phase (Avdeef 2003). For NCEs lacking an ionizable group, HPLC log P technique, also known as eLog P can be applied (Lombardo et al. 2000). A variety of techniques that are suitable for early discovery include liposome chromatography, immobilized artificial membrane (IAM) chromatography, capillary electrophoresis (CE) (Avdeef 2003), and artificial membrane preparations (Wohnsland and Faller 2001).

5.2.7 Permeability and Transporters

5.2.7.1 Gastrointestinal Absorption

Absorption of oral drugs is a complex phenomenon and it involves multiple mechanisms across the GI mucosa (Artursson and Tavelin 2003). In silico models are found useful to flag the potential permeability limitations of NCEs. Egan and coworker (Egan et al. 2000) have developed absorption model using polar surface area (PSA) and calculated log P (clog P). The data obtained by them show that almost all highly permeable marketed drugs are populated in an egg-shaped zone of the absorption model specified as "good." In contrast, poorly permeable drugs or those considered substrates for efflux transporters scatter outside the outer zone of the egg considered "poor" in the model. A number of in vitro models were developed to predict permeability as well as to assess the contributions of active transporters in the permeation process (Hämäläinen and Frostell-Karlsson 2004; Balimane et al. 2006).

5.2.7.2 PAMPA

Parallel artificial membrane permeability assay (PAMPA), pioneered by Kansy et al. (1998), utilizes a chemical membrane immobilized on a 96-well filter plate and samples are analyzed by a UV plate reader. Distinct membrane models were established by Kansy et al. (1998), Avdeef et al. (2001), Wohnsland and Faller (2001) and Sugano et al. (2001) to mimic passive diffusion of NCEs across the GI tract. Avdeef (2003) introduced the "double-sink" model that simulates the concentration and pH gradient across the GI membrane. Co-solvents (Ruell et al. 2004) or excipients (Kansy et al. 2001; Sugano et al. 2001; Liu et al. 2003; Bendels et al. 2006) were employed in PAMPA to overcome the low solubility issues frequently encountered in early discovery. Quantification using LCMS (Balimane et al. 2005; Wang and Faller 2007; Mensch et al. 2007) or HPLC (Liu et al. 2003) drastically improved the sensitivity and robustness of PAMPA by extending the limit of detection for NCEs with low solubility. It also prevents interference originating from impurity with high solubility and/or strong UV chromophore. PAMPA ultimately offers a fast and relatively cost effective method to estimate permeability for NCEs absorbed by passive diffusion mechanisms.

5.2.7.3 Caco-2

The human colon adenocarcinoma (Caco-2) cell permeability model exhibits morphological (e.g., tight junction and brush-border) and functional similarities (e.g., multiple transport mechanisms) to human intestinal enterocytes (Ungell and Karlsson 2003; Englund et al. 2006), thereby serving as the gold standard for *in vitro* permeability assessment in industry (Artursson and Tavelin 2003; Lennernas and Lundgren 2004). Caco-2 cells, which extensively express a variety of transport systems beyond P-glycoprotien (Pgp), are amenable to investigate the interplay among different transport systems and differentiate the relative contributions from passive and active transport mechanisms to the overall permeability across the human GI tract (Ungell 2004; Steffansen et al. 2004). While the conventional protocol (e.g., 21-day cell culturing) is essential to assure the full expression of transporters, the Caco-2 models using accelerated cell culturing procedures (e.g., 3–7 days) are not suitable for studying active transport of NCEs due to inadequate expression of transporters (Liang et al. 2000; Alsenz and Haenel 2003; Lakeram et al. 2007). Scientists (Yamashita et al. 2002; Miret et al. 2004) are still debating on if the short-term Caco-2 culturing system is appropriate to rank permeability, with the concerns on its poor correlation with human absorption data and inability to differentiate medium permeability compounds (Miret et al. 2004).

Today, the cumbersome long culturing procedure can be readily handled by automation (Saundra 2004; Wang and Faller 2007) and therefore Caco-2 has been successfully validated and widely implemented in the 24- or 96-well plate formats to assess the permeability and drug interaction with GI-related transporters (Ungell and Karlsson 2003; Kerns et al. 2004; Marino et al. 2005; Balimane et al. 2004). Similarly to PAMPA, precaution should be taken when dealing with low solubility compounds in discovery to choose agents inert to the permeability and transport process of the Caco-2 model (Yamashita et al. 2000). Introduction of bovine serum albumin (BSA) to the basolateral compartment is useful to mimic the *in vivo* sink condition and to help minimize the non-specific binding of NCEs to the cells and labware but the effect appeared to vary greatly upon the mechanism by which NCEs are transported across the monolayer (Saha and Kou 2002; Neuhoff et al. 2006).

"Bi-directional" approach is typically utilized to assess the transport mechanism in the Caco-2 model where permeability is measured from "apical" to "basolateral" compartments [absorptive permeability, $P_{app}(A-B)$] and in the reverse direction [secretory permeability, $P_{app}(B-A)$] (Artursson and Tavelin 2003; Lennernas and Lundgren 2004; Ungell 2004; Steffansen et al. 2004; Hochman et al. 2002; Varma et al. 2006). The NCEs that function as substrates for efflux transporters are one of the major concerns in discovery as they may significantly limit molecules from absorption into the enterocytes and GI membrane and eventually retard the exposure (Varma et al. 2006). Historically, efflux ratio (ER) has been utilized to identify the NCEs with potential efflux issues

$$ER = \frac{P_{app}(B-A)}{P_{app}(A-B)} \qquad (5.1)$$

while the classification boundary may vary by lab, NCEs with ER >> 1 are characteristic of potential efflux substrates and those with ER ~ 1 are dominated by passive mechanism(s). Once oral absorption of a drug candidate or scaffold is limited by efflux-dependent GI permeability, Caco-2 mechanistic study may help establish SAR, allowing for dialing out the efflux issue by optimization (Troutman and Thakker 2003a; Hochman et al. 2002). Experimentally, one can identify the transporters (e.g., Pgp) that NCEs may serve as substrates for and also the potential enhancement in oral absorption when primary transporters are inhibited. In addition, ER and $P_{app}(A-B)$ can be assessed at elevated NCE concentrations to investigate the potential of saturated transporters at high dose and eventually build in vitro-in vivo correlation (IVIVC) for highly soluble or formulated NCEs (Hochman et al. 2002).

As Pgp is, by far, the most prevailing one among transporters in human intestinal enterocytes, one frequently initiates the first transporter mechanistic study by applying potent Pgp inhibitors such as Cyclosporin (first generation), Verapamil and Valspodar (secondary generation), Zosuiquidar/LY335979, and Elacridar/GF120918 (third generation) (Kuppens et al. 2005; Nobili et al. 2006). Troutman and Thakker found ER is unable to properly characterize the Pgp-inhibition mediated enhancement of absorptive permeability due to the asymmetric behavior of Pgp substrates in the absorptive and secretory transport (Troutman and Thakker 2003a). Instead, absorption quotient (AQ) is recommended to better predict how Pgp-facilitated efflux activity attenuates intestinal permeability in vivo (Troutman and Thakker 2003b).

$$AQ = \frac{P_{Pgp}(A-B)}{P_{PD}(A-B)} = \frac{P_{PD}(A-B) - P_{app}(A-B)}{P_{PD}(A-B)} \qquad (5.2)$$

where $P_{Pgp}(A-B)$ is absorptive permeability (or apparent permeability from apical to basolateral direction) attributed to Pgp activity and $P_{PD}(A-B)$ is absorptive permeability measured in the presence of Pgp inhibitor(s).

AQ differentiates the absorptive permeability from secretory transport and offers a more relevant approach to quantify the functional activity of transporters such as Pgp observed during absorptive permeability. For a potent transporter (e.g., Pgp) substrate, inhibition of transporter (or Pgp) activity usually leads to drastically enhanced absorptive permeability $[P_{PD} (A - B) \gg P_{app} (A - B)]$ and thereby $AQ \approx 1$. On the other hand, comparable absorptive permeability values are anticipated (in the presence or absence of transporter inhibitor) for a weak transport substrate $[P_{PD}(A-B) \approx P_{app}(A-B)]$ or $AQ \approx 0$.

5.2.7.4 MDRI-MDCK

The Madin-Darby canine kidney (MDCK) model originating from dog kidney with different expression of transporters than human intestine (Balimane et al. 2000; Irvine et al. 1999) has commonly been utilized for permeability evaluation of NCEs using the passive diffusion mechanisms (Ungell and Karlsson 2003). Recent developments using Pgp-transfected multidrug resistance-1 (MDRI)-MDCK model (Bohets et al. 2001) allows for estimating the contributions of efflux transporters with reduced cell culturing cycle time (3–5 days). By regulating the level of Pgp expression, the sensitivity of NCEs to Pgp in the MDCK assay may be amplified, although the relevance of this to GI physiology has yet to be established.

Given the advantages and limitations of each approach, the latest consensus appears to favor a strategy that combines all three approaches with *in silico* models to ensure high quality assessment of permeability in early discovery (Kerns et al. 2004; Balimane et al. 2006; Faller 2008). A flow chart is proposed to collectively utilize all *in silico*, *in vitro* and *in vivo* tools to assess the absorption of NCEs (Fig. 5.2). PAMPA should serve as a fast and high-throughput permeability ranking tool in particular for scaffolds using passive diffusion mechanisms. Analysis of Novartis discovery compounds using Absorption model developed by Egan found that ~94% of NCEs in Absorption Model "good" zone fall in either "high" or "medium" ranks in PAMPA (LCMS approach), leading to decent predictivity. NCEs in "poor" or "Borderline" zones however showed mixed results in PAMPA, assuring the necessity of *in vitro* assessment. We found that NCEs fall into "low" permeable zone in PAMPA coincide with those in the same rank of Caco-2. Therefore, the Caco-2 model should be applied to challenging scaffolds involving active transport mechanisms or with higher molecular weight (e.g., >600). The former (potential substrates/inhibitors for efflux transporters), most likely, will exhibit "medium" to "high" permeability in PAMPA but poor *in vitro-in vivo* correlation. Caco-2 mechanistic studies are valuable to identify the major transporters such as Pgp, MRP2 and BCRP and to appraise the impact of shutting-down active transporters via either inhibitory (Varma et al. 2003) or saturation mechanisms (Bourdet and Thakker 2006). MDCK, expressed with a specific transporter, may be ideal to tackle the impact of an individual transporter subsequent to Caco-2 transporter assays (Varma et al. 2005).

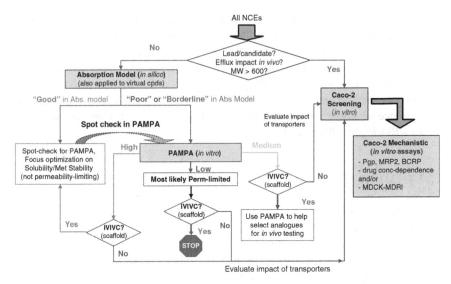

Fig. 5.2 Proposed flow chart to address permeability and transporters using collective *in silico*, *in vitro* and *in vivo* ADME tools in early discovery. All NCEs should be subject to evaluation using *in silico* tools such as absorption model (Egan et al. 2000). Whereas "good" predicted by absorption model should be spot-checked, most of "poor" and "borderline" given by the *in silico* model are recommended to follow up by *in vitro* permeability assay such as PAMPA. "Low" in PAMPA is found predictive to Caco-2 and also *in vivo* results. "Medium" and "high" in PAMPA should be spot-checked by *in vivo* data, where outliers should be tackled by the comprehensive suite of Caco-2 and MDCK screening and mechanistic assays

5.2.7.5 Blood–Brain Barrier

The blood–brain barrier (BBB) formed by the endothelial cells lining the cerebral microvessels has built an effective defense system for the brain parenchyma (Nicolazzo et al. 2006; Hitchcock and Pennington 2006). It however also causes severe burden for the development of central nervous system (CNS) and brain tumor drugs requiring penetration into brain tissue across BBB characterized by tight intercellular junctions. Efflux transporters, such as P-gp, BCRP and organic anion transporting polypeptide (OATP) introduced extra barrier for CNS exposure by actively propelling compounds back into the blood (de Lange 2004). For *in vivo* and *in situ* approaches to assess BBB permeability of NCEs, brain uptake by the intravenous (IV) injection technique was referred to as the "gold standard" (Smith 2003) due to its high sensitivity, availability of direct pharmacokinetic parameters as well as the ease of the experiment in comparison to other *in vivo* and *in situ* methods. A number of *in vitro* methods are utilized as well. Primary (bovine, human, mouse and rat) brain capillary endothelial cell cultures present the closest phenotypic resemblance to the *in vivo* BBB and have recently become standard assays for lead optimization and candidate selection processes in early drug discovery (Lundquist and Renftel 2002; Terasaki et al. 2003; Garberg et al. 2005). The application of these models, despite closely mimicking the physiology of the BBB and offering realistic expression of transport proteins (Hansen et al. 2002; Lundquist and Renftel 2002), is somewhat limited by the complicate culturing procedure as a result of "leaky"junction. Utility of non-cerebral cell models such as MDCK-MDR1 and Caco-2 appears to gain popularity lately but their predictivity to BBB is yet to be fully established due to distinct origin of cell genes and missing of other transporters in the model systems. The BBB models utilizing the parallel artificial membrane (Di et al. 2003) and immobilized artificial membrane (IAM) chromatography approaches (Reichel and Begley 1998) offer great potential for predicting BBB in early discovery phase by evading cumbersome cell culturing process.

5.2.8 Metabolism

5.2.8.1 Determinants to Assess Drug Exposure *In Vivo*

The absolute systemic availability, or bioavailability ($F\%$) of a drug can be described by (5.3)

$$F = F_a \cdot F_i \cdot F_h \qquad (5.3)$$

where F_a is the fraction of a drug dose absorbed across GI membrane that is proportional to solubility, dissolution, and intestinal permeability discussed in Sects 5.27 and 5.3. F_i and F_h represent the fraction escaping the first-pass intestinal (i) and hepatic (h) elimination, respectively.

Drug substances that permeate the GI tract are immediately subject to hepatic metabolism in the liver before they are exposed to systemic circulation and, most often, the therapeutic target. Though oral absorption of NCEs can be confirmed *in vivo* by monitoring drug concentrations in the portal vein, practical PK studies are in favor of sampling from the circulating plasma. This allows exposure of NCEs to be assessed systematically by comparing the maximum drug concentration (C_{max}) or area under curve (AUC) of the PK profile subsequent to a per os (PO) administration with those from the direct intravenous (IV) route, commonly referred to as bioavailability. Those PK parameters are essential to measure the availability of NCEs at the target site and for the SAR based optimization of leads and selection of development candidates. Without decent understanding of hepatic clearance and predictive *in vitro* tools to assess metabolic clearance, it is extremely difficult to properly interpret initial *in vivo* PK data, design further *in vivo* PK experiments or decipher the NCE properties that need to be optimized for maximum impact on bioavailability.

In general, a certain degree of metabolism is desired to help the clearance of NCEs from the human body. In early discovery, however, medicinal chemists are mostly concerned by the hepatic instability of NCEs and by the ultimate impacts on exposure, as a metabolically unstable NCE, albeit orally absorbed, might never reach the required therapeutic concentration. In addition, liver metabolism varies greatly between species therefore it is crucial to generate data in different species to be able to highlight possible PK-PD differences way before human data are available (Obach 2001; Naritomi et al. 2001). The currently available approaches to monitor metabolic clearance include the use of recombinant cytochrome P450 (CYP) enzymes, liver microsomes, S9 fraction (the 9000G supernatant of a liver homogenate), isolated hepatocytes and liver slices (Obach et al. 1997; Obach 1999; Lau et al. 2002a).

5.2.8.2 High Throughput Metabolic Clearance Assessment Using Liver Microsomes

Liver microsomes, enriched for CYP content, remain the accepted fast screening approaches in early discovery. CYPs (commonly referred to as "Phase I" metabolism) dominate the metabolic clearance of marketed drugs (Wienkers and Heath 2005). UGT (UDP-glucuronosyltransferase) activity can also be reconstituted in the microsomal assays to predict metabolism by glucuronidation (so called "Phase II") mechanisms. The concerted activation of the CYP and UGT enzymes sometimes may over-estimate the *in vivo* relevance of glucuronidation (Lin and Wong 2002). The analytical method of choice is LC-MS/MS to probe the disappearance of the test compound. The assays report either half-life and/or intrinsic clearance values (Obach 1999; Linget and du Vignaud 1999; Caldwell et al. 1999; Korfmacher et al. 1999) or percentage remaining of the parent compound at 1–2 time point(s) (Di et al. 2004). The co-solvent approach is applicable to minimize precipitation of insoluble NCEs and the subsequent false positives in metabolic clearance, although the risk to CYP inhibition must be carefully monitored (Di et al. 2006).

5.2.8.3 Hepatocyte Clearance (CL)

To account for the contributions of hepatic transporters and other metabolizing proteins, measurement of hepatocyte clearance, albeit at the expense of reduced throughput and increased cost (McGinnity et al. 2004; Reddy et al. 2005), indeed offers a number of advantages over the microsomal assays (see review article by Hewitt et al. 2007). First, it maintains the cell integrity of hepatocytes and avoids the destructive technique used in the microsomal preparation. This retains not only transporter protein systems which may be important for either the uptake or efflux of NCEs and/or metabolites but also cytosolic enzymes such as esterases, amidases, and alcohol dehydrogenases that may also significantly contribute to the metabolism of NCEs. In addition, utility of hepatocytes provides a more physiologically relevant model to measure hepatic clearance than the "artificial" microsomal systems. In the latter, one has to monitor cytochrome P450 metabolism by introducing cofactors (NADPH), or to predict metabolism by glucuronidation using reconstituted UGT.

All these, along with the missing contributions from hepatic transporters and other metabolizing enzymes, may cause additional challenges to reproduce the physiological sequential/parallel metabolism of a compound *in vivo*. In this aspect, hepatocyte clearance assays, as a self-consistent hepatic model, demonstrate improved predictivity to *in vivo* clearance data over microsomal assays (Hewitt et al. 2007). Hewitt et al. (2007) consolidated the cryopreserved human hepatocyte clearance data of 42 marketed drugs from four reports (Fig. 5.3), demonstrating a good correlation between predicted and observed *in vivo* clearance values. Soars et al. (2007a) also demonstrated a correlation between *in vivo* clearance and those predicted from *in vitro* hepatocyte experiments using discovery NCEs, where most of the "outliers" were from the chemical series with significant contributions from hepatic transporters or renal clearance.

The technical aspects of hepatocyte clearance assays were extensively examined. First, discrepancy was observed in measured *in vitro* CLint for conventional monolayer cultures when compared with suspensions, possibly attributing to the improved mixing in the reaction matrix and enhanced access of NCEs to the core structure in hepatocyte suspensions over cultured monolayer (Blanchard et al. 2005; Hewitt et al. 2007). Another critical comparison was focused on the choice of primary cultured vs. cryo-preserved hepatocyte systems. A number of studies (Soars et al. 2002; Shibata et al. 2002; Lau et al. 2002b; McGinnity et al. 2004; Blanchard et al. 2006; Hewitt et al. 2007; Jacobson et al. 2007; Li 2007) support the routine use of cryo-preserved human hepatocytes. Li (2007) concluded that

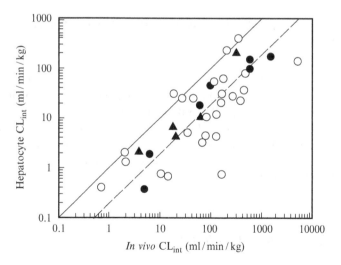

Fig. 5.3 Comparison between observed human *in vivo* CL$_{int}$ and CL$_{int}$ predicted using cryo-preserved hepatocytes for 41 substrates. *Open symbols* represent predictions from drug depletion studies by Soars et al. (2002), Shibata et al. (2002) and Lau et al. (2002b); *closed symbols* from metabolite formation studies by Hallifax et al. (2005). The *solid line* represents the line of unity, whereas the *dashed line* represents a 5.6-fold under-prediction. (The figure is taken from Hewitt et al. 2007, with permission.)

fresh and cryo-preserved human hepatocytes have comparable CYP, UGT and Glutathione (GSH) activity. Some lots of cryo-preserved human hepatocytes also retain hepatic transporter activities. In addition, the utility of cryo-preserved hepatocytes also exhibits several benefits over freshly prepared primary hepatocytes such as steady supply of the hepatocyte cells, pooling samples from different donors, improved cell variability and reduced operating expenses. Nonetheless, it requires a good practice to properly handle the thawing of cryo-preserved hepatocytes (Jacobson et al. 2007). Cryo-preserved human hepatocytes may not readily be cultured in monolayer cultures (Li 2007). Rather, they are more suitable to applications in suspension format where limited lifespan should be utilized. In addition, precaution should be taken in the GSH or transporter assays due to the reduced level of GSH and transporter expression in cryo-preserved human hepatocytes. Moreover it is advisable to minimize the utility of organic solvent such as DMSO due to its deleterious effects on the results (Soars et al. 2007a).

Numerous medium to high throughput hepatocyte clearance assays were developed lately in automated format (Reddy et al. 2005; Jacobson et al. 2007). It is suggested that usage of cryo-preserved human hepatocytes in suspension in microtiter plate format provide a satisfactory tool for predicting hepatic clearance in early discovery. In addition, the "media loss" approach was reported to monitor the interplay of all enzymes and transporters within hepatocytes and to predict the *in vivo* clearance of NCEs with significant hepatic uptake (Soars et al. 2007b).

5.3 High Throughput Techniques for Characterization of NCEs

High throughput technology can be utilized for multiple applications at the drug discovery/development interface. In this section, we will focus on pharmaceutics/drug delivery related techniques. High throughput screening (HTS) is preferred to maximize the information obtained in the fastest possible time with the least amount of material (Balbach and Korn 2004). Following applications of High throughput technology have been used in pharmaceutical industry for the physico-chemical characterization of NCEs.

5.3.1 Solubility and Dissolution

Aqueous solubility of an NCE and it's dissolution rate are two critical parameters that may affect both *in vitro* and *in vivo* attributes. Additionally, interpretation of *in vitro* efficacy data and profiling assays such as permeability, clearance, hERG, or other toxicity and drug interaction assays can be difficult and misleading without solubility information. Solubility is one of the two key parameters for waiver of *in vivo* bioequivalence studies on the well known Biopharmaceutics Classification System (BCS) recommended by the FDA (Amidon et al. 1995), as the reproducibility and predictability of absorption may depend on the drug substance solubility; poor aqueous solubility can lead to failure in drug development.

A number of approaches have been utilized to assess solubility. Softwares have been developed to assess the solubility based on *in silico* approach, utilizing the structural information. However, the predictability of thermodynamic solubility of molecules is complex due to an interplay among hydrogen-bond acceptors and donors, conformational effects, and crystal packing energy. Therefore, limited *in silico* models were developed on the pharmacophore basis that are extremely valuable for chemical optimization of solubility via the structure-property relationship (SPR) approach. Rather, many solubility prediction models are mathematically based such as quantitative SAR (QSAR). In this case, deviations from the predicted values for dissolution is highly dependent of the training set and whether such data are for non-drug-like chemicals in the non-physiologically relevant solubility range (10^{-12}–10^{3} g/L) (Bergström et al. 2004; Wang and Urban 2004). Therefore, it should not be surprising if a commercial *in silico* tool works reasonably well only for non-drug-like molecules but not for drug discovery compounds. Furthermore, some *in silico* predications may require other physico-chemical parameters such as partition coefficient (Log P) or melting points (MP) (Liu and So 2001; Chen et al. 2002) of the test compounds that are not always available in the early stage of drug discovery.

Experimentally, thermodynamic solubility is commonly employed to monitor the exposure of test compounds, to predict *in vitro–in vivo* correlations, as well as to understand the interplay among absorption, distribution, metabolism, excretion, and toxicity (ADMET) parameters. Generally, the Kinetic or apparent solubility is determined for characterizing and qualifying biological and ADMET profiling assays owing to similar experimental conditions used. Intrinsic solubility can be used to assess the solubility of NECs in their neutral species and predict their impact of permeability, which is believed to correlate better with the availability of neutral molecules in solution.

5.3.1.1 Apparent or Kinetic Solubility

A variety of high throughput approaches have been developed to scrutinize the kinetic events of solubilization of NCEs and rapidly estimate aqueous solubility in early drug discovery. The approach, first introduced by Lipinski, generally featured direct introduction of NCEs in DMSO solution into the designated media, followed by an immediate readout using optical detection such as turbidimetry (Lipinski et al. 1997; Kibbey et al. 2001), nephelometry (Bevan and Lloyd 2000; Kariv et al. 2002), laser flow cytometry (Stresser et al. 2004), or direct-UV (using a UV plate reader) (Pan et al. 2001), or HPLC (Zhou et al. 2007). In most cases, multiple samples are prepared with increasing concentrations of the test compound so that one can define the solubility by monitoring the concentration where particles are falling out of the solution. For direct UV or HPLC approaches, a procedure to secede the solution from undissolved solid is needed. The advantage of kinetic solubility is fast turn-around time, cost-effectiveness, and the use of relatively universal detection modes. Challenges include the interference of low soluble impurities with the readout, unless an analytical separation is involved such as HPLC.

5.3.1.2 Thermodynamic or Equilibrium Solubility

The conventional thermodynamic solubility, also referred to as "saturation shake-flask" method, starts with solid material, reflecting not only the molecular interactions of NCEs in solution phase but also their packing information in solid phase (polymorph) (James 1986; Yalkowsky and Banerjee 1992; Grant and Brittain 1995). The assay typically involves a long (24–72 h) agitated incubation of pre-weighed solid samples in a designated buffer at a specific temperature followed by phase separation (of saturated solution from solute) and quantification (Avdeef et al. 2007; Zhou et al. 2007). This method is considered the "gold standard" for solubility determination by industry and the FDA. This approach, however, is time-consuming and labor-intensive, and it also requires significant sample material and resource, and therefore, can not accommodate the need in early discovery. For example, filtration and centrifugation approaches are commonly employed to separate the saturated solution from excess solid subsequent to incubation. Solubility of NCEs with poor physicochemical properties could be underestimated by filtration procedures due to adsorption or non-specific binding to the filtration membrane. For the centrifugation approach, overestimation commonly occurs for "fluffy" compounds. Frequently, one tends to use one approach as the primary tool and the other as a complementary method for challenging compounds. Alternatively, one can run both separation approaches in parallel, to assure the accuracy of the determination. In addition, LC-UV, LC-MS and LC-chemiluminescent nitrogen detector (CLND) are utilized to quantify solubility, which may require development of special analytical methods or an accurately measured calibration curve. Therefore, the time consuming and resource intensive conventional shake-flask approach is ideal for compounds in late discovery stage or in development, when the number of compounds are limited and sufficient amount of compound is available in crystalline forms.

5.3.1.3 High-Throughput (HT) Equilibrium Solubility Using Miniaturized Shake-Flask Method

In early discovery, alternative viable high-throughput solubility approaches are employed during lead optimization and candidate selection, and the information is used to establish *in vitro/in vivo* correlation and structure-property based relationship (SPR). Kinetic solubility, albeit useful in qualifying biological and *in vitro* ADMET profiling results, generally fails in predicting thermodynamic solubility for drug discovery compounds (Zhou et al. 2007) and in establishing SPR (Avdeef et al. 2007). This is mostly due to the simplified protocol with insufficient incubation

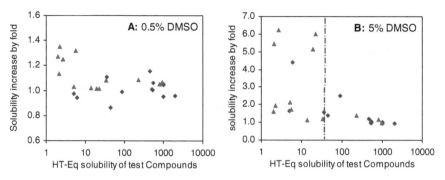

Fig. 5.4 Increase of HT-Eq solubility values caused by the presence of 0.5% (**a**) and 5% DMSO. Marketed drugs (*filled diamond*) and in-house NCEs (*filled triangle*) were used for the evaluation. Data show that the DMSO-induced solubility enhancement is more significant for compounds with solubility lower than 40 M as marked in *dotted line*, and/or for in-house discovery NCEs (Figure is taken from Zhou et al. 2007, with permission.)

time, improper handling of the phase separation process, presence of DMSO (Balakin 2003; Zhou et al. 2007), and indirect readout of solubility data. Wang and coworkers found that presence of a small amount of DMSO in aqueous buffers (0.5–5%) led to drastic overestimation of equilibrium solubility of NCEs and marketed drugs (Fig. 5.4). Unfortunately, it is unlikely to predict the magnitude of the DMSO-dependent solubility estimation based on NCE's structure.

It should be noted that NCEs in discovery phase show distinct dependence on the incubation time (Fig. 5.5). For instance, compounds with high equilibrium solubility such as Norfloxacin (pH 1.0) tend to reach equilibrium in a very short period of time (accomplishing >90% equilibrium within minutes). Some compounds of medium to high solubility (~100's μM) such as Carbamazepine (pH 6.8) show medium dissolution rate in the initial phase but can still reach equilibrium (e.g., >80%) within hours. However, the dissolution process appears to rise monotonously during the 24-h incubation. A number of compounds (such as Spironolactone at pH 6.8) with medium to low solubility (~10's μM) behave distinctly in that they possibly supersaturate and attain higher kinetic solubility quickly but then precipitate before eventually reaching equilibrium solubility. The discrepancies, in 1–3 orders of magnitude as a collective consequence of utility of DMSO and insufficient incubation time and variable readouts, indeed become problematic for the drug property assessment as either false positives (unpleasant surprise in late stage) or false negatives (mistaken termination of promising chemistry series) are fatal and costly to the programs.

Several miniaturized platforms with different throughput have been evaluated to assess equilibrium solubility using Uni-Prep filter chamber (Kerns 2001; Glomme et al. 2005), circulation pump (Chen and Venkatesh 2004) and vials (Bergström et al. 2002, 2004; Tan et al. 2005) and 96 well microplate (Wang 2009). Meanwhile, modifications of the "kinetic" approach were also attempted to improve the predictability to equilibrium solubility without sacrificing the throughput (Avdeef 2003). Recently, Zhou et al. (2007) reported a true HT and miniaturized equilibrium solubility approach utilizing mini-prep vials (MPV) and fast HPLC. This novel approach

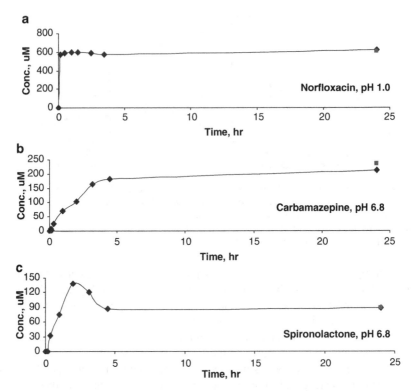

Fig. 5.5 Dissolution of selected compounds using the current MPV platform (PVDF filter) determined within 24 h at room temperature (diamond). (**a**) Norfloxacin, pH 1.0 (Pattern 1); (**b**) Carbamazepine, pH 6.8 (Pattern 2); and (**c**) Spironolactone, pH 6.8 (Pattern 3). Data annotated in *green square* were collected at 24 h with the additional 30-min incubation after phase separation by pushing filter plunger down to the bottom of the MPV. (Figure is taken from Zhou et al. 2007, with permission.)

addresses most of the caveats encountered by the kinetic or "semi-equilibrium" solubility approaches (no DMSO, 24-h incubation, minimal adsorption due to the "dynamic filter" in MPV and fast HPLC quantification), thereby exhibiting an excellent agreement to the conventional shake-flask approach for not only commercial drugs (Zhou et al. 2007) but also exigent NCEs in early drug discovery. It also offers the capacity and streamlined logistics feasible for supporting lead identification and optimization demands. Recently this HT equilibrium method has been successfully established in the 96 well microplate format (Wang 2009).

5.3.1.4 Solubility and Dissolution in Biorelevant Media

Thermodynamic or equilibrium solubility in aqueous buffers offers simplified determinants to flag solubility risks of NCEs to oral absorption. To better predict the *in vivo* performance of a drug after oral administration, it was proposed to identify and model the dissolution limiting factor to absorption using biorelevant media such as simulated intestinal fluid (SIF), simulated gastric fluid (SGF), fasted-state simulated intestinal fluid (FaSSIF) and fed-state simulated intestinal fluid (FeSSIF) (Galia et al. 1998; Nicolaides et al. 1999; Dressman and Reppas 2000; Kostewicz et al. 2002, 2004; Vertzoni et al. 2007). FaSSIF and FeSSIF contain a mixture of bile salts and surfactants (such as sodium taurocholate) mimicking the physiological media in human stomach and intestine and promoting wetting process in GI tract (Galia et al. 1998). Detailed characterizations of human or animal intestinal fluid as well as simulated biorelevant media were reported (Levis et al. 2003; Vertzoni et al. 2004). In addition, the species difference should also be considered in selecting simulated biorelevant media due to distinction in the composition of gastric/intestinal fluid between human and animal models (Kalantzi et al. 2006).

Dressman, Reppas, and coworkers reported a number of case studies demonstrating an excellent correlation of solubility and dissolution in the biorelevant media with drug's exposure in the clinical studies (Nicolaides et al. 1999; Dressman and Reppas 2000). Determination of solubility in the biorelevant media is essential for the accurate classification of BCS class II (and also class I) NCEs where solubility or dissolution is rate-limiting to absorption and also for lead optimization via establishing SPR and IVIVC in early discovery (Obata et al. 2004; Wei and Loebenberg 2006). Comparison of solubility in FaSSIF and FeSSIF can help foresee the food effects *in vivo* (Nicolaides et al. 1999; Dressman and Reppas 2000; Kostewicz et al. 2004; Persson et al. 2005).

5.3.1.5 Intrinsic Solubility and Solubility pH Profile

Intrinsic solubility, the equilibrium solubility of NCEs in their neutral form, is potentially beneficial for properly interpreting data from permeability as well as other cellular assays. It can be attributed to the general consensus that the neutral species of an NCE tends to preferably penetrate the lipophilic GI membrane more than its ionized counterparts.

Intrinsic solubility for ionizable NCEs may be determined using the potentiometric titration that was initiated by Avdeef et al. (2000). A comparable approach with increased throughput was reported recently (Stuart and Box 2005). The methodology has the advantage of eliminating the separation procedure of the filtrate from the precipitate, as the pH-electrode measures proton exchange of the fraction of compound which is in solution, in comparison to the conventional shake-flask approach. Another attractive feature of the potentiometric method is the capacity to "extrapolate" the solubility pH profile of an NCE from its intrinsic solubility using an accurately determined pKa. This method, however, is time consuming and requires skilled operators for experiment design and data interpretation, and thereby is not amenable to high-throughput. In addition, it is not applicable to NCEs lacking an ionization center. It should be noted that significant discrepancies were reported between thermodynamic solubility data collected using the conventional "shake-flask" approach and those of solubility products (e.g., salt and ionized species) extrapolated by the current approach at the same pH (Bergström et al. 2004). In these cases, the

Henderson–Hasselbach equation fails to describe the solubility pH profile for some low-solubility compounds, and the nature and concentration of the counter ions can influence solubility significantly.

5.3.1.6 Monitoring Dissolution Process in Early Discovery

The HT equilibrium set up using mini-prep vial offers a novel and unique platform to monitor drug dissolution in early discovery (Zhou et al. 2007). With utilization of mimi-prep vial approach, the effectiveness of the dissolution process was evaluated by Zhou et al., on the extent of 24-h dissolution for nine marketed drugs. Most compounds reached ~90% dissolution within the initial phase of the 24-h incubation and approached plateaus in 24 h, as observed in other miniaturized assays (Chen and Venkatesh 2004).

Miniaturized dissolution apparatus has been invented by Novartis Technical Research and Development scientists in collaboration with $_p$ION in 2004. This apparatus is used to determine the dissolution rates using a small volume of dissolution medium with small amount of material. The data are very useful to compare dissolution rates between different physical forms, formulations, salts, etc., and the information obtained is useful in selecting a developable physical form, salt, and/ or formulation principles before NCE enters into development.

5.3.2 Formulation Screening for Preclinical Studies

Traditionally, formulation development was done in development to optimize physico-chemical and biopharmaceutical properties, and to support clinical development. However, with a significant push to improve candidate success in R&D and to promote developable candidates from discovery into development, formulation development and screening have become an integral part of the preclinical programs. While in development, formulation activities are more focused on assessing a candidate's developability and supporting *in vitro* and *in vivo* preclinical assessment. Since the effort at this early stage is to screen various preclinical candidates and amount of material available is generally small, it is essential to manage such activities differently than in development. As a result, HT formulation has become an integral part of drug discovery efforts in preclinical development (Chaubal 2004; Neervannan 2006). Early formulation of compounds supports *in vivo* studies, obtaining more clinically relevant PK-PD information and allows for conducting toxicology and safety pharmacology experiments to move into early drug development (Neervannan 2006; Pouton 2006). With this new strategy, discovery teams can prioritize their limited resources on promising candidates selected for the proof of concept (POC) experiments and eliminate risky NCEs flagged by early *in vitro* and *in vivo* toxicology assays. Suboptimal formulations can critically affect the fate of compounds in early stages by either leading to underestimated exposure or by allowing over-optimistic liability assessment for toxic candidates. These aspects are important as a large proportion of early drug discovery phase compounds have poor solubility and permeability that limit the use of *in vitro* ADMET assays. Since many additives and surfactants affect the biological assay conditions, and therefore the readout of *in vitro* assays, formulation scientists involved at this stage must understand the role that such additives can play in the assessment of preclinical candidates (e.g., Caco-2, PgP, Cytochrome P450s, etc.) (Yamashita et al. 2000; Mountfield et al. 2000; Wagner et al. 2001). In summary, HT-formulation efforts enable choice of vehicles to be selected for preclinical studies using small amount of material in short duration.

Typical formulation approaches include pH adjustment, co-solvents, surfactant, salt selection, particle size reduction (Merisko-Liversidge et al. 2003; Dubey 2006; Kocbek et al. 2006), self emulsifying drug delivery systems, lipid complexation (Pouton 2006), and selection of various solubilizing agents. Several HT-formulation platforms (Chen et al. 2003; Gardner et al. 2004; Chaubal 2004; Chan and Kazarian 2005, 2006; Hitchingham and Thomas 2007), have addressed the challenges in discovery phase such as high capacity, limited amount of material, fast turnaround, and insufficient development time.

The potential impacts and limitations of all solubility and dissolution tools should be fully understood and attention should be focused on their interplay with other key parameters. For example, identification and understanding of the factors limiting systemic exposure are a pre-requisite for seeking appropriate formulations principles. In addition, formulations principles can be employed most effectively for NCEs with solubility and dissolution issues, but are less effective when the molecule is permeability or clearance-limited. Indeed, sometimes HT-formulation approaches may help screen quickly mechanisms for bioavailability enhancement for candidates that have permeability and metabolism issues. Such improvement, however, may be achieved at the expense of solubility, deferring the solubility issues to later stages. Formulations of poorly soluble NCEs can be difficult, costly, not always successful, and possibly lead to late stage failures, and therefore, optimizing the structure is a preferred route to improve solubility and exposure *in vivo* than to expand a significant amount of time during preclinical development to overcome exposure issues.

5.3.3 *Pharmaceutical Salts and Polymorphs*

Salt selection is a strategy that is commonly employed to improve properties of pharmaceutical compounds (Berge et al. 1977; Gould 1986). Crystalline salts can confer useful attributes such as improved physical properties, aqueous solubility, dissolution rate, chemical stability, and bioavailability relative to those of the free base or acid of the NCEs. Drug delivery options, such as oral, topical, parenteral solutions, inhalation, and controlled release technologies, often depend on the ability to modulate physicochemical properties of NCEs. Furthermore, salts and other crystal forms (e.g., hydrates, solvates, and co-crystals) of NCEs are patentable compositions that enhance the intellectual property and give freedom to operate. Due to its central placement in pharmaceutical and chemical development and its impact on biopharmaceutical properties and pharmaceutical research, salt selection is becoming increasingly automated to meet the need for rapid identification of crystalline salt forms in early development (Gould 1986; Morris et al. 1994; Tong et al. 1998). Salt selection at earlier stages has also been demonstrated using high-throughput crystallization (Morissette et al. 2003). The sudden appearance of a more stable polymorph of an NCE/API is an unpredictable and challenging aspect of drug development. Generally, the most stable crystal form of a given compound is desired, and much effort is expended to find it in the early stages of development. When a new form appears late in development, the product must be carefully evaluated to ensure that performance and quality (e.g., content dissolution rate, intrinsic solubility, bioavailability) have not changed. In some case, formulation may need to be adapted or changed entirely to restore the original performance. The change in physical form of the drug in the product may have a significant impact on product performance that it could delay development, drug approval, and market entry. A bigger problem is the potential emergence of such a polymorph in a marketed product (Bauer J; Spanton et al. 2001); In such scenario, the product must be withdrawn from the market for reformulation, if specifications of the original product cannot be met, as was the case of ritonavir (Bauer J; Spanton et al. 2001). As a result, in recent years pharmaceutical solid polymorphism has received much scrutiny throughout various stages of drug development, manufacturing, and regulation. Extensive experimentation is needed to ensure that all polymorphs and solvates of the selected neutral or salt forms are found early in development. In the case of compounds that can make crystalline salts, it may be possible to find salt forms with lesser propensity for polymorphism.

Using high throughput screening technology, maximum information can be obtained with limited amount of compound in a short period of time. However, the more time and resources spent on polymorphism, the higher the likelihood of discovering more polymorphs, and therefore, a careful assessment must be made on each candidate on the extent of screening that should be performed during preclinical development to find a stable polymorph.

Complete characterization of an NCE/API and their salts is extremely important in order to develop a drug product successfully. A number of methods have been employed for the physical characterization of solid materials such as crystallography, microscopy, thermal analysis, dynamic vapor sorption (DVS), and spectroscopy (FTIR, Raman, and nuclear magnetic resonance (NMR)). Additionally, solubility/intrinsic dissolution studies are the most useful methods for characterization of polymorphs and solvates (Brittain 1999a, b). Single crystal X-ray diffraction studies lead to the structural elucidation of small molecules within a crystal lattice and provide the most valuable information on the polymorphic solid. In fact, the definitive criterion for the existence of polymorphism is demonstration of a nonequivalent crystal structure. However, a limiting requirement of this technique is the necessity of obtaining appropriate single crystals for analysis. X-ray powder diffraction is another powerful technique suited for distinguishing solid phases with different internal crystalline structures. However, unlike single crystal X-ray diffraction,

X-ray powder diffraction does not require single crystals and is also an effective tool for the routine analysis of powdered samples. In some instances, X-ray powder diffraction has been used in the determination of the unit cell parameters and space group as well as in molecular structure determination (Stephenson 2000). X-ray powder diffraction may also be used for the determination of degree of crystallinity, quantitative analysis of polymorphic solids, and kinetic determination of solid-state reactions (Suryanarayanan 1995). A different X-ray powder diffraction pattern could be misleading in differentiating salt vs. free form. During the crystallization of a salt in presence of the counter ions, an NCE may crystallize out as a different physical form of a free form due to an acidic or basic microenvironment without forming a salt. Therefore, it is very important to confirm the salt formation by Raman, NMR, Elemental analysis, or by some other techniques. Once the existence of polymorphism is established, other methods such as solid-state spectroscopy, microscopy, and thermal analysis may be used for further characterization. Microscopy (light and electron) characterizes polymorphs through the optical and morphologic properties of the crystal. Microscopy is especially useful when only a limited amount of the drug substance is available. A full microscopic examination can reveal possible differences in crystal habit or structural class. Hot-stage microscopy is a useful tool for discovering polymorphs and determining their stability relationship, as illustrated through the study of fluoroquinolone. Thermal analysis, such as differential scanning calorimetry (DSC) and thermogravimetric analysis (TGA), distinguishes polymorphs and salts with different counter ions on the basis of their phase transitions during heating. These techniques can be used to obtain additional information regarding these phase transitions including melting, desolvation, crystallization, and glass transition. They can also be used to determine the relative stability among polymorphs and to differentiate enantiotropic and monotropic systems. For an enantiotropic system, the relative stability of a pair of solid forms inverts at some transition temperature beneath the melting point, whereas in a monotropic system, a single form is always more stable beneath the melting point (1). Solid-state spectroscopy (IR, Raman, and NMR) has become an integral part of the physical characterization of pharmaceutical solids (Bugay 2001). Both IR and Raman spectroscopy measure the fundamental molecular vibrational modes and provide a fingerprint of the pharmaceutical solid (7). Both of these techniques complement one another in that IR spectroscopy measures vibrational modes that change in dipole moment, whereas Raman spectroscopy measures vibrational modes that change in polarizability. Solid-state NMR is a powerful technique used to measure the magnetic environment around a nucleus. Solid-state NMR can be used not only to differentiate between solid-state forms in the bulk drug substance and in the drug product but also to probe the molecular structure of each solid-state form. The NMR technique is finding increasing utility in deducing the nature of these solid-state polymorphic variations, such as variations in hydrogen bonding, molecular conformation, and molecular mobility (Vippagunta et al. 2001).

The solubility or physical form screening by HTS technology needs to be confirmed for its scalability and reproducibility by manual techniques. HTS technology is a good tool to obtain maximum information using a small amount of material in early drug development phase. Drug-like new chemical entities can be selected after screening of multiple candidates by rank ordering the molecules based on the data from HTS as well as scale up and reproducibility by manual process.

5.4 Importance of Physical Form in Suspension

5.4.1 Required/Desired Form: Crystalline or Amorphous

Physical form of a new chemical entity can have a significant impact on the physical, chemical and biopharmaceutical attributes of a suspension. Since in early preclinical studies, the physical form is often not characterized, the implication of the physical form on the preclinical results must be assessed. Ideally, it is best to use a stable crystalline form in the preclinical studies so that there is no change in attributes of the material with the time. Suspension of amorphous material drug will give better exposure in preclinical studies, but due to physical stability of amorphous material and inability to predict when such suspension will undergo transformation into a stable crystalline form can lead to non-reproducible results. Therefore, it is important to characterize drug substance before its use in important preclinical studies. If the new chemical entities contains acidic or basic functional group(s), attempts should be made to evaluate if the salt form leads to better biopharmaceutical behavior than the parent molecule. If the NCE is a neutral molecule, formulation approaches can yield improved solubility/dissolution rate or even use of co-solvents, surfactant or special drug delivery system should be evaluated in the preclinical studies. Therefore, if new chemical entities give promising preclinical results, and it has high potential to enter into development, it is very important to collect all the preclinical data with the developable physical form otherwise extensive work would be required during clinical development (Matteucci et al. 2007).

In preclinical studies such as pharmacological dose range finding and telemetry, it is usually preferred to dose new chemical entities in a simple vehicle to avoid undesirable effects from the excipients/vehicles. Therefore, often materials used in such studies are suspensions which are formulated at concentration below 1% celluloses with or without presences of a surfactant. The physical form of the NCE used in these suspensions should be determined, and the impact of the physical form on the outcome of the study must be assessed. The authors highly recommend that the final physical form must be selected as early as possible, and preferably while the drug is still in preclinical development. By selecting a final form early in preclinical studies, one can obtain reliable and reproducible data, which can aid in the selection of good candidates (developable candidates) into development portfolio. Also, the reliable data allow giving discovery chemists good feedback on their molecule sooner than later so that they can modify molecules depending on liabilities that the leads or scaffolds carry. This cascading mechanism helps pharmaceutical companies to reduce the attrition rate and enrichment of product pipeline with promising candidates (Venkatesh and Lipper 2000).

5.4.2   Techniques to Characterize the Physical Form

There are a number of techniques used to characterize the physical form. However, as mentioned earlier during preclinical studies, the available quantity of material is small. Therefore, it is advisable to use non-destructive techniques such as polarized light microscopy and X-ray powder diffraction. In both techniques, the amount of material required in less than 1.0 mg.

5.4.2.1   Polarized Optical Microscopy

Light microscope consists of a linear polarizer located below the condenser, a second polarizer mounted on the top of eye piece and a rotating stage. The main goal is to identify whether the sample is amorphous or crystalline or a mixture of amorphous and/or crystalline. The light from the source is rendered linearly polarized by the initial polarizer and the second polarizer (analyzer) is oriented so that its axis of transmission is orthogonal to that of the initial polarizer. During the condition of crossed polarizer, no transmitted light can be perceived by the observer. If the sample gives rise to birefringence, then it is classified as crystalline. If there is no birefringence, then the sample is classified as amorphous (Brittain 1999a, b).

5.4.2.2   X-Ray Powder Diffraction

X-ray powder diffraction (XRPD) is mainly used to understand a polycrystalline material and solving of a chemical structure by collecting data on a single crystal (Chang and Smith 2000). However, in preclinical formulation, such as suspensions, it is mainly used to understand whether the NCE used in preclinical studies is amorphous/crystalline or semi-crystalline. The XRPD pattern for crystalline material consists of a series of peaks detected at characteristic scattering angles, whereas the amorphous material does not show peaks, and instead shows large one or two hollows (what do you mean by hollows?).

5.5   Excipients Used in Suspension for Preclinical Studies

In this book, there is a special chapter on excipients used in suspension. However, in this chapter, a brief summary is given on excipients. The excipients used in pre-clinical suspension are "GRAS" (generally regarded as a safe) listed. The most of the GRAS listed excipients are well described in the hand book of excipients (Kibbe 2000) and characterized (Bugay and Findlay 1999). As described below, it is very important to understand physico-chemical attributes of an NCE before selecting excipients for preclinical formulation.

- Chemical nature of NCE, such as neutral, acidic, basic, or a salt
- Pro-drug, co-crystal (chemically and physically unstable respectively in an aqueous environment)
- Bulk density
- Aqueous solubility
- Solubility in pH modifiers, co-solvents, surfactant and combination, etc.

For basic compounds (weak and strong bases) and their salts (made with acidic counter ions), it is recommended to use celluloses which are neutral. And for acidic compounds or their salts (made with basic counter ions), it is good to use polymer solutions that result in alkaline pH. Often it is observed that when a suspension of a basic compound is prepared in sodium carboxymethyl cellulose (Na CMC), free base of the compound precipitates out as a different crystalline or amorphous form. It is due to hydrolysis of Na CMC at 0.5 or 1% aqueous solution giving rise to basic pH. Hence for basic compounds and their salts, it is recommended to use neutral polymers as a viscosity modifier. Therefore, it is important to evaluate different suspending agents with or without surfactant before initiation of any long term studies (Zietsman et al. 2007).

5.5.1 Viscosity Modifiers

Selection of viscosity modifiers completely depends on the physico-chemical properties of new chemical entities. Very commonly used viscosity modifiers are Na CMC, methyl celluloses (MC) of low, med, and high viscosity grades, hydroxypropyl cellulose (HPC-Klucel), hydroxypropyl methyl cellulose (HPMC)-low viscosity grade. These celluloses are used as aqueous solutions ranging in concentrations from 0.5 to 2%. Generally, the high bulk density materials require higher viscosity modifier concentrations to yield viscous solution to reduce settling rate.

5.5.2 Surface Modifiers

Surface modifiers are also called as wetting agents or surfactants. There are some basic principles that can be used in selecting a surface modifier depending on physico-chemical properties of molecules. The most commonly used principle is the "like dissolves like." It essentially means that a surfactant that is structurally similar to the compound of interest could be a good choice. Surfactants are classified as per their hydrophilic/lipophilic balance (HLB), which can be helpful in selection of a surfactant. It has been observed that for the surfactant molecules that are zwitter ions, with both acidic and basic functionalities, and those that are cationic are good wetting agents. The effectiveness of a surfactant in solubilizing a molecule depends on the surfactant concentration up to a limit; the higher the concentration of surfactant, the higher the solubility enhancement. This could be due to formation of micelle in the solution, which can entrap the compound to be dissolved on a molecular level, and thus prevent aggregation (Maeda et al. 1984).

During preclinical suspension evaluation, the selection of surface modifiers is based on the results such as agglomeration, sedimentation rate, particle size distribution, caking, re-dispersability, and physico- chemical stability.

The commonly used surface modifiers include ionic surfactants such as sodium lauryl sulfate, dioctyl sulfosuccinate (DOSS), cationic surfactants such as cetyltrimethyl ammonium bromide, and amphiphilic surfactants such as tweens, pluronics, cremophor, and solutol. They are usually used in the concentration range of 0.1–2% w/v. The final percentage depends on the concentration required to obtain a stable suspension and maximum dose used in preclinical studies.

5.5.3 pH Modifiers

A number of new chemical entities have pH dependant solubility and stability, and therefore, it is important to understand the pH solubility and stability profile for new chemical entities which are intended to be dosed as a suspension.

For example, when a suspension of an acidic salt of a basic NCE is prepared at concentrations exceeding 10 mg/mL, the pH of suspension may be as low as 1.5–2.0, and such a suspension is not suitable for dosing in animals. If the pH of such suspensions, containing weakly basic compounds with pKa<4, is adjusted to a satisfactory range (around pH=5–7), most of the salt would convert to free base. The poor solubility and dissolution rate of the free base may limit exposure of such a compound at higher doses. Therefore, sometimes it is advisable to optimize the pH of the suspension to avoid or minimize the conversion. However, one should not neglect the gastric irritation caused by repeated dosing of an acidic suspension during long team studies, especially in rodent species.

Chemical stability of an NCE may also be impacted by pH, especially those that have the potential to undergo hydrolysis or oxidation. For such NCEs, it is equally important to adjust the pH of suspension to achieve maximum stability. Commonly used pH modifiers are citric, tartaric acid, hydrochloric acid, sodium hydroxide, potassium hydroxide, potassium and sodium bi carbonates, sodium hydrogen phosphate, tris buffers. All these agents are usually used within 0.1 M to 1.0 molar range. Both the nature of the buffer and the buffer concentrations may have impact on the stability of a compound, and thus, must be optimized.

5.6 Preparation and Characterization of Suspension for Preclinical Evaluation

5.6.1 Batch Size 1–10 mL

For preclinical studies, the availability of material is always constrained, and often suspensions are prepared at a small scale. The most commonly used methods are described as follows (Niazi 2004).

1. Mortar and pestle: It is a very old technique, however it is very efficient, and even today, it is still widely used to make suspension at 1–10 mL scale. In this method, an active pharmaceutical ingredient is triturated gently with solution of cellulose and/or surfactant at a concentration between 0.5 and 2% w/v. After a few minutes of triturating the mortar contents, it forms homogeneous paste/viscous flock. Mortar content are transferred to the measuring cylinder or volumetric flask to make up the volume with remaining suspending agent/surfactant. The bottle or vial is always shaken well before administration because usually suspensions tend to settle with time.

2. Homogenizer: This technique is very commonly used in pharma industry to break agglomerate of particles when formulated as a dispersed system. In this method, an active pharmaceutical ingredient is transferred to a vial containing cellulose and/or surfactant solution. Vial contents are mixed well by manual shaking or vortexing followed by homogenization.

3. Probe sonication: In this method, an active pharmaceutical ingredient is weighed into the vial. A cellulose and/or surfactant containing solution is added to the vial. The vial is kept under probe sonicator/ultrasonic homogenizer to reduce particle size at known sonication amplitude. Heat may generate during the probe sonication, and hence care must be taken to prevent decomposition of the active. It is probably the most effective technique to reduce particle size significantly (Fig. 5.6).

4. Wet ball milling: A ball mill grinds material in a rotating cylinder with steel grinding balls, causing the balls to fall back into the cylinder and onto the material to be ground. A Planetary Ball Mill (the PM 100) utilizes Coriolis forces to

Fig. 5.6 Probe sonicator/ultrasonic homogenizer

Fig. 5.7 Ball mill (wet and dry milling)

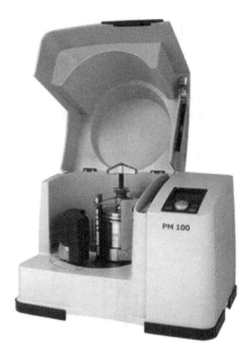

attain a high-energy, fine-grind of samples. The sample and grinding balls are placed in a grinding jar and clamped into the jar station. The milling process is initiated; the sun-wheel rotates clock-wise, while the jar station rotates counter clock-wise. This opposing spin causes the Coriolis force, which results from the grinding balls traversing from one end of the jar to the opposing end at high speeds. The high speed impact causes a high energy collision with sample particles. In addition, the centripetal motion of the grinding action causes frictional forces that aid in fine grinding applications (Fig. 5.7).

5.6.2   Batch Size 10 mL to 1 L

Preparation of suspensions at liter scales during preclinical studies is very common for pharmacological and toxicity studies. Commonly used techniques to make suspension at a large scale are described as follows.

5.6.2.1   High Speed Homogenizer

A number of high speed homogenizers are available in the market with RPM ranging from 50 to 50,000. The main purpose to use homogenizers is to obtain uniform

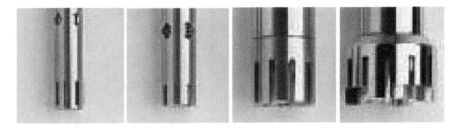

Fig. 5.8 Propellers for homogenization

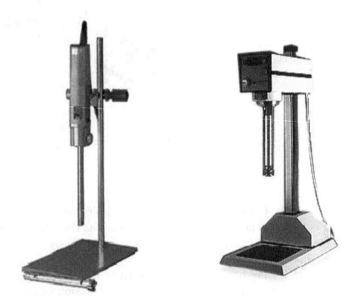

Fig. 5.9 Laboratory homogenizer

distribution of particles in the suspension at a larger scale. These homogenizers give continuous dispersability of solid particles into liquid and are designed for solids which have a poor wetability (Li et al. 2006). Homogenizers consist of motor, stand, beaker, clamps, and speed controller as show in the diagram below (Figs. 5.8 and 5.9).

5.6.2.2 High Shear Microfluidizer® Processors

Microfluidizer processors (http://www.microfluidics.com) are used for particle size reduction of emulsions and dispersions. The reduction of particle size in a microfluidizer is a result of high shear impact as the fluid flows through microchannels of a fixed geometry, which differentiates microfluidizer processors from standard high pressure homogenizers. Also, it is the fixed geometry aspect of the technology that generates fairly uniform shear fields and processes the fluid under uniform conditions.

The advantages of Microfluidizer processor technology are as follows:

- Smaller particle and droplet size (from nm to μm range)
- More uniform particle and droplet size distribution
- Faster processing times (>2 orders of magnitude in some applications)
- Scalability from small batches to continuous production
- No moving parts in the interaction chamber
- Easy to clean and sterilize
- Little or no contamination
- Uniform and stable dispersions
- High batch to batch reproducibility
- Used in cGMP manufacturing environments

Microfluidizers are very useful for producing economically very small and very uniform particles in a short time. Alternative approaches such as ball milling would require weeks to yield particles of similar size and consistency. In pharmaceuticals, the lack of contamination and the ability to uniformly disperse an active agent throughout its delivery media (e.g., water, buffers, cellulose, surfactant) often makes microfluidizer processor technology the method of choice.

5.6.2.3 Mechanism of Processing

Microfluidizer processors depending on the model have either a pneumatic or electric-hydraulic intensifier pump. During processing, the product stream accelerates to very high velocities, creating shear rates within the product stream that are orders of magnitude greater than by any other conventional means. The entire batch experiences an identical processing condition, producing the desired results, including uniform particle and droplet size reduction (often submicron) and deagglomeration.

5.6.2.4 Other Advantages

A constant pressure, when applied uniformly, results in particles of uniform size distribution. Very often multiple process/mechanism such as high pressure/force and velocity, all working at once on the same fluid stream results in very small particles. A broad pressure range (up to 40,000 psi) opens a wide variety of applications and materials to the advantages of small and uniform particle size distribution.

The principles that enable microfluidizer processor technology remains the same, whether the equipment is for large or small scale manufacturing. An application that has been developed, optimized, and validated on a smaller machine can be easily migrated to incrementally larger machines.

Additional advantages are lower processing cost, increased speed of operation, and more flexible manufacturing capability in the face of evolving market opportunities (Figs. 5.10 and 5.11).

5.6.3   Particle Size/Physical Form

5.6.3.1   Particle Size

During and after high pressure homogenization/microfluidization, it is important to confirm the particle size distribution after every 5–10 min. *Timely analysis on particle size measurement is useful in optimization of processing parameters during fluidization.* For the new chemical entities which belong to biopharmaceutical classification system (BCS) class II and IV (Yu et al. 2002), particle size distribution has the major impact on their solubility/dissolution rate. A number of studies using less than 1 µm particle size for biopharmaceutical classification system (BCS) II and IV suggest that there was a 1.1- or 1.5-fold increase in exposure. For the salts of BCS II, compound had a significant improvement in dissolution rate/exposure in animal models by particle size reduction as compared to salts of BCS class IV molecules. Because the NCE belongs to BCS class IV category, their exposure is not only limited by permeability but also by solubility/dissolution rate. Therefore, improving the solubility and dissolution rate may not help to increase the exposure for BCS class IV molecules.

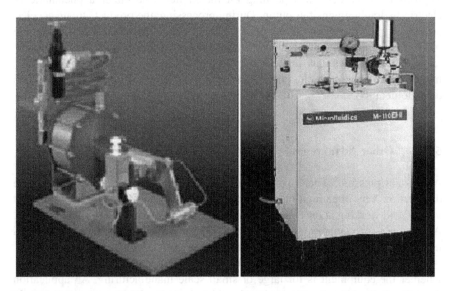

Fig. 5.10 Laboratory scale Microfluidizer processors (with permission from Microfludics, Newton, MA)

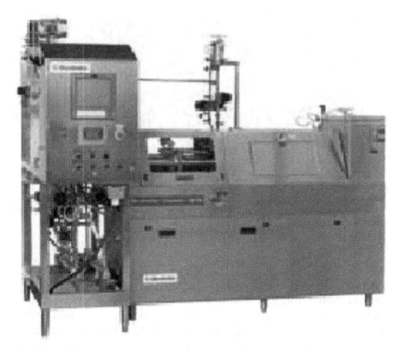

Fig. 5.11 Industrial scale Microfluidizer processor with steam sterilization (with permission from Microfludics, Newton, MA)

Particle size distribution is measured using a very commonly used instrument called by Malvern and Horiba particle size analyzer. A very small amount of sample is required, as low as 0.1 mL could be sufficient for measurement. It gives not only an average particle size, but also helps to understand the entire particle size distribution range.

5.6.3.2 Physical Form

During preparation of suspensions especially by microfluidization, the possibility of physical form change is very high. Physical form conversion depends on crystal lattice energy of new chemical entities. NCEs, which are strongly crystalline and have high enthalpy greater than 100 J/g, remain stable (crystalline) during particle size reductions. However, weakly crystalline forms usually covert to an amorphous form and during processing upon storage revert back to the same or different crystalline forms. The converted amorphous form may be used to perform initial pharmacokinetic studies. The physical form change can have a significant impact on *in vivo* study, and therefore, it is very important to confirm the physical form prior to *in vivo* studies. Methods to characterize physical form are described in Sect. 5.2.2.

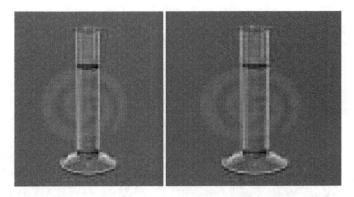

Fig. 5.12 Graduated measuring cylinders for determining sedimentation volumes initial height (Hi) for example 30 cm height after settling (Hs) 20 cm after 24 h. Therefore the ratio is 20/30=0.666

5.6.4 Sedimentation Volume

Suspensions tend to settle with time, and therefore, they are designed to be readily redispersed by gentle shaking or stirring resulting in a homogeneous suspension. Redispersability depends on particle size of the dispersed phase and the type of suspension (floccules, non-floccules or coagule). The sedimentation rate can be measured by a simple method. Sedimentation volume is the ratio of the height of the sediment after settling (Hs) and initial height of suspension (Hi). The larger the ratio, the better is the suspendability (Patel et al. 1976). Sedimentation volume determination is explained in Fig. 5.12.

5.7 Stability

5.7.1 Physical Stability

Physical stability of suspension is very important to ensure uniform dosing in preclinical studies. Physical parameters that play an important role in determining stability of suspensions include physical form and particle size of the dispersed phase, and particle–particle interactions. These properties determine if the suspension will form a sediment that will form a cake or a sediment that can be easily dispersed. It is critical to ensure that there are no changes in physical form due to polymorphic changes or increase in particle size due to Ostwald ripening (Stella and Yoshioka 2002).

5.7.1.1 Change in Physical Form

During preclinical studies, often the physical form is not well characterized because of the lack of sufficient quantity of material at an early stage of drug discovery, and sometimes due to a lack of understanding of how the form can impact preclinical results. Changes in physical form can result in significant changes in dissolution properties of the drugs, which can lead to changes in exposure. For example, Carbamazepine anhydrates converts to dihydrate when formulated as a suspension (Zheng et al. 2005). Dihydrate has lower solubility and dissolution which affects the overall exposure. In addition, a number of weakly crystalline compounds and salts of weakly basic compounds with pKa less 3.5 can convert into amorphous basic form or crystalline free base in suspension. This could also affect the solubility/dissolution rate and the exposure significantly for drugs post oral administration.

To assess the behavior of a physical form in a suspension formulation, typically 5–6 combinations of polymers and surfactants together or alone are evaluated at different concentrations for the suspensions at different pH. The main goal in selection of suspending agents (polymer) and surfactant is to confirm that the physical form does change during the duration of study.

5.7.1.2 Change in Particle Size

When suspensions are used in preclinical studies, it is a very common procedure to prepare them at the start of study and use over the entire study duration (2–3 weeks). Depending on the crystal habit and thermodynamic solubility of the particles, either particle size or distribution of the suspended drug may change. Even during this short duration studies that employ suspensions of low dose (10 mg/kg) NCEs, particle size of drugs can change. Such changes can impact attribute of suspension. Only in formulation containing high drug doses (>100 mg/kg), the overall performance of the product may not be impacted over a short duration of a few weeks, as the small changes in particle size will not be meaningful given the high dose.

Ostwald ripening refers to the growth of larger crystals from those of smaller size. It occurs due to the faster rate of solubilization of smaller particles due to their large surface area relative to large particles. Faster solubilization leads to supersaturation, and over time this leads to crystallization. As the smaller particles dissolve and the crystallization occurs around large particles, there is a tendency to have particle size growth over time. In the process, many small crystals disappear, and the larger ones grow larger and larger at the expense of the small crystals. The smaller crystals act as fuel for the growth of large crystals. Molecules on the surface are energetically less stable than the ones already well ordered and packed in the interior. Large particles, with their greater volume to surface area ratio, therefore represent a lower energy state. Hence, many small particles will attain a lower energy state if transformed into large particles and this is what is known as Ostwald ripening. It is possible to inhibit the Ostwald ripening by use of second component which has extremely low aqueous solubility. To inhibit ripening, the drug/inhibitor mixture (in the particles) must form a single phase. The drug/inhibitor mixture can be characterized by the interaction parameter using the Bragg-Williams theory, where the parameter can be calculated from the (crystalline) solubility of the drug in the inhibitor, provided the inhibitor is a liquid, and the melting entropy and temperature of the drug are known (Lindfors et al. 2006). Redispersability and caking in suspension can be avoided by use of certain polymers and surfactant. Use of electrolyte also helps to redispersability of the particles in suspension.

5.7.2 Chemical Stability

Since reactions in solution state take place more rapidly than in solid state, drug in suspended forms with limited solubility are far more stable than the drug in solution. However, chemical stability of drugs in preclinical suspension should still be monitored, especially if NCE's fall in the following categories:

1. Aqueous solubility >1 mg/mL
2. Drug suspensions are stored for extended period
3. When the drug concentration in suspension is low

Often the chemical reactions in suspensions involve hydrolysis, oxidation, complexation, etc. Chemical functionalities such as ester and amides are susceptible to hydrolytic degradation and amino functions may undergo oxidative degradation. One way to minimize the impact of such degradation is to prepare suspensions of high concentrations (>100 mg/mL).

If an NCE is not stable in suspension, decomposing more than 2% over the duration of study, it is recommended that a fresh batch can be prepared daily prior to dosing. Alternatively, the suspension can be stored under refrigerated conditions, or freezed and thawed for a few hours prior to use, provided the uniformity of the drug is not compromised. While preparation of a fresh suspension each time prior to dosing is an option when significant physical or chemical changes occur during storage, it is not very convenient and could potentially result in lot-to-lot variability. Therefore, for short term studies, it may be a good strategy to prepare a single large batch and freeze/thaw prior to dosing. However, for long term studies it is not advisable to do so. While the physical and chemical instability challenges can be overcome during preclinical assessment, development of such products for a long-term clinical studies may not be feasible. For such drugs, it may be worthwhile to invest in the identification of an alternative new chemical entity, which will not be burdened with such limitations.

5.8 Applications

5.8.1 Oral Suspensions

Oral suspensions are widely used during preclinical development due to the ease of preparation low likelihood of stability issues, and ease of dosing. If a suspension formulation yields satisfactory exposure in a pharmacokinetic study and good linearity over the dose range of interest, it can be used for efficacy and toxicity studies. However, often because of poor solubility, a non linear pharmacokinetic profile is observed at higher doses. In such cases, it is important to understand and overcome such issues as early as possible. A large number approaches including non aqueous vehicles, as well as other excipients such as cylcodextrin, surfactants, etc., have been used in multiple species to overcome absorption issues due to poor solubility when dosed as suspension (Gad and Cassidy 2006).

Pharmaceutical suspensions are also very useful commercially for infants, children, and older patients who have difficulty in swallowing. Such products are marketed either as suspensions or often as powders for reconstitution. The latter, when reconstituted, have a limited shelf life, typically in the range of a few days to weeks.

5.8.2 Parenteral Suspension

Parenteral suspensions are manufactured with lots of consideration, such as prolonged effect of new chemical entities at the site of administration, i.e., intramuscular or subcutaneous. All the agents used in the parenteral formulations must be safe, should not have pyorgenic or antigenic or irritant/hemolytic effects. Heat sterilization can not be used with the final powder because its physical stability can be affected such as crystal growth and protecting colloids modification are potentially possible. Therefore, all the operations must be done in sterile environments with aseptic techniques.

5.8.3 Topical Suspension: Ophthalmic, Dermal

5.8.3.1 Ophthalmic

There are a number of products commercially available as suspensions for ophthalmic use. While solutions are preferred over suspension, the physico-chemical properties of molecule may not allow formulating them as a solution (Lieberman et al. 1998). Suspensions may provide a prolonged therapeutic effect and may be preferred over solutions for certain applications.

Diclofenac ophthalmic (0.1%) solution and Nepafenac suspensions (0.03 and 0.1%) were compared with each other for analgesic efficacy and safety for treatment of pain and photophobia from after excimer laser surgery in patients. Nepafenac suspension at both the concentrations was effective in pain and photophobia after excimer photoreactive keratectomy (Colin and Paquette 2006).

Ibuprofen and cloricromene were studied for ophthalmic indication using Eudragit retard polymer nanoparticles technology (Bucolo et al. 2002; Pignatello et al. 2006). Eudragit based polymer nano suspension showed very fine particle distribution and a positive surface charge along with the good physico-chemical stability. Nano suspension with a positive surface charge can allow a longer residence time on the cornea surface with the consequence of slower drug release and higher drug concentration in aqueous humor with classical eye drops. Eudragit based ophthalmic solution (placebo) are found to be safe when administered topically to the rabbit eyes (Pignatello et al. 2002). There was no irritation to cornea, iris and conjunctiva.

Gelatin based nano particles have been studied for the ophthalmic indication using pilocarpine and hydrocortisone (Vandervoort and Ludwig 2004). Sustained release profiles for both the drugs were observed using gelatin based nanoparticles. However, for hydrocortisone, use of cyclodextrin was a must to increase the solubility and the release profile of hydrocortisone was dependant on type of cyclodextrin. In the case of pilocarpine, gelatin based nanoparticles showed a zero order release profile.

Hydrocortisone suspension in polymers demonstrated 55% increase in bioavailability of hydrocortisone relative to hydrocortisone solution with cyclodextrin in aqueous humor and 75% in cornea in rabbits (Bary et al. 2000).

Formulation of ophthalmic emulsion is another approach to deliver water insoluble drugs to the eye. Lipid emulsion containing an anti-inflammatory drug difluprednate, synthetic glucocorticoids was formulated in 5% caster oil and 4% polysorbate 80% with 0.05% of drug. When difluprednate 0.05% suspension was compared with lipid emulsion, 5–7 times higher concentration was observed in aqueous humor 1-h post installation (Yamaguchi et al. 2005).

5.8.3.2   Dermal

Very limited use of suspensions for dermal application has been observed. Generally, the ointments or gels are the preferred dosage form for dermal applications as these approaches provides better drug delivery and thus efficacy. However, for dermal suspension, it is important for the drug particles to be uniformly mixed with the carrier/excipients. The homogeneity/miscibility is very important for uniform application of drug on the skin. Therefore, hydrophilic or hydrophobic based ointments/pastes containing active ingredients are available for dermal application. For example, neomycin and gentamycin antibacterial ointments are available.

A study was performed using four formulations, including solid, paste, aqueous suspension, and solution in ethanol in animal models to evaluate systemic delivery of a highly permeable molecule following dermal application. It was observed that the chemical in solid form, aqueous suspension and paste was comparable with a solution in ethanol (Hughes et al. 1992). The chemicals investigated were [14C]-2-sec-butyl-4,6-dinitrophenol (DNBP, 4.2 µmol), 2,4,5,2′,4′,5′-[14C]-hexachlorobiphenyl (HCB, 2.3 µmol), and 3,4,3′,4′-[14C]-tetrachlorobiphenyl (TCB, 0.5 µmol). Absorption of HCB applied as a solid was significantly higher (p less than or equal to 0.05) as compared to HCB applied in ethanol. There were no other significant differences in the absorption. This data indicates that the chemicals examined in this study can penetrate the skin as readily when applied either as a solid, aqueous paste, or suspension, as when applied in the volatile vehicle ethanol.

5.8.4 Aerosol Suspensions

Aerosols are mainly used to deliver NCEs for topical administration, by inhalation through the respiratory tract (Offener et al. 1989). Aresol suspensions have given through transdermal or inhalation routes advantages because they show rapid therapeutic action, avoiding first pass effect and gastrointestinal degradation when given orally. Aerosol products have mainly two components, the concentrate containing active compound(s) and the propellant gas mixture. The concentrate can be an emulsion, solution or semisolid paste, however, in suspension the active compound is directly dispersed in the propellant vehicles (Sciarra et al. 1989). Although manufacturing aerosol suspension can be difficult, methods to achieve stable and efficient suspension require the following: (a) to decrease the rate of settling and flocculation adding surfactant or dispersing agents (lecithin, oleic acid, ethyl alcohol), (b) reducing particle size to less than $5\,\mu m$, (c) matching the densities of new chemical entities or active pharmaceutical ingredient with propellant mixtures, (d) minimizing the moisture content because the respirable fraction is strongly influenced by the moisture content.

There are a number of aerosol suspensions available, such as epinephrine bitartarate, steroids, and bronchodilators (salbutamol, phenylephrine, and disodium chromoglycate) (Gallardo et al. 2000).

References

Alsenz J, Haenel E: Development of a 7-day, 96-well Caco-2 permeability assay with high-throughput direct UV compound analysis. *Pharm. Res.* (2003) **20**(12):1961–1969.

Amidon GL, Lennernas H, Shah VP, Crison JR: A theoretical basis for a biopharmaceutics drug classification: the correlation of in vitro drug product dissolution and in vivo bioavailability. *Pharm. Res.* (1995) **12**:413–420.

Artursson P, Tavelin S: Caco-2 and emerging alternatives for prediction of intestinal drug transport: a general overview. In: van de Waterbeemd H, Lennernas H, Artursson P (eds.) *Drug Bioavailability*. Wiley-VCH, Germany (2003):72–89.

Atkinson A, Kenny JR, Grime K: Automated assessment of the time-dependent inhibition of human cytochrome P450 enzymes using liquid chromatography-tandem mass spectrometry analysis. *Drug Metab. Dispos.* (2005) **33**(11):1637–1647.

Avdeef A, Berger CM, Brownell C: pH-Metric solubility. 2: Correlation between the acid–base titration and the saturation shake-flask solubility pH methods. *Pharm. Res.* (2000) **17**(1):85–89.

Avdeef A, Strafford M, Block E, Balogh MP, Chambliss W, Khan I: Drug absorption in vitro model: filter-immobilized artificial membranes. 2. Studies of the permeability properties of lactones in piper methysticum forst. *Eur. J. Pharm. Sci.* (2001) **14**:271–280.

Avdeef A: In: *Absorption and Drug Development, Solubility, Permeability and Charge State*. Wiley-Interscience, Hoboken, NJ, USA (2003).

Avdeef A, Voloboy D, Foreman A: Dissolution and solubility. In: Testa B, van de Waterbeemd H (eds.) *Comprehensive Medicinal Chemistry*, Volume 5, second edition. ADME-Tox Approaches. Elsevier Ltd, Oxford, UK (2007):399–423.

Balakin KV: DMSO solubility and bioscreening. *Curr. Drug Discov.* (2003) **Aug**:27–30.

Balimane PV, Chong S, Morrison RA: Current methodologies used for evaluation of intestinal permeability and absorption. *J. Pharmacol. Toxicol.* (2000) **44**:301–312.

Balimane PV, Patel K, Marino A, Chong S: Utility of 96 well Caco-2 cell system for increasing throughput of P-gp screening in drug discovery. *Eur. J. Pharm. Biopharm.* (2004) **58**:99–105.

Balimane PV, Pace E, Chong S, Zhu M, Jemal M, Van Pelt CK: A novel high-throughput automated chip-based nanoelectrospray tandem mass spectrometric method for PAMPA sample analysis. *J. Pharm. Biomed. Anal.* (2005) **39**(1–2):8–16.

Balimane PV, Han YH, Chong S: Current industrial practices of assessing permeability and P-glycoprotein interaction. *AAPS J.* (2006) **8**(1):E1–E13.

Bary AR, Tucker IG, Davies NM: Considerations in the use of hydroxypropyl-b-cyclodextrin in the formulation of aqueous ophthalmic solutions of hydrocortisone. *Eur. J. Pharm. Biopharm.* (2000) **50**:237–244.

Bauer J, Spanton S, Henry R, Quick J, Dziki W, Porter W, Morris J: Ritonavir: an extraordinary example of conformational polymorphism. *Pharmaceut. Res.* (2001) **18**(6):859–866.

Bendels S, Tsinman O, Wagner B, Lipp D, Parrilla I, Kansy M, Avdeef A: PAMPA – excipient classification gradient map. *Pharm. Res.* (2006) **23**(11):2525–2535.

Berge SM, Bighley LD, Monkhouse DC: Pharmceutials salts. J. Pharmaceut. Sci. (1977) **66**(1):1–19.

Bergström CAS, Norinder U, Luthman K, Artursson P: Experimental and computational screening models for prediction of aqueous drug solubility. *Pharm. Res.* (2002) **19**(2):182–188.

Bergström CAS, Luthman K, Artursson P: Accuracy of calculated pH-dependent aqueous drug solubility. *Eur. J. Pharm. Sci.* (2004) **22**(5):387–398.

Bevan CD, Lloyd RS: A high-throughput screening method for the determination of aqueous drug solubility using laser nephelometry in microtiter plates. *Anal. Chem.* (2000) **72**:1781–1787.

Blanchard N, Alexandre E, Abadie C, Lave T, Heyd B, Mantion G, Jaeck D, Richert L, Coassolo P: Comparison of clearance predictions using primary cultures and suspensions of human hepatocytes. *Xenobiotica* (2005) **35**(1):1–15.

Blanchard N, Hewitt NJ, Silber P, Jones H, Coassolo P, Lave T: Prediction of hepatic clearance using cryopreserved human hepatocytes: a comparison of serum and serumfree incubations. *J. Pharm. Pharmacol.* (2006) **58**(5):633–641.

Bohets H, Annaert P, Mannens G, Beijsterveldt VL, Anciaux K, Verboven P, Meuldermans W, Lavrijsen K: Strategies for absorption screening in drug discovery and development. *Curr. Top. Med. Chem.* (2001) **1**:367–383.

Bourdet DL, Thakker DR: Saturable absorptive transport of the hydrophilic organic cation Ranitidine in Caco-2 cells: role of pH-dependent organic cation uptake system and P-glycoprotein. *Pharm. Res.* (2006) **23**(6):1165–1177.

Box K, Bevan C, Comer J, Hill A, Allen R, Reynolds D: High-throughput measurement of pKa values in a mixed-buffer linear pH gradient system. *Anal. Chem.* (2003) **75**:883–892.

Brittain HG: Methods for characterization of polymorphs and solvates. In: Brittain HG (ed.) *Polymorphism in Pharmaceutical Solids*, Volume 95. Marcel Dekker, New York (1999a):240.

Brittain HG: Methods for characterization of polymorphs and solvates. In: Brittain HG (ed.) *Polymorphism in Pharmaceutical Solids*, Volume 95. Marcel Dekker, New York (1999b):264.

Bucolo C, Maltese A, Puglisi G, Pignatello R: Enhanced ocular anti-inflammatory activity of ibuprofen carried by an Eudragit RS 100 nanoparticle suspension. *Ophthalmic Res.* (2002) **34**:319–323.

Bugay DE: Characterization of the solid state: spectroscopic techniques. *Adv. Drug. Delivery Rev.* (2001) **48**(1): 43–65.

Bugay DE, Findlay WP: *Pharmaceutical Excipients Characterized by IR, Raman and NMR Spectroscopy*, Marcel Dekker, New York (1999): **94**:669.

Caldwell GW, Masucci JA, Chacon E: High throughput liquid chromatography-mass spectrometry assessment of the metabolic activity of commercially available hepatocytes from 96-well plates. *Comb. Chem. High Throughput Screen.* (1999) **2**:39–51.

Chan KLA, Kazarian SG: Fourier transform infrared imaging for high-throughput analysis of pharmaceutical formulations. *J. Comb. Chem.* (2005) **7**(2):185–189.

Chan KLA, Kazarian SG: ATR-FTIR spectroscopic imaging with expanded field of view to study formulations and dissolution. *Lab Chip* (2006) **6**(7):864–870.

Chaubal MV: Application of drug delivery technologies in lead candidate selection and optimization. *Drug Discov. Today* (2004) **9**(14):603–609.

Chen H, Zhang Z, McNulty C, Olbert C, Yoon HJ, Lee JW, Kim SC, Seo MH, Oh HS, Lemmo AV, Ellis SJ, Heimlich K: A high-throughput combinatorial approach for the discovery of a cremophor EL-free paclitaxel formulation. *Pharm. Res.* (2003) **20**:1302–1308.

Chen XQ, Cho SJ, Li Y, Venkatesh S: Prediction of aqueous solubility of organic compounds using a quantitative structure-property relationship. *J. Pharm. Sci.* (2002) **91**(8):1838–1852.

Chen XQ, Venkatesh S: Miniature device for aqueous and non-aqueous solubility measurements during drug discovery. *Pharm. Res.* (2004) **21**(10):1758–1761.

Chang FH, Smith DK (eds.): Industrial application of X-ray diffraction in pharmaceuticals, by Joel Bernstein and Jan-Olav Henck " development and formulation", CRS Press (2000) 527.

Cleveland Jr JA, Benko MH, Gluck SJ, Walbroehl YM: Automated p*K*a determination at low solute concentrations by capillary electrophoresis. *J. Chromatogr. A* (1993) **652**:301–308.

Colin J, Paquette B: Comparison of the analgesic efficacy and safety of nepafenac ophthalmic suspension compared with diclofenac ophthalmic solution for ocular pain and photophobia after excimer laser surgery: a phase II, randomized, double-masked trial. *Clin. Ther.* (2006) **28**(4):527–536.

de Lange ECM: Potential role of ABC transporters as a detoxification system at the blood–CSF barrier. *Adv. Drug Deliv. Rev.* (2004) **56**:1793–1809).

DeWitte RS: Avoiding physicochemical artefacts in early ADME-Tox experiments. *Drug Discov. Today* (2006) **11**:855–859.

Di L, Kerns EH: Biological assay challenges from compound solubility: strategies for bioassay optimization. *Drug Discov. Today* (2006) **11**:446–451.

Di L, Kerns EH, Fan K, Mcconnell OJ, Carter GT: High throughput artificial membrane permeability assay for blood–brain barrier. *Eur. J. Med. Chem.* (2003) **38**:223–232.

Di L, Kerns EH, Gao N, Li SQ, Huang Y, Bourassa JL, Huryn DM: Experimental design on single-time-point high-throughput microsomal stability assay. *J. Pharm. Sci.* (2004) **93**(6):1537–1544.

Di L, Kerns EH, Li SQ, Petusky SL: High throughput microsomal stability assay for insoluble compounds. *Int. J. Pharm.* (2006) **317**:54–60.

Dressman JB, Reppas C: *In vitro–in vivo* correlations for lipophilic, poorly water-soluble drugs. *Eur. J. Pharm. Sci.* (2000) **11**(Suppl. 2):S73–S80.

Dubey R: Impact of nanosuspension technology on drug discovery and development. *Drug Deliv. Technol.* (2006) **6**(5):67–71.

Ecanow B, Gold B, Ecanow C: *Am Cosmet. Perfumer* (1969) **84**:27

Egan WJ, Merz KM, Baldwin JJ: *J. Med. Chem.* (2000) **43**:3867–3877.

Englund G, Rorsman F, Roennblom A, Karlbom U, Lazorova L, Grasjoe J, Kindmark A, Artursson P: Regional levels of drug transporters along the human intestinal tract: co-expression of ABC and SLC transporters and comparison with Caco-2 cells. *Eur. J. Pharm. Sci.* (2006) **29**(3–4):269–277.

Faller B: Artificial membrane assays to assess permeability. *Curr. Drug Metab.* (2008) **9**:886–92.

Fernando GC, Jaun de DG, Lopes Duran, suspension formulation. In: Nielloud F, Marti-Mestres G (eds.) *Pharmaceutical Emulsions and Suspensions*, Volume 105. Marcel Dekker, New York (2000):165.

Gad SC, Cassidy CD, Aubert N, Spainhour B, Robbe H: Nonclinical vehicle use in studies by multiple routes in multiple species. *Int. J. Toxicol.* (2006) **25**:499–521.

Galia E, Nicolaides E, Horter D, LoÈ benberg R, Reppas C, Dressman JB: Evaluation of various dissolution media for predicting *in vivo* performance of class I and II drugs. *Pharm. Res.* (1998) **15**:698–705.

Gallardo V, Ruiz MA, Delgado AV: Pharmaceutical suspension and their application. In: Nielloud F, Marti-Mestres G (eds.) *Pharmaceutical Emulsion and Suspension*, Volume 105. Marcel Dekker, New York (2000):444–445.

Garberg P, Ball M, Borg N, Cecchelli R, Fenart L, Hurst RD, Lindmark T, Mabondzo A, Nilsson JE, Raub TJ, Stanimirovic D, Terasaki T, Öberg JO, Österberg T: In vitro models for the blood–brain barrier. *Toxicol. In Vitro* (2005) **19**:299–334.

Gardner CR, Almarsson O, Chen H, Morissette S, Peterson M, Zhang Z, Wang S, Lemmo A, Gonzalez-Zugasti J, Monagle J, Marchionna J, Ellis S, McNulty C, Johnson A, Levinson D, Cima M: Application of high throughput technologies to drug substance and drug product development. *Comput. Chem. Eng.* (2004) **28**(6–7):943–953.

Glomme A, Marz J, Dressman JB: Comparison of a miniaturized shake-flask solubility method with automated potentiometric acid/base titrations and calculated solubilities. *J. Pharm. Sci.* (2005) **94**(1):1–16.

Gould PL: Salt Selection for basic drugs. *International J. Pharmaceut.* (1986) **33**(1–3):201–217.

Grant DJW, Brittain HG: Solubility of pharmaceutical solids. In: Brittain HB (ed.) *Drugs and the Pharmaceutical Sciences*, Volume 70. Physical Characterization of Pharmaceutical Solids. Marcel Dekker Inc., New York, USA (1995):321–386.

Hallifax D, Rawden HC, Hazook N, Houston JB: Prediction of metabolic clearance using cryopreserved human hepatocytes: kinetic characteristics for five benzodiazepines. *Drug Metab. Dispos.* (2005) **33**:1852–1858.

Hämäläinen MD, Frostell-Karlsson A: Predicting the intestinal absorption potential of hits and leads. *Drug Discov. Today Technol.* (2004) **1**:397–406.

Hansen DK, Scott DO, Otis KW, Lunte SM: Comparison of in vitro BBMEC permeability and in vivo CNS uptake by microdialysis sampling. *J. Pharm. Biomed. Anal.* (2002) **27**:945–958.

Hewitt NJ, Lechón MJ, Houston JB, Hallifax D, Brown HS, Maurel P, Kenna JG, Gustavsson L, Lohmann C, Skonberg C, Guillouzo A, Tuschl G, Li AP, LeCluyse E, Groothuis GM, Hengstler JG: Primary hepatocytes: current understanding of the regulation of metabolic enzymes and transporter proteins, and pharmaceutical practice for the use of hepatocytes in metabolism, enzyme induction, transporter, clearance, and hepatotoxicity studies. *Drug Metab. Rev.* (2007) **39**(1):159–234.

Hitchingham L, Thomas VH: Development of a semi-automated chemical stability system to analyze solution based formulations in support of discovery candidate selection. *J. Pharm. Biomed. Anal.* (2007) **43**(2):522–526.

Hitchcock SA, Pennington LD: Structure–brain exposure relationships. *J. Med. Chem.* (2006) **49**(26):7559–7583.

Hochman JH, Yamazaki M, Ohe T, Lin JH: Evaluation of drug interactions with P-glycoprotein in drug discovery: in vitro assessment of the potential for drug-drug interactions with P-glycoprotein. *Curr. Drug Metab.* (2002) **3**(3):257–273.

Hughes MF, Shrivastava SP, Sumler MR, Edwards BC, Goodwin JH, Shah PV, Fisher HL, Hall LL: Dermal absorption of chemicals: effect of application of chemicals as a solid, aqueous paste, suspension, or in volatile vehicle. *J. Toxicol. Environ. Health* (1992) **37**(1):57–71.

Irvine JD, Takahashi L, Lockhart K, Cheong J, Tolan JW, Selick HE, Grove JR: MDCK (Madin-Darby canine kidney) cells: a tool for membrane permeability screening. *J. Pharm. Sci.* (1999) **88**:28–33.

Ishihama Y, Nakamura M, Miwa T, Kajima T, Asakawa N: A rapid method for pKa determination of drugs using pressure-assisted capillary electrophoresis with photodiode array detection in drug discovery. *J. Pharm. Sci.* (2002) **91**:933–942.

Jacobson L, Middleton B, Holmgren J, Eirefelt S, Fröjd M, Blomgren A, Gustavsson L: An optimized automated assay for determination of metabolic stability using hepatocytes: assay validation, variance component analysis, and in vivo relevance. *Assay Drug Dev. Technol.* (2007) **5**(3):403–415.

James KC: Solubility and related properties. In: *Drugs and the Pharmaceutical Sciences*, Volume 28. Marcel Dekker Inc., New York, USA (1986):36–52.

Kalantzi L, Persson E, Polentarutti B, Abrahamsson B, Goumas K, Dressman JB, Reppas C: Canine intestinal contents vs. simulated media for the assessment of solubility of two weak bases in the human small intestinal contents. *Pharm. Res.* (2006) **23**(6):1373–1381.

Kansy M, Senner F, Gubernator K: Physicochemical high throughput screening: parallel artificial membrane permeation assay in the description of passive absorption processes. *J. Med. Chem.* (1998) **41**:1007–1010.

Kansy M, Fischer H, Kratzat K, Senner F, Wagner B, Parrilla I: High-throughput artificial membrane permeability studies in early lead discovery and development. In: Testa B, van de Waterbeemd H, Folkers G, Guy R (eds.) *Pharmacokinetic Optimization in Drug Research.* Verlag Helvetica Chimica Acta (2001):447–464.

Kariv I, Rourick RA, Kassel DB, Chung TDY: Improvement of "hit-to-lead" optimization by integration of in vitro HTS experimental models for early determination of pharmacokinetic properties. *Comb. Chem. High Throughput Screen.* (2002) **5**:459–472.

Kerns EH: High throughput physicochemical profiling for drug discovery. *J Pharm. Sci.* (2001) **90**:1838–1858.

Kerns EH, Di L: Physicochemical profiling: overview of the screens. *Drug Discov. Today Technol.* (2004) **1**(4):343–348.

Kerns EH, Di L, Petusky S, Farris M, Ley R, Jupp P: Combined application of parallel artificial membrane permeability assay and Caco-2 permeability assays in drug discovery. *J. Pharm. Sci.* (2004) **93**(6):1440–1453.

Kerns EH, Di L: Chemical stability. In: Testa B, van de Waterbeemd H (eds.) *Comprehensive Medicinal Chemistry*, Volume 5, second edition. ADME-Tox Approaches. Elsevier Ltd, Oxford, UK (2007):489–507.

Kibbey CE, Poole SK, Robinson B, Jackson JD, Durham D: An integrated process for measuring the physicochemical properties of drug candidates in a preclinical discovery environment. *J. Pharm. Sci.* (2001) **90**:1164–1175.

Kocbek P, Baumgartner S, Kristl J: Preparation and evaluation of nanosuspensions for enhancing the dissolution of poorly soluble drugs. *Int. J. Pharm.* (2006) **312**(1–2):179–186.

Korfmacher WA, Palmer CA, Nardo C, Dunn-Meynell K, Grotz D, Cox K, Lin CC, Elicone C, Liu C, Duchoslav E: Development of an automated mass spectrometry system for the quantitative analysis of liver microsomal incubation samples: a tool for rapid screening of new compounds for metabolic stability. *Rapid Commun. Mass Spectrom.* (1999) **13**:901–907.

Kostewicz ES, Brauns U, Becker R, Dressman JB: Forecasting the oral absorption behavior of poorly soluble weak bases using solubility and dissolution studies in biorelevant media. *Pharm. Res.* (2002) **19**:345–349.

Kostewicz ES, Wunderlich M, Brauns U, Becker R, Bock T, Dressman JB: Predicting the precipitation of poorly soluble weak bases upon entry in the small intestine. *J. Pharm. Pharmacol.* (2004) **56**(1):43–51.

Kuppens IELM, Breedveld P, Beijnen JH, Schellens JHM: Modulation of oral drug bioavailability: from preclinical mechanism to therapeutic application. *Cancer Invest.* (2005) **23**(5):443–464.

Kibbe AH: *Handbook of Pharmaceutical Excipients.* American Pharmaceutical Association, Pharmaceutical Press (2000).

Lachman L, Lieberman H, Kanig J: *The Theory and Practice of Industrial Pharmacy.* Lea & Febiger, Philadelphia (1970).

Lakeram M, Lockley DJ, Sanders DJ, Pendlington R, Forbes B: Paraben transport and metabolism in the bio-mimetic artificial membrane permeability assay (BAMPA) and 3-day and 21-day Caco-2 cell systems. *J. Biomol. Screen.* (2007) **12**(1):84–91.

Lau YY, Krishna G, Yumibe NP, Grotz DE, Sapidou E, Norton L, Chu I, Chen C, Soares AD, Lin CC: The use of in vitro metabolic stability for rapid selection of compounds in early discovery based on their expected hepatic extraction ratios. *Pharm. Res.* (2002a) **19**:1606–1610.

Lau YY, Sapidou E, Cui X, White RE, Cheng K-C: Development of a novel in vitro model to predict hepatic clearance using fresh, cryopreserved and sandwich-cultured hepatocytes. *Drug Metab. Dispos.* (2002b) **30**:1446–1454.

Lennernas H, Lundgren E: Intestinal and blood–brain drug transport: beyond involvement of a single transport function. *Drug Discov. Today Technol.* (2004) **1**:417–422.

Levis KA, Lane ME, Corrigan OI: Effect of buffer media composition on the solubility and effective permeability coefficient of ibuprofen. *Int. J. Pharm.* (2003) **253**(1–2):49–59.

Li AP: Human hepatocytes: isolation, cryopreservation and applications in drug development. *Chem. Biol. Interact.* (2007) **168**(1):16–29.

Li N, Degennaro MD, Liebenber W, Tied LR, Zahr AS, Pishko MV, de Villires MM: Increased dissolution and physical stability of micronized nifedipine particles encapsulated with a bio-compatible polymer and surfactants in a wet ball milling process. *Pharmazie* (2006) **61**(8):659

Liang E, Chessic K, Yazdanian M: Evaluation of an accelerated Caco-2 cell permeability model. *J. Pharm. Sci.* (2000) **89**(3):336–345.

Lieberman HA, Rieger MM, Banker GS: *Pharmaceutical Dosage Forms: Disperse Systems*, Volume 3, revised and expanded. Marcel Dekker, New York/Basel (1998):488.

Lindfors L, Skantze P, Skantze U, Rasmusson M, Zackrisson A, Olsson U: Amorphous drug nanosuspensions. Inhibition of Ostwald ripening. *Langmuir* (2006) **22**(3):906–910.

Lin JH, Wong BK: Complexities of glucuronidation affecting in vitro in vivo extrapolation. *Curr. Drug Metab.* (2002) **3**:623–646.

Linget JM, du Vignaud P: Automation of metabolic stability studies in microsomes, cytosol and plasma using a 215 Gilson liquid handler. *J. Pharm. Biomed. Anal.* (1999) **19**:893–901.

Lipinski CA, Lombardo L, Dominy BW, Feeney PJ: Experimental and computational approaches to estimate solubility and permeability in drug discovery and development settings. *Adv. Drug Deliv. Rev.* (1997) **23**:3–25.

Liu H, Sabus C, Carter GT, Du C, Avdeef A, Tischler M: In vitro permeability of poorly aqueous soluble compounds using different solubilizers in the PAMPA assay with liquid chromatography/mass spectrometry detection. *Pharm. Res.* (2003) **20**(11):1820–1826.

Liu R, So SS: Development of quantitative structure-property relationship models for early ADME evaluation in drug discovery 1 aqueous solubility. *J. Chem. Inf. Comput. Sci.* (2001) **41**:1633–1639.

Lombardo F, Shalaeva M, Tupper KA, Gao F, Abraham MJ: ElogPoct: a tool for lipophilicity determination in drug discovery. *J. Med. Chem.* (2000) **43**:2922–2928.

Lundquist S, Renftel M: The use of in vitro cell culture models for mechanistic studies and as permeability screens for the blood–brain barrier in the pharmaceutical industry – background and current status in the drug discovery process. *Vascul. Pharmacol.* (2002) **38**:355–364.

Maeda H, Kato H, Ikeda S: Effect of cationic surfactants on the conformation and aggregation of poly(L-glutamic acid). *Biopolymers* (1984) **23**(7):1333.

Marino AM, Yarde M, Patel H, Chong S, Balimane PV: Validation of the 96 well Caco-2 cell culture model for high throughput permeability assessment of discovery compounds. *Int. J. Pharm.* (2005) **297**(1–2):235–241.

Matteucci ME, Brettmann BK, Rogers TL, Elder EJ, Williams RO III, Johnston KP: Design of potent amorphous drug nanoparticles for rapid generation of highly supersaturated media. *Mol. Pharm.* (2007) **4**(5):782–793.

McGinnity DF, Soars MG, Urbanowicz RA, Riley RJ: Evaluation of fresh and cryopreserved hepatocytes as in vitro drug metabolism tools for the prediction of metabolic clearance. *Drug Metab. Dispos.* (2004) **32**:1247–1253.

Mensch J, Noppe M, Adriaensen J, Melis A, Mackie C, Augustijns P, Brewster ME: Novel generic UPLC/MS/MS method for high throughput analysis applied to permeability assessment in early drug discovery. *J. Chromatogr. B Analyt. Technol. Biomed. Life Sci.* (2007) **847**(2):182–187.

Merisko-Liversidge E, Liversidge GG, Cooper ER: Nanosizing: a formulation approach for poorly-water-soluble compounds. *Eur. J. Pharm. Sci.* (2003) **18**(2):113–120.

Miret S, Abrahamse L, de Groene EM: Comparison of in vitro models for the prediction of compound absorption across the human intestinal mucosa. *J. Biomol. Screen.* (2004) **9**(7):598–606.

Morris KR, Fakes MG, Thakur AB, Newman AW, Singh AK, Venit JJ, Spangnulo CJ, Serajuddin ATM: An integrated approach to the selection of optimal salt form for a new drug candidate. *International J. Phamaceut.* (1994) **105**(3):209–217.

Morissette SL, Read MJ, Soukasene S, Tauber MK, Scoppettuolo LA, Apgar JR, Guzman HR, Sauer J-M, Collins DS, Jadhav PK, Engler T, Gardner CG: High throughput crystallization of poly-morphs and salts: Applications in early lead optimization: 225th ACS National Meeting, New Orleans, LA, United States, March 23–27, (2003).

Mountfield RJ, Senepin S, Schleimer M, Walter I, Bittner B: Potential inhibitory effects of formulation ingredients on intestinal cytochrome P450. *Int. J. Pharm.* (2000) **211**(1–2):89–92.

Naritomi Y, Terashita S, Kimura S, Suzuki A, Kagayama A, Sugiyama Y: Prediction of human hepatic clearance from in vivo animal experiments and in vitro metabolic studies with liver microsomes from animals and humans. *Drug Metab. Dispos.* (2001) **29**:1316–1324.

Nash RA: Pharmaceutical suspensions. In: Lieberman HA, Rieger MM, Banker GS (eds.) *Pharmaceutical Dosage Form. Dispersed System*, Volume 1. Marcel Dekker, New York (1988):151.

Neervannan S: Preclinical formulations for discovery and toxicology: physicochemical challenges. *Expert Opin. Drug Metab. Toxicol.* (2006) **2**(5):715–731.

Neuhoff S, Artursson P, Zamora I, Ungell A-L: Impact of extracellular protein binding on passive and active drug transport across Caco-2 cells. *Pharm. Res.* (2006) **23**(2):350–359.

Niazi S: *Handbook of Pharmaceutical Manufacturing Formulations Uncompressed Solid Products*, Volume 2. CRC Press, Boca Raton (2004):41.

Nicolaides E, Galia E, Efthymiopoulos C, Dressman JB, Reppas C: Forecasting the in vivo performance of four low solubility drugs from their *in vitro* dissolution data. *Pharm. Res.* (1999) **16**:1876–1882.

Nicolazzo JA, Charman SA, Charman WN: Methods to assess drug permeability across the blood–brain barrier. *J. Pharm. Pharmacol.* (2006) **58**:281–293.

Nobili S, Landini I, Giglioni B, Mini E: Pharmacological strategies for overcoming multidrug resistance. *Curr. Drug Targets* (2006) **7**(7):861–879.

Obach RS: Prediction of human clearance of twenty-nine drugs from hepatic microsomal intrinsic clearance data: an examination of in vitro halflife approach and nonspecific binding to microsomes. *Drug Metab. Dispos.* (1999) **27**:1350–1359.

Obach RS: The prediction of human clearance from hepatic microsomal metabolism data. *Curr. Opin. Drug Discov. Dev.* (2001) **4**:36–44.

Obach RS, Baxter JG, Liston TE, Silber BM, Jones BC, MacIntyre F, Rance DJ, Wastall P: The prediction of human pharmacokinetic parameters from preclinical and in vitro metabolism data. *J. Pharmacol. Exp. Ther.* (1997) **283**:46–58.

Obata K, Sugano K, Machida M, Aso Y: Biopharmaceutics classification by high throughput solubility assay and PAMPA. *Drug Dev. Ind. Pharm.* (2004) **30**(2):181–185.

Offener III CM, Schmaare RL, Schwartz JB: Reconstitutable suspensions. In : Lieberman HA, Rieger MM, Banker GS (eds.) *Pharmaceutical Dosage Form. Dispersed System*, Volume 2. Marcel Dekker, New York (1989):317–334.

Pan L, Ho Q, Tsutsui K, Takahashi L: Comparison of chromatographic and spectroscopic methods used to rank compounds for aqueous solubility. *J. Pharm. Sci.* (2001) **90**:521–529.

Patel NK, Kennon L, Levison RS: Pharmaceutical suspension. In: Lachman L, Lieberman HA, Kanig JL (eds.) *The Theory and Practice of Industrial Pharmacy*, Volume 3. Lea and Febiger, Philadelphia (1976):484.

Persson EM, Gustafsson A-S, Carlsson AS, Nilsson RG, Knutson L, Forsell P, Hanisch G, Lennernaes H, Abrahamsson B: The effects of food on the dissolution of poorly soluble drugs in human and in model small intestinal fluids. *Pharm. Res.* (2005) **22**(12):2141–2151.

Pignatello R, Bucolo C, Puglisi G: Ocular tolerability of Eudragit RS100 and RL100 nanosuspensions as carriers for ophthalmic controlled drug delivery. *J. Pharm. Sci.* (2002) **91**(12).

Pignatello R, Ricupero N, Bucolo C, Maugeri F, Maltese A, Puglisi1 G: Preparation and characterization of Eudragit retard nanosuspensions for the ocular delivery of cloricromene. *Pharm. Sci. Technol.* (2006) **7**(1):E27.

Pouton CW: Formulation of poorly water-soluble drugs for oral administration: physicochemical and physiological issues and the lipid formulation classification system. *Eur. J. Pharm. Sci.* (2006) **29**(3–4):278–287.

Reddy A, Heimbach T, Freiwald S, Smith D, Winters R, Michael S, Surendran N, Cai H: Validation of a semi-automated human hepatocyte assay for the determination and prediction of intrinsic clearance in discovery. *J. Pharm. Biomed. Anal.* (2005) **37**:319

Reichel A, Begley DJ: Potential of immobilized artificial membranes for predicting drug penetration across the blood–brain Barrier. *Pharm. Res.* (1998) **15**:1270–1274.

Ruell JA, Tsinman O, Avdeef A: PAMPA – a drug absorption in vitro model. 12. Cosolvent method for permeability assays of amiodarone, itraconazole, tamoxifen, terfenadine, and other very insoluble molecules. *Chem. Pharm. Bull.* (2004) **52**:561–565.

Saha P, Kou JH: Effect of bovine serum albumin on drug permeability estimation across Caco-2 monolayers. *Eur. J. Pharm. Biopharm.* (2002) **54**(3):319–324.

Saunders KC: Automation and robotics in ADME screening. *Drug Discov. Today Technol.* (2004) **1**:373–380.

Sciarra JJ, Cutie AJ: Aerosol suspension and emulsion. In: Lieberman HA, Rieger MM, Banker GS (eds.) *Pharmaceutical Dosage Form. Dispersed System*, Volume 2. Marcel Dekker, New York (1989):417–460.

Shibata Y, Takahashi H, Chiba M, Ishii Y: Prediction of hepatic clearance and availability by cryo-preserved human hepatocytes: an application of serum incubation method. *Drug Metab. Dispos.* (2002) **30**:892–896.

Smith QR: A review of blood–brain barrier transport techniques. In: Nag S (ed.) *The Blood–Brain Barrier: Biology and Research Protocols.* Humana Press Inc., Totowa (2003):193–208.

Soars MG, Burchell B, Riley RJ: In vitro analysis of human drug glucuronidation and prediction of in vivo metabolic clearance. *J. Pharm. Exp. Ther.* (2002) **301**:382–390.

Soars MG, McGinnity DF, Grime K, Riley RJ: The pivotal role of hepatocytes in drug discovery. *Chem. Biol. Interact.* (2007a) **168**(1):2–15.

Soars MG, Grime K, Sproston JL, Webborn PJ, Riley RJ: Use of hepatocytes to assess the contribution of hepatic uptake to clearance in vivo. *Drug Metab. Dispos.* (2007b) **35**(6): 859–865.

Suryanarayanan R: X-ray powder diffractometry. *Drugs Pharmaceut. Sci.* (1995) **70**:187–221.

Steffansen B, Nielsen CU, Brodin B, Eriksson AH, Andersen R, Frokjaer S: Intestinal solute carriers: an overview of trends and strategies for improving oral drug absorption. *Eur. J. Pharm. Sci.* (2004) **21**(1):3–16.

Stella VJ, Yoshioka S: Physical stability of drug substances. In: Stella VJ, Yoshioka S (eds.) *Stability of Drugs and Dosage Forms.* Springer Publisher (2002):139.

Stephenson GA: Structure determination from convetional powder diffraction data: application to hydrates, hydrochloride salts, and metastable polymorphs. *J. Pharmaceut. Sci* (2000) **89**(7):958–966.

Stresser DM, Broudy MI, Ho T, Cargill CE, Blanchard AP, Sharma R, Dandeneau AA, Goodwin JJ, Turner SD, Erve JCL, Patten CJ, Dehal SS, Crespi CL: Highly selective inhibition of human CYP3A in vitro by azamulin and evidence that inhibition is irreversible. *Drug Metab. Dispos.* (2004) **32**:105–112.

Stuart M, Box K: Chasing equilibrium: measuring the intrinsic solubility of weak acids and bases. *Anal. Chem.* (2005) **77**(4):983–990.

Sugano K, Hamada H, Machida M, Ushio H, Saitoh K, Terada K: Optimized conditions of bio-mimetic artificial membrane permeability assay. *Int. J. Pharm.* (2001) **228**:181–188.

Tan H, Semin D, Wacker M, Cheetham J: An automated screening assay for determination of aqueous equilibrium solubility enabling SPR study during drug lead optimization. *JALA* (2005) **10**(6):364–373.

Terasaki T, Ohtsuki S, Hori S, Takanaga H, Nakashima E, Hosoya K: New approaches to in vitro models of blood–brain barrier drug transport. *Drug Discov. Today* (2003) **8**:944–954.

Tong W-Q, Whitesel G: In situ salt screening-a useful techique for discovery support and perfor-mulation studies; *Pharmaceut. Dev. Tech.* (1993) **3**(2):215–223.

Troutman MD, Thakker DR: Efflux ratio cannot assess P-glycoprotein-mediated attenuation of absorptive transport: asymmetric effect of P-glycoprotein on absorptive and secretory transport across caco-2 cell monolayers. *Pharm. Res.* (2003a) **20**:1200–1209.

Troutman MD, Thakker DR: Novel experimental parameters to quantify the modulation of absorptive and secretory transport of compounds by P-glycoprotein in cell culture models of intestinal epithelium. *Pharm. Res.* (2003b) **20**:1210–1224.

Ungell AL, Karlsson J: Cell culture in drug discovery: an industrial perspective. In: van de Waterbeemd H, Lennernas H, Artursson P (eds.) *Drug Bioavailability*. Wiley-VCH, New York (2003):90–131.

Ungell A-LB: Caco-2 replace or refine? *Drug Discov. Today Technol.* (2004) 1:423–430.

Van de Waterbeemd H: Physicochemistry, in pharmacokinetics and metabolism in drug design, 2nd edn (eds D.A. Smith, H. van de Waterbeemd and D.K Walker), Wiley-VCH, Verlag GmbH, Winheinm, (2006), 1–18

Varma MVS, Ashokraj Y, Chinmoy SD, Panchagnula R: P-glycoprotein inhibitors and their screening: a perspective from bioavailability enhancement. *Pharmacol. Res.* (2003) 48(4): 347–359.

Varma MVS, Sateesh K, Panchagnula R: Functional role of P-glycoprotein in limiting intestinal absorption of drugs: contribution of passive permeability to P-glycoprotein mediated efflux transport. *Mol. Pharm.* (2005) 2(1):12–21.

Varma MVS, Perumal OP, Panchagnula R: Functional role of P-glycoprotein in limiting peroral drug absorption: optimizing drug delivery. *Curr. Opin. Chem. Biol.* (2006) 10(4):367–373.

Vertzoni M, Fotaki N, Kostewicz E, Stippler E, Leuner C, Nicolaides E, Dressman JB, Reppas C: Dissolution media simulating the intralumenal composition of the small intestine: physiological issues and practical aspects. *J. Pharm. Pharmacol.* (2004) 56(4):453–462.

Vertzoni M, Pastelli E, Psachoulias D, Kalantzi L, Reppas C: Estimation of intragastric solubility of drugs. *Pharm. Res.* (2007) 24(5):909–917.

Venkatesh S, Lipper RA. Role of the development scientist in compound lead selection and optimization. *J. Pharm. Sci.* (2000) 89:145–154.

Vandervoort J, Ludwig A. Preparation and evaluation of drug-loaded gelatin nanoparticles for topical ophthalmic use, Eur J Pharm Biopharm. (2004) Mar; 57(2):251–61.

Vippagunta SR, Brittain HG, Grant DJW: Crystalline solids. *Adv. Drug Delivery Rev.* (2001) 48(1):3–26.

Wagner D, Spahn-Langguth H, Hanafy A, Koggel A, Langguth P: Intestinal drug efflux: formulation and food effects. *Adv. Drug Deliv. Rev.* (2001) 50(Suppl. 1):S13–S31.

Wan H, Ulander J: High-throughput pKa screening and prediction amenable for ADME profiling. *Expert Opin. Drug Metab. Toxicol.* (2006) 2(1):139–155.

Wang J: Comprehensive Assessment of ADMET Risks in Drug Discovery. *Curr. Pharmaceut. Design* (2009) 15:2195–2219.

Wang J, Urban L: The impact of early ADME profiling on drug discovery and development strategy. *Drug Discov. World* (2004) 5:73–86.

Wang J, Faller B: Progress in bioanalytics and automation robotics for ADME screening. In: Testa B, van de Waterbeemd H (eds.) *Comprehensive Medicinal Chemistry*, Volume 5, second edition. ADME-Tox Approaches. Elsevier Ltd, Oxford, UK (2007):341–356.

Wang J, Urban L, Bojanic D: Maximising use of in vitro ADMET tools to predict in vivo bioavailability and safety. *Expert Opin. Drug Metab. Toxicol.* (2007b) 3(5):641–665.

Wei H, Loebenberg R: Biorelevant dissolution media as a predictive tool for glyburide a class II drug. *Eur. J. Pharm. Sci.* (2006) 29(1):45–52.

Wienkers LC, Heath TG: Predicting *in vivo* drug interactions from *in vitro* drug discovery data. *Nat. Rev. Drug Discov.* (2005) 4:825–833.

Wohnsland F, Faller B: High-throughput permeability pH profile and high-throughput alkane/water Log P with artificial membranes. *J. Med. Chem.* (2001) 44:923–930.

Yalkowsky SH, Banerjee S: *Aqueous Solubility Methods of Estimation for Organic Compounds*. Marcel Dekker Inc., New York, USA (1992):149–154.

Yamashita S, Furubayashi T, Kataoka M, Sakane T, Sezaki H, Tokuda H: Optimized conditions for prediction of intestinal drug permeability using Caco-2 cells. *Eur. J. Pharm. Sci.* (2000) 10(3):195–204.

Yamashita S, Konishi K, Yamazaki Y, Taki Y, Sakane T, Sezaki H, Furuyama Y: New and better protocols for a short-term Caco-2 cell culture system. *J. Pharm. Sci.* (2002) 91(3):669–679.

Yamaguchi M, Yasueda S, Isowaki A, Yamamoto M, Kimura M, Inada K, Ohtori A: Formulation of an ophthalmic lipid emulsion containing an anti-inflammatory steroidal drug, difluprednate. *Int. J. Pharm.* (2005) 301:121–128.

Yu LX, Amidon GL, Polli JE, Zhao H, Mehta MU, Conner DP, Shah VP, Lesko LJ, Chen ML, Lee VHL, Hussain AS: Biopharmaceutics classification system: the scientific basis for bio-waiver extensions. *Pharm. Res.* (2002) **19**(7)

Zheng W, Kim H, Garad S: Effect of Polymeric Excipients on the Solid Phase Transformation in Aqueous Suspensions. APPS, 2005, Annual Meeting Poster Presentation.

Zhou L, Yang L, Tilton S, Wang J: Development of high throughput equilibrium solubility assay using miniaturized shake-flask method in early drug discovery. *J. Pharm. Sci.* (2007) **98**(11): 3052–3071.

Zietsman S, Kilian G, Worthington M, Stubbs C: Formulation development and stability studies of aqueous metronidazole benzoate suspensions containing various suspending agents. *Drug Dev. Ind. Pharm.* (2007) **33**(2):191–197.

Chapter 6
Analytical Tools for Suspension Characterization

Alan F. Rawle

Abstract Particle shape, size, and density play a fundamental role in determining both primary and bulk powder and suspension properties. For example, dissolution is related to the primary particle size in the system and rheological properties to the bulk particle size distribution (agglomerates and all). There are a large number of techniques both qualitative and quantitative that can be used to characterize particulate suspensions and use these generated data to predict the behavior of the materials in formulations or for quality control purposes. This chapter explores the various particle size, morphological, and other properties that can be used in the characterization of suspensions.

6.1 Introduction

The answer to the question "Why?" is too infrequently tackled within analytical measurement and the reasons for the determination forgotten or ignored. Harold Heywood (1966), in the closing speech at the first major particle size congress in 1966, answered this question as follows:

> "However, it must be realised that particle size analysis is not an objective in itself but is a means to an end, the end being the correlation of powder properties with some process of manufacture, usage or preparation"

Thus we need to examine the real scientific reasons behind taking the particular measurement that we will be attempting. Are we concerned with bulk suspension properties such as flowability or viscosity (rheological properties), filter blockage, etc? Or is our measurement related to primary particle properties that are usually reacted to the available surface area? Properties such as dissolution (Noyes and

A.F. Rawle (✉)
Malvern Instruments Inc., 117 Flanders Road, Westborough, MA 01581-1042, USA
e-mail: alan.rawle@malvern.com

A.K. Kulshreshtha et al. (eds.), *Pharmaceutical Suspensions: From Formulation Development to Manufacturing*, DOI 10.1007/978-1-4419-1087-5_6,
© Springer Science + Business Media, LLC 2010

Whitney 1897), chemical reactivity and absorption of molecules onto the surface are governed by the size of the individual particulate element in the system. We do recognize that in a solid-in-liquid suspension, dissolution would be expected to be minimal (the solid may be in saturated solution, of course) – unless dilution or chemical change e.g. in the gastric juices occurs.

If, as too often, the answer to the question "Why?" is "Because my boss says so" or "For Q.A.", then we must take time out to go back and explore the scientific reasons for the measurement. This is to prevent the provision of data becoming meaningless as the measured result may be taken in a regime that is not appropriate to the needs and objectives of the investigation. It can be noted at this stage that the energies found in any particle size "dispersion" unit, wet or dry, are considerably higher than which can be found in most conventional plant processing conditions. Thus, if we have a problem related to inability of a suspension to be pumped adequately, then it is ridiculous to try to correlate this to a primary dispersed size achieved by sonication to stability in a wet particle size dispersion unit.

In the pharmaceutical scenario, we may be trying to relate our suspension properties to how quickly an oral formulation is likely to act or trying to relate the numbers to drug bioavailability. Shape characterization is vital here as different crystal faces dissolve at different rates, as well as the fact that the less-rounded or platy or particles of lower apparent density particles will settle out of suspension much more slowly than the ideal spherical equivalent. At the preformulation stage, we may need to identify the techniques we would need to use further down the line in production in order to produce a formulation with the desired efficacy in line with optimized and economic production techniques.

A little time at the start thinking through the answers to the question "Why?" will pay dividends further down the analytical road. On this journey, this chapter may contribute hopefully to some essential and useful knowledge, but it is not expected to make the reader an expert. Indeed the author is still an open-minded learner in this field. In the words (Nelligan 2000) of Chris Leslie (referring to violin making):

"These are the basics for a lifetime's improvement"

Hopefully we will not be giving the reader enough information to get him or her into serious trouble!

6.2 Basic Concepts and Assumptions

A typical suspension could be considered that of a solid in a liquid phase or a liquid in a liquid phase, the latter normally given the term emulsion. We will consider these systems only, even though other two-phase suspensions (e.g. solid in gas) could be deemed pedantically correct. A suspension is kept in the state of "suspension" by input of energy – this would be Brownian motion for systems typically under 1 μm or of "low density". For larger systems typically larger than say 1 μm, then agitation (stirring) would be needed to prevent the particles from settling. We assume standard gravity conditions as found on earth!

We also should assume that a stable suspension or formulation has been achieved by the standard processes known to formulation chemists in industries such as ceramics and paints as well as pharmaceuticals:

- Wetting – obtaining intimate contact between the solid phase and the liquid phase. If wetting is needed, then use of a surfactant (=surface active agent) or other liquid that reduces the contact angle and lowers the interfacial tension between the phases is appropriate. Generally speaking, it can be argued that if a material wets in its interaction with the bulk liquid phase, then no surfactant is needed. This is unlikely in the case of most organic solids unless they have some ionic make-up (and then are likely to have water solubility). If the suspension is of colloidal dimensions, it is likely to have been made by a bottom-up process

- Separation/dispersion – the creation of primary, individual particles rather than agglomerates by the input of energy. In the laboratory experiment, this can be achieved with ultrasound energy. In the production context, a high shear mixer may be the only possibility and this can never deliver the energy input that ultrasound is capable of achieving. In manufacture, then ball-milling combined with high ionic surfactant (usually anionic) can achieve separation down to 100 nm or so.

- Stabilization – this would be required in situations under 5 μm, say, where the attractive forces between particles can cause the agglomeration of particles. In the aqueous phase, then the need for stabilization can be assessed by means of the measurement of zeta potential (the charge on the particle's slipping plane) in the system.

Will the particles stay in suspension? This is a key question for the chapter topic. The minimum settling time can be predicted by the application of Stokes' Law. Any deviations from sphericity (increased surface in relation to volume) or systems where the Brownian motion becomes significant will increase the settling times in comparison to the ideal spherical particle settling as Stokes predicts. A predictive settling table (taken from an ASTM standard ASTM E2490-09 work item; Ref (ASTM E56 Committee on Nanotechnology 2007)) is easily constructed from the basic Stokes' equation (Table 6.1):

These calculations indicate that if any settling is seen over a period of several hours then the material cannot be considered truly nano (based on a <100 nm definition of "nano": ASTM E-2456). Thus, the table indicates the nature of a true colloidal suspension.

Incorporation of fluid into the particulate matrix will reduce the apparent density of the material and again prolong settling making the material appear smaller than the diameter obtained from, say, a microscope measurement.

Table 6.1 Stokes' law Predictions for spheres of differing sizes and density

Diameter (μm)	Diameter nm	ρ (Material) kg/m^3	ρ (Water) kg/m^3	η (Water) 298k, Poise	Time to settle 1 cm (1×10^{-2}m) in water		
					Minutes	Hours	Days
0.01	10	2,500	1,000	0.0008905	1,815,494.39	30,258	1,261
0.1	100	2,500	1,000	0.0008905	18,154.94	302.58	12.61
1	1,000	2,500	1,000	0.0008905	181.55	3.03	0.126
10	10,000	2,500	1,000	0.0008905	1.82	0.03	0.001
100	100,000	2,500	1,000	0.0008905	0.02	0.00	0.000
0.01	10	3,500	1,000	0.0008905	1,089,296.64	18,154.94	756
0.1	100	3,500	1,000	0.0008905	10,892.97	181.55	7.56
1	1,000	3,500	1,000	0.0008905	108.93	1.82	0.076
10	10,000	3,500	1,000	0.0008905	1.09	0.02	0.001
100	100,000	3,500	1,000	0.0008905	0.01	0.00	0.000
0.01	10	4,200	1,000	0.0008905	851,013.00	14,183.55	591
0.1	100	4,200	1,000	0.0008905	8,510.13	141.84	5.91
1	1,000	4,200	1,000	0.0008905	85.10	1.42	0.059
10	10,000	4,200	1,000	0.0008905	0.85	0.01	0.001
100	100,000	4,200	1,000	0.0008905	0.01	0.00	0.000
0.01	10	5,500	1,000	0.0008905	605,164.80	10,086.08	420
0.1	100	5,500	1,000	0.0008905	6,051.65	100.86	4.20
1	1,000	5,500	1,000	0.0008905	60.52	1.01	0.042
10	10,000	5,500	1,000	0.0008905	0.61	0.01	0.000
100	100,000	5,500	1,000	0.0008905	0.01	0.00	0.000

6.3   The Effect of Particle Size on Product Performance

We need not point out to the reader that there are many facets and parameters associated with the term "product performance". At the end of the day, we really should only be concerned with the efficacy of the treatment rather than any generated numbers or even the specification per se. However, we do need to remember, paraphrasing Lord Rayleigh (and many others) that "If we can't measure it, then we can't manage it".

We will tackle an overview of some of the more important parameters that may be dealt with in the formulation laboratory. These include size, shape, zeta potential (related to stability) and rheological as well as other physical properties. The effectiveness or otherwise of the preparation is outside the scope of this chapter and the knowledge of the author.

6.3.1   Particle Size and Dissolution

There is a basic (FDA) requirement to study the dissolution enshrined in USP XXIII <711>. In the simplest analogy, dissolution should broadly correlate with bioavailability. It is a simple and inexpensive indicator that something may be awry with a formulation.

The basic equation linking particle size and dissolution is attributed to Noyes–Whitney, (Noyes and Whitney 1897), and dates back to 1897.

$$\delta A/\delta T = kS\ (Cs - C)$$

where A = amount of drug in solution, t = time, k = intrinsic dissolution rate constant, S = Surface area, Cs = concentration of the drug-solvent boundary on the surface of the particle (approximately equal to the solubility of the drug in the solvent), C = concentration of the drug in the dissolution medium at time t.

Playing with dimensional analysis indicates that we would expect the diameter of a dissolving particle to decrease linearly with time.

It is certain that different crystal faces dissolve at different rates and Carstensen (Lai and Carstensen 1978) indicates a generic plot for different aspect ratios (in length/width format) for needle and plate-like crystals (Fig. 6.1).

Thus, knowledge of the shape of the material is a prerequisite, as we will see later. Apparatus for testing the dissolution of a solid material is commonly seen in pharmaceutical analytical laboratories in the form of paddle testers. A general review article on dissolution methodology by Shiu working for the FDA is contained within (Shiu 1996).

6.3.2 Effect of Particle Size on Content Uniformity

If a suspension will be used as the basis of, say a tabletting formulation, then it is fairly obvious that the mass of the active ingredient in relation to the overall tablet

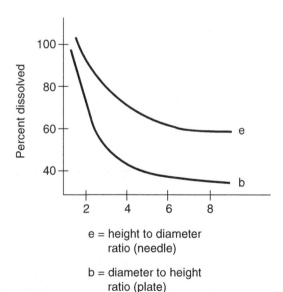

e = height to diameter
ratio (needle)

b = diameter to height
ratio (plate)

Fig. 6.1 Generic plot of dissolution of different crystal faces (after Lai and Carstensen 1978)

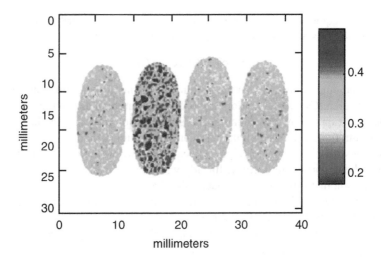

Fig. 6.2 Homogeneity of tablets viewed with NIR chemical analysis (SyNIRgi, Malvern Instruments)

volume will give some potential measure of the amount of drug delivered in a single tablet. Mainly, it is the top end size and the width of the distribution of the active ingredient that will govern the final uniformity of the system and the potential tablet-to-tablet variation. The uniformity issue is basically a problem in sampling theory. Experimental indications of homogeneity or otherwise can be explored by means of chemically imaging the appropriate material by means of NIR or Raman spectroscopy. These techniques allow stunning visual depictions, similar to science-fiction planetary bodies, to be displayed as above (Fig. 6.2).

The plot shows four tablets with same active ingredients processed on the SyNIRgi NIR chemical imaging equipment. The active and inactive ingredient are easily identified and the homogeneity or otherwise of the tablets can be clearly seen. The rogue (actually counterfeit) tablet is easily identified by this "fingerprinting" technique.

The move towards smaller sized, more potent (so less needed) active ingredients will assist the goal of tablet and batch homogeneity.

6.4   Particle Size Analysis – The Theory

Consider the dimensional changes in the development of a butterfly. It could be considered to begin life as an approximately spherical egg, modelled by a single dimension – the diameter. As it evolves to the caterpillar stage, then this creature could be modelled by a length and diameter – two dimensions – considering it resembling a cylinder. Once the butterfly emerges from the chrysalis then we are in an impossible position even to describe it dimensionally with three numbers.

The last situation is akin to the challenges facing the analyst attempting to describe the irregularly shaped particles.

Hindsight is a wonderful ally and it may seem obvious to state that a particle is a three-dimensional entity. We may easily accept that any image of a particle is only a two-dimensional representation of the same particle. However, the next stage of accepting that any particle sizing technique provides only a one-dimensional descriptor of the particle becomes virtually impossible to accept or understand. However, it is this dimensional reduction (to absurdity?) that all particle characterization techniques must battle with. This is because single numbers are desired for process control and specification purposes. In these days of Computer Hindered Design (CHD), then it may be possible or even desirable to describe a particle with a large number of parameters but this will be too many degrees of freedom for the Production or Quality Control Manager. It is generally the case especially for regulatory bodies such as the FDA that control and specification revolve around a single number and the accepted tolerances on this selected number.

6.4.1 The Equivalent Sphere

All particle-sizing techniques measure some one-dimensional property of the particle (e.g. weight, volume, settling rate, projected area, minimum sieve aperture through which the particle will fall) and relate this to a spherical particle of the same property as that of our irregular particle (Rawle 1993). This then allows a single number ("equivalent spherical diameter") to be reported. It is clear that for a cylinder, only two numbers are required to specify the particle exactly (height and diameter) and for a cuboid (e.g. a matchbox) then three-numbers (length, breadth, height) are required. Thus a single number obtained from any particle sizing technique cannot be used to fully specify the material and some form of imaging of the particle is required and essential in order to obtain an understanding of the shape of the material. This finds expression in USP General Test <776> (2000):

> "For irregularly shaped particles, characterization of particle size must include information on particle shape"

Thus shape and size go hand in hand like strawberries and cream.

The end result of this linkage is that we have a number of correct one-dimensional answers as to the particle size of a single three-dimensional particle. We also have a number of one-dimensional "answers" taken from two-dimensional information of the "size" of the particle when imaged (Fig. 6.3 and 6.4).

We thus have the seemingly unbelievable situation in Fig. 6.3 below that we have seven different but all correct answers for the reported size of the indicated irregular particle. The two important consequences of these being irregular particles are:

- Different techniques will measure different properties of the particle and will thus generate different answers. Therefore there is a whole gamut of international standards on particle size measurement reflecting the different techniques

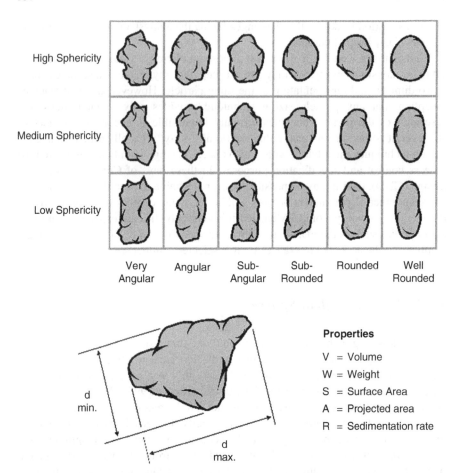

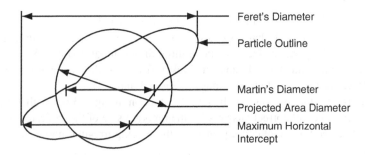

Fig. 6.3 Particle shape descriptions – adapted from Powers (1953) and most commonly used one-dimensional descriptors

Fig. 6.4 Most frequently used image analysis diameters – adapted from Yamate and Stockham (1977)

of measurement. Result emulation (sometimes called "shape correction") is not advisable as it will identically map the two results obtained by the two different techniques, with the big assumption that the two samples offered to the two techniques are identical in both particulate composition (sampling) and dispersion. Should the sample change markedly (e.g. introduction of enormous particles from some production disaster. A good example is a wrench falling into a powder) then this result emulation cannot be obviously now valid (a real wrench in the works!) as these large particles were never emulated.

- Different techniques can only be strictly compared by the use of spherical particles – only for these particles should identical answers (the "true" diameter of the spheres) be obtained. The corollary to this consequence is that particle size standards should be spherical in order that there is no dispute as to the size of the particle. Nonspherical particles can be characterized with one technique and should read the same on this particular technique but not necessarily the same with other techniques. A good example of this is the irregular quartz standard BCR66 produced within the EEC. This material has been thoroughly characterized by sedimentation in the range 0.5–2.5 µm roughly. Thus it is a good test of sedimentation equipment. Laser diffraction will read a larger result than the certified (sedimentation) values, as particles in this size range do not strictly behave as predicted by Stokes' Law. They do not settle in a straight line due to Brownian motion and have a bigger drag on them than an equivalent spherical volume particle (the sphere compresses the most volume or mass into any shape). Thus, the particles will be measured larger on laser diffraction equipment, which measures volumes of particles (in the derived sense – actually scattering is mapped to identical scattering behaviour in the same way that sedimentation has measured settling rate and not "size"). Indeed agreement between techniques on such a system would be suspicious. Again we must exercise caution as the preparative technique may have induced errors. In the early 1960s, spherical monodisperse particle size "standards" were characterized by electron microscopy and found to have a median diameter around half that expected and to be extremely broad in distribution. Puzzled researchers soon realized that this was a consequence of dispersing the particles in epoxy resin and microtoming to produce thin sections. A nice distribution of spheres had been produced by slicing through all diameters of the sphere from "0" units in diameter to the expected size with a preponderance of slices tending towards the $d/2$ value.

As an example, let us take (at least for mathematicians!) a set of cylinders on different heights but the same radii and calculate how these would appear in terms of equivalent volume and equivalent surface (Table 6.2).

Note that d_v falls much more rapidly than d_{sa} for this system because $d_v \propto V^{1/3}$ but $d_{sa} \propto S.A.^{1/2}$. This again shows how the sphere packs in more materials than any other shape. As volume considers the amount of material any particle possesses, it is desirable that any technique should at least generate this as first reported diameter and then report the particular diameter that the technique measures (Scarlett 1985).

Table 6.2 Equivalent volume and surface area diameters of various cylinders of constant diameter

Size of cylinder						
Height	Diameter	Aspect ratio	Volume	Surface area	d_v	d_{sa}
400	20	20:1	125,663.6	25,796.38	62.14	90.62
200	20	10:1	62,831.8	13,212.78	49.32	64.85
100	20	5:1	31,415.9	6,920.98	39.15	46.94
40	20	2:1	12,566.36	3,145.9	28.84	31.64
20	20	1:1	6,283.18	1,887.54	22.89	24.51
10	20	1:2	3,141.59	1,258.36	18.17	20.01
4	20	1:5	1,256.636	880.852	13.39	16.74
2	20	1:10	628.318	755.016	10.63	15.50
1	20	1:20	314.159	692.098	8.43	14.84
0.4	20	1:50	125.6636	654.3472	6.21	14.43

d_v = Equivalent spherical diameter by volume (= mass when constant density)
d_{sa} – Equivalent spherical diameter by surface area

6.4.2 Fundamental Measures of Central Tendency

For a nonmathematician, it may have been easier to title this section "Averages" but we will see shortly why this is likely to be ambiguous.

Consider a simple system consisting of three spherical particles of diameters one, two, and three units respectively. It is likely that the statistician would want the average of the system, which they may call the mean size. What would this average be? The most likely answer to be shouted out in a class new to particle size is two. This is a value that anyone without time-share on a neuron can fathom out the origin:

$$(1 + 2 + 3)/3 = 2.00$$

This is the value that would probably be isolated by simple microscopy on the system – add up all the diameters of the measured particles and divide by the number of particles thus obtaining the average value 2.00. As the mean contains the number of particles in the system it is referred to as a number mean. As the numerator in the equation contains diameter terms expressed as simple lengths, then this is called a number-length mean diameter, d_{nl}. This is a bit of a mouthful so one mathematical shorthand for reducing this mouthful is to note that there are no diameters terms on the denominator of the equation, but diameter terms (lengths) to the power 1 in the numerator – a D[1, 0] is thus generated.

However, consider the following scenarios:

• The dissolution rate of a material is governed amongst the other parameters by the surface area of the material – the more surface area then the more rapid the dissolution rate.
• The surface area exposed to the reactants governs the catalytic activity of a substance. More finely divided metals, for example Raney nickel, speed up the hydrogenation of fats. Having finer constituents – gunpowder being a prime example, in general, speeds up chemical reactions.

- The hydration rate of cement (yet another chemical reaction) is related to the surface area (usually expressed in Blaine units) of the cement.
- The burning rate of fuels is governed by the rate at which oxygen can get to the surface of the fuel. Too large a particle and coke formation results from too slow a burn whilst finely divided fuel is explosive (the internal combustion engine) as can be all finely divided oxidizable materials.

Thus we may wish to consider calculating an average based on the surface areas of the three particles. We have a slightly more complicated calculation. The surface areas are calculated by taking the squares of the diameters (because S.A. of a sphere is πd^2). We then calculate an average surface area by dividing by the number of particles. Taking the square root of this average recovers an average diameter by surface area.

$$[(1^2 + 2^2 + 3^2)/3]^{0.5} = 2.16$$

For those who wonder what has happened to the π's in the equation, then a little thought will show that one first multiplies by the π's to get surface areas and then divides by them again when the diameter is recovered.

The above average diameter is again a number mean as the number of particles appears in the equation. It is though a number-surface area diameter, d_{ns}. And in the simplest mathematical shorthand, D[2, 0]. The could be the first mean isolated by an image analysis technique by scanning a beam of light (laser or white) across a photograph and processing the particle to recover the equivalent spherical surface area.

The unwary may not consider the next step! A chemical engineer will consider the pumping fluids around a plant or weighing materials that are to be sold or value added to them in the process. This individual may therefore consider averaging the particles on the basis of volume or weight/mass. Volumes or masses are obtained by cubing the diameters, averaging for an average volume or mass, and taking a cube root to recover an average diameter:

$$[(1^3 + 2^3 + 3^3)/3]^{1/3} = 2.29$$

On the basis of the preceding comments, we can see that this is a number-weight or number-volume mean, d_{nv}, more conveniently expressed as D[3, 0]. An electrozone-sensing instrument may measure the volumes of particles and divide by the number of particles to obtain such a mean.

It should be noted at this stage that there is an international (ISO) norm (ISO9276-2: 2001), which recommends alternative ways of showing the above means. For example, the D[1, 0] shown above can be represented as:

$$x_{1,0} = D_{1,0} = M_{1,0} = M_{-2,3}/M_{-3,3}$$

Although more mathematically correct for the pedants, it is less easy to understand from the simplistic forms first proposed by the Muegle and Evans (1953) – based on earlier work by Hatch and Choate (1929) and well-described in Heywood (1947) – and which we have derived above.

Other common statistical ways of looking at the central tendency of any distribution are:

- Median – this is the point at which exactly half (50%) of the distribution lies above and exactly half (50%) below. Normally we would specify which type of distribution we are dealing with by the appropriate suffix e.g. D[v, 0.5] for a volume median.
- Mode. This is the most common or frequent point in the distribution – "the summit of the mountain"

6.4.3   The Moment Means

We now need to take a journey in which the reader must have some faith or wishes to delve deeper into a textbook such as Allen's (1997) or ISO9276-2 (2001). The simple number means shown above possess one major difficulty and that is that they contain the number of particles. A simple calculation will show that 1 g of SiO_2 ($\rho=2.5$) will contain around 750×10^6 particles at $10\,\mu m$ size and around 750×10^9 particles when $1\,\mu m$ in size. It is generally a difficult technical task to count such huge numbers of particles. Particle counters generally deal with very small numbers of particles and this will be discussed further in a later section. At this stage only the example of Federal Standard 209D (we will ignore the metric version – 209E – at this stage) will be quoted. This requires for a Class 100 clean room that no more than 100 particles of $0.5\,\mu m$ and above are present in a cubic foot of air. This equates to ppb (usually accepted to be 10^{-9} nowadays). So particle counters are equipped to deal with very low levels of particles whereas particle size analysers are equipped to measure powders, suspensions, emulsion etc in raw form (albeit at high dilutions too).

The problem of having to deal with huge unfeasible quantities of particles is statistically dealt with in particle size analysis by what are known as the moment means, the two most commonly encountered being:

- The Surface Area Moment Mean or Sauter Mean Diameter (SMD) calculated as D[3,2] for the Muegle and Evans method (Muegle and Evans 1953) of displaying particle moments. For our 3-sphere system, the calculation is thus:

$$D[3, 2] = (1^3 + 2^3 + 3^3)/(1^2 + 2^2 + 3^2) = 2.57$$

- The Volume or Mass Moment Mean less commonly known as the De Brouckere Mean calculated as follows for our 3-sphere system:

$$D[4, 3] = (1^4 + 2^4 + 3^4)/(1^3 + 2^3 + 3^3) = 2.72$$

The big advantage of the above means is that they do not involve the need to know the total number of particles in the system in order that particle statistics can be calculated. Laser diffraction which in the Mie deconvolution of the scattering Intensity – Angle measurement considers total volumes of particles on an ensemble basis will thus usually extract the D[4,3] at an early stage. The laser diffraction instrument does not therefore need to know or calculate the total numbers of particles present in size classes.

The disadvantage for the student to particle size analysis is that these means seem less easily understood than the numbers means present in the preceding section. They arise from considering the frequency plot of surface area or volume diameters and the centre of gravity or moment of inertia of such a distribution. The average weighted diameters represent the abscissa of the centre of gravity of the relevant frequency distribution (See (Heywood 1947)). This is the point at which the entire distribution will rotate freely. As a moment of inertia involves an extra term, (Force = Mass × Distance) then in the particle size analysis an extra diameter term gets introduced. Thus surface area (αd^2) "moves up" to a d^3 and a mass or volume (αd^3) moves up to the 4th power of the diameter. At this stage it is better just to accept these mean diameters; the pedant can consult the literature for more rigorous treatments.

So where are we now? For our seemingly simple system containing only three spheres, we have these mean or average diameters:

$$D[1, 0] = (1 + 2 + 3)/3 = 2.00 \quad (= d_{nl})$$

$$D[2, 0] = [(1^2 + 2^2 + 3^2)/3]^{0.5} = 2.19 \quad (= d_{ns})$$

$$D[3, 0] = [(1^3 + 2^3 + 3^3)/3]^{1/3} = 2.26 \ (= d_{nv})$$

$$D[3, 2] = (1^3 + 2^3 + 3^3)/(1^2 + 2^2 + 3^2) = 2.57 \ (= d_{sv})$$

$$D[4, 3] = (1^4 + 2^4 + 3^4)/(1^3 + 2^3 + 3^3) = 2.72 \ (= d_{mv})$$

Note how as we progress up the scale from number through surface and then to volume then the values of the means increases. Although this may seem confusing at this stage, this increase is a clue as to when such means should be employed.

It is possible (mathematically, at least) to convert between the various means listed above but it is usually desirable not to do so. The conversions are known as the Hatch–Choate transformations (Hatch and Choate 1929) and rely on the assumption of the data being log-normal and mapping parallel lines to this data. See (Heywood 1947; Allen 1997). Errors in converting number to volume will be increased markedly as volume is proportional to the cube of the diameter. It is always best to rely on the fundamental output of any instrument type rather than try to bend it to simulate another technique or type of result.

6.4.4 Which Mean is the Most Useful?

The real question is which mean should we use? The 3-sphere system was obviously too complicated for the average manager, so let us consider a more simple system consisting of only two spheres of diameters 1 and 10 units respectively. We will change the rules slightly and assume that these are particles of gold. The simplest number means considers each particle as equivalent (equal weighting but maybe this is a confusing term in this context and maybe equal validity is better) and equally important from a statistical point-of-view:

$$D[1, 0] = (1 + 10)/2 = 5.50$$

Now if we have a choice of which particle we would rather have then most would opt for the larger diameter particle as this contains more mass of gold and is thus worth financially more. It can be easily seen that this larger particle has a mass of 10^3 or 1,000 units and constitutes 1,000/1,001 parts of the complete system. The volume moment mean expresses this well:

$$D[4, 3] = (1^4 + 10^4)/(1^3 + 10^3) = 9.991$$

It tells us where the mass of the system is located and is thus treating the larger particle as "more important" (we can see why higher weighting now sounds confusing) from a statistical perspective.

Perhaps another example illustrates this better. The most common difficulty encountered in particle size for the novice relates to the difference between number means (usually obtained by a microscope or imaging technique) and a weight or volume based technique (e.g. sieves or light scattering). Because the eye cannot judge volumes or masses in the same way as numbers, we tend to feel that "seeing is believing" and this clouds our statistical judgement on the many occasions when mass or volumes are required as opposed to numbers of particles. The example below is that of artificial objects in space classified by scientists on the basis of size from the largest (satellites and rocket motors) to the smallest (flakes of paint, dust etc). See Table 6.3 below:

If this represented a gold mine, then we would want to collect the 7,000 easy nuggets rather than the dust in the system. To design a spacesuit though would require protection against the many tiny objects, the larger satellites being easy to avoid (one would think).

The person with a headache does not request 4.3 billion particles of paracetamol from the pharmacist. Likewise, the supplier of titanium dioxide to the paint industry supplies tonnes rather than numbers of particles. Mass balances combined with particle size data on mine classifier circuits in order to generate Tromp plots (classifier efficiency curves) are exactly that – masses not numbers.

Number and volume are also applicable to literature according to a text that the author read in an Indian hotel:

"Words should be weighed and not counted"

Table 6.3 Volume-Number relationship for man-made objects in space (Adapted and recalculated from article in New Scientist (13th October 1991). See (Rawle 1993)

Size (cm)	Number of objects	% by number	% by mass	% by $/£/€
10–1,000	7,000	0.2	99.96	99.96
1–10	17,500	0.5	0.03	0.03
0.1–1	35,00,000	99.3	0.01	0.01
	Number Mean ~0.6cm			
	Mass Mean ~500cm			

So when do we need to know or count the numbers of particles?

6.5   Particle Counting

Particle counting is the basis of a number of techniques, which will merit further discussion in the context of the pharmaceutical industry. These include:

* Microscopy (manual or electron)
* Image analysis
* Light obscuration or shadowing techniques for air or liquid applications
* Electrozone sensing

The use of counting is when we wish to know about any of the 3 C's in a system:

* Concentration
* Contamination
* Cleanliness

We have already seen the small number of permitted particles that need to be counted in air for a Class 100 clean room. Indeed for air cleanliness in the semiconductor or pharmaceutical industries according to Fed Stan 209D, then we only need to consider particles at one size only (0.5 μm) although intuitively we would realize that the number of particles will increase exponentially as the size decreases. The number of size classes in the first three techniques above is usually very limited. The electrozone sensing device will be treated later in this paper and is a "special case" of counter. The other pertinent pharmaceutical example is that of parenteral (or intravenous) fluids as opposed to intralipid emulsions. These parenteral fluids (small or large volume) have BP and USP specifications related to them.[1] For example, particles measuring 50 microns or larger can be detected by visual inspection. Specialised equipment is needed to detect particles less than 50 microns in size. The USP 24/NF19 Section <788> sets limits on the number and size of particulates that are permissible in parenteral formulations. This standard seems to be undergoing constant revision, so up-to-date information should be sought (See footnote 1). This information was adapted from a recent application note (July 2007) posted on the website of Particle Measuring Systems, Boulder, CO and probably represents the latest position (Table 6.4).

There appears to be a consistency across the pharmacopoeias except for that of dry powders within EP and one assumes that sample preparation is key here.

Microscopy has been removed from the pharmacopoeias as a quantitative form of particle sizing. For example, Appendix XIII of BP2000 "Microscope Method" based on Ph. Eur. Method 2.9.21 states "This test is intended to provide a qualitative method for identifying any particles that may be present in a solution." Normally a light

---

[1] For example: http://pharmlabs.unc.edu/parenterals/equipment/ch16.htm and http://www.pmeasuring.com

Table 6.4 Particulate specifications for various pharmacopeias

	Type of parenteral		
	Small Volume	Large Volume	Dry powders
USP 10 μm	<6,000	<25/mL	None
USP 25 μm	<600	<3/mL	
EP 10 μm	<6,000	<25/mL	<10,000
EP 25 μm	<600	<3/mL	<1,000
JP 10 μm	<6,000	<25/mL	None
JP 25 μm	<600	<3/mL	

obscuration counter is used for particle counting purposes. Using either a laser or white light source the typical range of such a counter used in the pharmaceutical industry is quite limited (e.g. 1–100 μm for liquid, 0.3–10 μm or so for air). It need not do more than this as both BP now and USP call for measurement at 10 and 25 μm for parenterals and 0.5 μm only for air counting (Fed Stan 209D/E). Note the limited range and limited number of size classes that a counter operates with. Also the range of a counter is limited – the larger the size of particle to be counted, the lower the absolute count or concentration can be. Typically a 1–100 μm liquid counter cannot deal with more than around 50,000 particles/cm^3 of fluid – well below the 1 g of SiO_2 particles that would be no trouble to an ensemble technique such as diffraction.

A particle counter operating on the obscuration or shadowing principle is a secondary form of measurement. A reference standard is delivered through the measurement zone and the response adjusted until the signal sits in the required or "correct" channel. Note that it is quite difficult for the standard latex material larger than a micron or so to be nebulized so that it can pass through the measurement zone of an air counter. Furthermore, although the x-axis (size) can be calibrated quite easily, it is another matter to confirm that the absolute count is correct.

So a particle counter will answer the question "How many?"

6.6 Particle Size Analysis

A particle size analyzer in contrast to a particle counter is interested in the answer to the question "How much?"(Bendick 1947).[2] This can be, for example:

- How much (volume or weight) material >x μm?
- How much (volume or weight) material <y μm?
- How much material between two size bands?
- Mill efficiency calculations (Tromp curves)
- What is the average size? How has this changed on a quality control chart?
- How much is the value of my material?

---

[2]Some languages other than English can seem to deal with this distinction in an apparently superior manner. For example we have in Finnish "Kuinka monta?" or "Montako?"(How many?) in contrast to "Paljonko?" (How much?)

These are the situations where we have a powder, emulsion, suspension or spray and we are concerned with obtaining a particle size distribution. There is a subtle difference between "How many?" and "How much?" This is sometimes not always appreciated by the average purchaser of an instrument. Relating the question back to USP General Test <776> (2000), then the question "How many?" is either answered by some form of imaging or a light obscuration counter (counting single particles) and the question "How much?" by a laser particle size analyser (ensemble method) in a modern laboratory.

6.6.1 Methods of Particle Size Analysis – Techniques Overview

We will discuss briefly the most common means of particle size analysis found in the pharmaceutical laboratory – sieves, electrozone sensing, and laser diffraction. BET measurement of powder surface area is extremely valuable, as it will probably correlate to the amount of available surface for dissolution. A mean particle size is only inferred with surface area measurements and no distribution obtained we will only point the reader to the standard text in this field (Gregg and Sing 1967) for further details. This is not to belittle surface area measurements – indeed they could be considered to be providing more information in a practical sense (dissolution) than those methods measuring particle size, so to say. But particle sizing is the theme of this script and therefore we must stick to it. Other chapters within this volume will be devoted to the other techniques related to surface area.

6.6.2 Sieves/Sieving/Sifting/Screening

There are a lot of excellent texts and standards relating to this technique e.g. Allen (1997), Cadle (1965) and Wills (1988), but the author has gratefully made use of material used by Dominic Rhodes of BNFL at a number of UK Royal Society of Chemistry (RSC) meetings. In the case of suspensions, we should normally be considering wet-sieving which, in theory, is capable of measurement down to 2 or 5 μm but blockage would be a regular feature of any screening measurement if this was considered an achievable situation!

Reference texts (Heywood 1970) suggest that this technique has been around since ancient Egyptian times, although many depictions stated as sieving appear to be separating wheat from chaff ("winnowing"). However, amongst the hieroglyphics, we can see foodstuffs being sifted with sieves made from reeds and holes punched in metal plates. So we can accept that the technique has been with us a long time. It is normally encountered in production environments where separation or rejection/recycling of material is to be made. In the pharmaceutical area, this is likely to be an Alpine Jet Sieve as well as the standard type of screens found in the laboratory. Specifications are often based around a single point (e.g. <0.5% material >38 μm or >99.5% material <38 μm), which, as Kaye points out, is prone to be exceedingly dangerous (Kaye 1998). He refers to two separate micrographs of ibuprofen made to the "same specification":

> "In Fig. 6.3 we show two different ibuprofen powders which both met the specification that a small amount of residue on a sieve constitutes an adequate definition of the fineness of the product. The specification made no mention of the shape of the powder grains and the two powders differ enormously in their grain shape subsequently creating large differences in the way that they packed in a storage device and responded to pressure".

The basis of the sieve separation of particles is around both size and shape although a weight distribution is usually postulated as the sieved fractions are weighed. Of all techniques, this one as well as producing a crude particle size distribution has the advantage that a material can be separated into distinct size classes, which can be advantageous on occasions. For large material (>2 mm/2,000 μm) there are no real competitive techniques, although for a single material of known density then counting and weighing can produce a mean size and some form of distribution if every particle is laboriously recorded. Imaging techniques can also be used in this large size area but again these will not be routinely encountered in the preformulation stage of a pharmaceutical.

Interestingly, there are similarities between sieving and image analysis techniques. Shape is combined with orientation. Both these techniques deal with geometrical area or cross-sectional perimeter. With a microscope or image analysis, the object or particle will always present a preferred two-dimensional axis to the direction of view. With sieving a preferred orientation of particle will be passed by the sieve – this can be a time-dependent process too.

The shape dependence of a screen has been known for many years. To quote a text that has already celebrated its 100th anniversary (Bleininger 1904):

> "In considering the results of a sieve analysis, it must be remembered that the particles passing through the sieve are not necessarily round or symmetrical in shape, but may often be needle-like in form, and though they pass through the sieves, they are apt to be larger in bulk than particles retained, which approach the more ideal spherical shape"

The sieve equivalent diameter is defined as the diameter of a sphere, which will just pass through the mesh of a particular size of screen. Hence, a particle must have two dimensions less than the sieve gap diameter in order to pass through. Consequently, sieving is often said to size particles according to their second largest dimension, although if the smallest dimension allows penetration of the mesh without falling, it seems that the second smallest dimension would allow the particle passage. Care should be taken with the equivalent sphere concept in sieving as particle length may prevent passage due to stability preventing an end-on orientation. Sieves will not cope well with elongated particles and it is difficult to see what it actually being measured in such a case.

The types of sieve routinely encountered are:

• Punched hole or perforated screens normally coping with the larger material and rarely seen in a pharmaceutical laboratory but more common in mining. These can have round or square-shaped holes.

- Wire cloth sieves, which are woven to produce nominally square apertures with a defined tolerance (see later!). These are generally brass or stainless steel although plastic is also available. One should avoid carrying out a heavy metal analysis by AAS or ICP on a screened fraction after screening through metallic sieves!
- Micro-sieves generally from 150 μm downward (even 1 μm sizes are specified!) but used extensively for wet sieving below 38 μm or so. These are generally made by electro-etching nickel alloys in a process identical to that of forming semi-conductor components (photoresist, (UV) expose, develop, etch, strip). In theory any possible shape could be produced by a different photomask but again, in reality, square and round holes are the norm. In practical terms for organic materials and water-based systems, these can be a real pain to use as blockage is common and tightly bound material (agglomerated) resists all attempts to "disperse" it through the screen. At small sizes ultrasonic sieving will considerably speed up the process.

It is common to specify sieves in terms of mesh number (the number of wires per unit length e.g. inch) but this is ambiguous as a variety of National test standards are in use world-wide:

- ASTM E11-04
- American Tyler series
- German DIN 4188
- French Standard AFNOR (on a $^{10}\sqrt{10}$ separation basis as opposed to the more common $^{4}\sqrt{2}$ progression between sizes. The advantage of this is that it gives a 2:1 volume change between adjacent screens rather than a 2:1 change in specific surface area as the BS sieves may be considered having).
- British Standard BS410 (now replaced by the ISO standard 3310-1:2000, as one assumes the French and German standards will also be superseded). This refers to the metal wire cloth sieves. The −2 and −3 of this standard refer to perforated metal plate and electroformed sheets respectively).

The technique is generally fairly low resolution as a limited number of size classes will usually be determined. It is not the case that large numbers of sieves are stacked on top of one another to form a "nest". Very often, only four or five sieves are used which can make comparison ("result emulation") with other techniques that are problematic at best.

In terms of accuracy, sieves are calibrated less frequently than they ought to be and at this stage it is useful to examine permitted tolerances for screens of the smallest dimension, which are those to be found routinely in the pharmaceutical laboratory. Taking the set of Tyler mesh sieves from Cadle's text (Allen 1997), we see that the 37 μm screen (400# in this designation but this would be 38 μm in others), is permitted to have a maximum opening of 90% of the nominal opening if no more than 5% of the holes exceed this value! Thus 70 μm holes would be permitted $(37 + 33 = 70 \,\mu m)$ in such a screen. If we sieved long enough on such a screen, 50 μm beads would eventually pass through and could be sized as <37 μm! It should not be necessary to indicate that the same listing states that ±7% is permitted on the average size of hole against the nominal. In practice, though, sieves are made to considerably tighter tolerances than this although the example does show the importance of regularly calibrating the screens in use in the laboratory aside from the possibility of operator-induced damage, which also increases the size of the aperture. Even in the modern ISO standard, the mean aperture tolerance of a 250 μm sieve is 240–260 μm. The maximum acceptable aperture is 308 μm!

In terms of methodology, dry materials are rarely sieved below 63 μm or so and this is the region where wet sieving finds its main application. Traditionally the region below 38 μm was (and is) referred to as the "sub-sieve size" region.

Although hand sieving is considered to be the reference point, it is clearly susceptible to difficulties relating to standardization. Thus some mechanical form of assistance (e.g. The Ro-Tap) is preferable and the duration and frequency can be regulated.

The Chairman of the Institution of Mining and Metallurgy speaking in 1903 and quoted by Leschonski in 1977 gave this riposte (Leschonski 1977):

"Screening is not a scientific means of measurement"

The author disagrees with this statement, but it takes a little care (like all techniques) to understand the basis of this deceptively simple form of measurement and to put it on a scientific and logical platform.

Perhaps the words of Edward Lear's poem "The Jumblies" are appropriate to end on a note of humor but not intended to be scathing of sieving as a particle sizing technique:

"They went to sea in a sieve....."

6.6.3   Electrozone Sensing

Users of this technique often know this as the Coulter Counter even though other devices from other manufacturers exist in the marketplace. Many researchers over the age of 50 or so will have been brought up on this technique, as it was one of the few reputable electronic techniques available in the 1960's and 1970's for particle sizing. First developed by Wallace Coulter in the 1950's for blood cell counting (after optical methods were found to be problematical!), the technique quickly showed its versatility for large numbers of applications where witchcraft had been the norm before. Even in areas where its use was complicated (e.g. lack of conductivity), researchers found interesting routes to bypass these (e.g. measurement in ethanol + ammonium thiocyanate, NH_4SCN).

The principle of measurement is simple. Particles are forced to flow through an aperture or orifice situated between two electrodes (Fig. 6.5).

Nonconductive particles are suspended in the electrically conductive diluent. When a particle passes through the aperture during the counting process, the particle decreases the flow of current, increasing the resistance between the electrodes. The resistance causes a voltage change between the electrodes proportional to the volume of the particle.

Electrical Sensing Zone Method

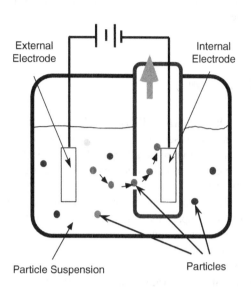

External Electrode

Internal Electrode

Particle Suspension

Particles

Particles, which are non-conductors, are suspended in the electrically conductive diluent. When a particle passes through the aperture during the counting process, the particle decreases the flow of current, increasing the resistance between the electrodes. The resistance causes a voltage change between the electrodes proportional to the volume of the particle.

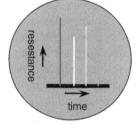

Fig. 6.5 Principle of the electrical zone sensing method

When the particle traverses the measurement zone there is a change in voltage, current, resistance, and capacitance all of which could be measured and used to count and size the particle. The method is a secondary one in that the instrument is calibrated against known NIST-traceable latex standards and all measurements of our material interpolated against these standards. Given its birth just before the time of the Information Age, then there is a wealth of literature to be found on practical examples of the technique.

Orifices are available in a variety of sizes and it is recommended that the particles are not less than 10% of the aperture size and not more than 30% of the same size for efficient processing. Like most rules this can be (and has been) stretched. This restricts the dynamic range of the system to a maximum of around 100:1 although it is possible to combine analyses from different aperture sizes.

The main difficulties in using the technique for pharmaceutical usage are:

- The solubility's of many pharmaceuticals in aqueous and alcoholic media
- The broad size range of pharmaceutical preparations leading to blockage of the orifices and slow speed of measurement
- In emulsion work the 0.9% NaCl electrolyte can promote agglomeration
- The inadequate lower size range for a number of formulations

This is not to be negative against an extremely high-resolution technique capable of dividing a monodisperse latex material into 256 or more size channels. ISO13319 can be consulted for further detail on the method.

    In practice, though another electronic technique based on the laws of light
scattering has virtually superseded the usage of electrozone sensing for powder,
emulsion, and suspension measurement.

6.6.4   Light Scattering ("Diffraction")

Here we will deal with both laser light scattering (ISO13320-1 specification is
0.1–3,000 μm) and photon correlation spectroscopy (PCS), another light based
technique exclusively for the sub-100nm region, although capable of dealing with
most submicron suspensions. Again we have a technique that appears extremely
simple in concept. We quote Rose (1954):
    "I wish now to mention a method of mean size determination which is charming
in its simplicity and in the low cost of the apparatus which is but a few shillings,
but I am sorry to have to damp enthusiasm at the outset by stating that the method
is applicable to but a limited range of materials."
    He then goes on to describe classic diffraction!
    The principle of measurement is that large particles scatter light into narrow
angles (specifically the forward scattering direction) and small particles scatter
light into wider angles. A simple rough (but helpful tool for linking size and
diffraction angle is (Davies 1966):

$$\theta \sim 35/d$$

where θ is the scattering in degrees and d the size of the particle in microns (μm).
    This formula gives the total extent of scattering but an important note is that one
angle does not correspond to one size. This is easily seen by measurement of the
scattering that occurs with different sizes of sphere (Fig. 6.6):
    Note the broad scattering occurring with the smallest particle shown and also the
secondary scattering arising from passage of light through rather than around the
particle. The crude formula shows roughly that (Fig. 6.6, Table 6.5):
    Thus small particles need very wide angular detection and conversely large
particles will require detectors to be crammed into very small space. Bond pad limitations and cross-talk prevent detectors being spaced closer than around 75 μm for
conventional silicon technology. Charge Coupled Devices (CCD's) operate in a
different fashion and permit closer spacing of detectors as these are in an X–Y matrix
form (see Image Analysis). The scope of laser diffraction is given in the ISO13320-1
(1999) document as 0.1–3000 μm broadly fitting in with the scattering of particles.
At the smaller size the scattering becomes weak (I $\alpha\, d^6$ or V^2 in the Rayleigh region
; $d < \lambda/10$ typically). At the larger sizes the detector has to be mounted so far away
from the particles that temperature and vibration compromise the measurement.
    The advent of three developments after the writing of Rose's book mentioned
above allowed this "simple" technique to become a practical reality – the laser, the
silicon slice allowing precise photodetectors to be built and the computer. This also

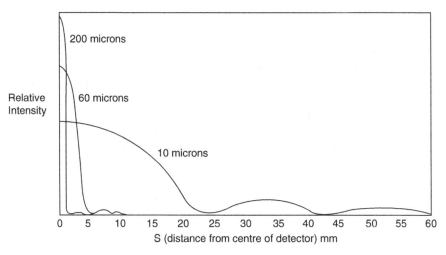

Fig. 6.6 Scattering angle and form in relation to different sized spheres

Table 6.5 Approximate range of scattering angles for various sizes of particle

Size of particle (μm)	Approximate total scattering angle (deg)
3,000	0.012
1,000	0.035
10	3.5
1	35
0.1	350

increased the cost beyond the few shillings that he and some present-day customers envisaged!

Although developed in the 1970's, this method has pushed forward the frontiers of particle size measurement for a number of reasons:

- Flexibility – ability to measure dry, suspensions, emulsions, and even sprays. The latter was the first application of the technique to military jet and diesel engines to optimize the performance
- Wide range – 0.02 to 2,000 μm or better in a single unit. We must be a little suspicious at anything outside the quoted scope stated within ISO13320-1, but limits are there to test and stretch
- Rapid – single measurement in 0.001 s or quicker. This allows sprays to be "sliced" and huge amounts of interesting information generated
- Excellent repeatability – integration of many single measurements. A 5 s measurement will have 5,000 individual measurements registered and integrated
- No calibration is necessary but easy for verification. This is sometimes a problem for QA inspectors and users alike but relates to fundamental, first principle mea-

surements as opposed to secondary or comparative measurements. There is no formal way to "calibrate" any laser diffraction instrument. That is, it does not operate by comparing a signal to that of a known material and adjusting the performance to read the designated value. Rather, if the wavelength of the light is known together with the optical properties of the material and the angular scattering is measured, then the particle size is derived from first principles as the only unknown remaining in the equations linking the light scattering with size. However, challenging the equipment with a verifiable material is a regular occurrence to prove performance (within the tolerance of the standard and the manufacturing tolerances) and this is termed verification. As seen earlier, particle size standards (verify the x-axis, "*d*") should be both spherical (no debate as to the "size") and polydisperse (verify the y-axis, quantity) to correctly challenge the performance

The disadvantages must also be seen as well:

- Very new in relative terms (mid-1970's), so limited number of formal methods (but now ASTM-1458-92 and ISO13320-1)
- Not applicable to material over 3 mm or so where sieves take over
- Is an optical technique so the optical properties of materials are required (ISO13320-1 indicates that when $D < 40\lambda$ (~25 μm for He–Ne laser at 632.8 nm), then the optical properties are essential for accuracy

The basic layout of a state-of-the-art instrument (Malvern Instruments Mastersizer 2000) is used as an illustration of the basic construction of a unit (Fig. 6.7).

Usually, operation is remarkably and deceptively simple and many users lull themselves into a false sense of security because of this.

Method development is simple for any unknown material. This method development must be carried out before a Standard Operating Procedure (SOP) is formalized, or the latter is just a means to getting the wrong result more precisely. Method development follows the guidelines of ISO13320-1, in particular Section 6.2.3:

- Ensure that the equipment has formal traceablility by verification with NIST-traceable standards (ISO 13320-1 Section 6.4) appropriate to the specific material.
- Ensure that a representative sample of the bulk material is taken – this almost will necessitate the use of a spinning riffler if any material >75 μm is present.

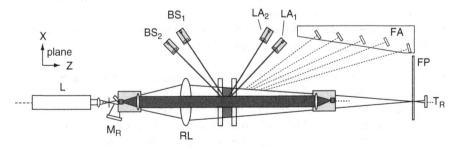

Fig. 6.7 Schematic outline of a laser diffraction instrument (Malvern Mastersizer 2000)

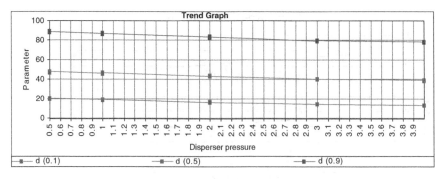

Fig. 6.8 Dry measurement. Pressure-size titration as per ISO13320-1 (Rawle 2000)

- Dry: pressure – size titration followed by wet to decide on an appropriate pressure (ISO 13320-1 Section 6.2.3.2) or to reject dry if significant attrition is occurs (like with many organics).
- Wet: before, during, and after sonication in order to obtain a plateau of stability. For organics the formation of a stable dispersion will follow the steps wetting, separation, stabilization. This may need some work with solvents and perhaps surfactants. These are only vehicles in which to circulate the powder (or slurry/suspension etc).
- SOP formulated around the stable dispersed region for primary size or against input of minimum energy for bulk size (possibly only one measurement may be possible in the latter if dispersion is occurring in the wet.

The key features of any method must be:

- Repeatability
- Reproducibility
- Robustness

These three R's are explored further in (Rawle 2000) to which the reader is referred. As an illustration, the following particle size plots and comparisons involved in method development for an organic compound should be easy to understand without further comment (Figs. 6.8–6.11).

Interpretation of the raw angular scattering pattern (Intensity – Angle plot) can either be undertaken with approximations (Fraunhofer, Anomalous) or more rigorous and exact theories (Mie).

6.6.5 Fraunhofer Approximation and Mie Theory

The interested reader is advised to read the Appendix in ISO13320-1.

We note that for many years, computing power was not adequate in order to deal with the complex calculations inherent in Mie theory. Thus, the simpler Fraunhofer

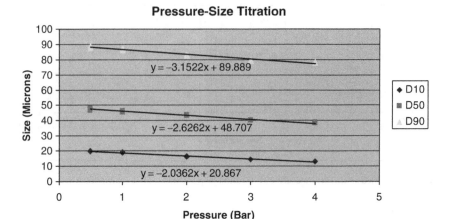

Fig. 6.9 Dry. This is an initial check on method. Note the magnitude of the slopes is in the order $D_{90} > D_{50} > D_{10}$. The magnitude of the slopes (−2 to −3 μm/bar increased pressure here indicate little or no attrition but this needs confirmation by comparison against wet

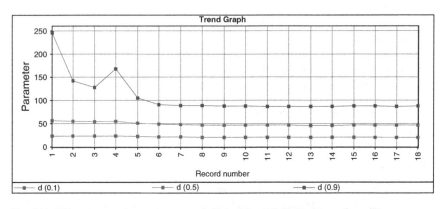

Fig. 6.10 What pressure do we measure at in Figs. 6.8 and 6.9? The above figure illustrates a wet measurement of the same material with sonication (dispersion) to stability

approximation was used for larger material where $d \gg \lambda$. Both interpretations (mathematical "engines") rely on given shapes (disc or slit – Fraunhofer; sphere – Mie) and sizes from which the angular scattering intensities can be calculated. The reverse step of angular light intensities to particle size distribution is the challenge for the light scattering computer. Mie accounts for the optical properties of the particle including polarization (and hence the need for the refractive index) while Fraunhofer assumes a completely absorbing particle. The division in ISO13320-1 between these two interpretations is stated to be 40λ (with Fraunhofer stated to be similar to Mie above this level), which is around 25 μm for a He–Ne

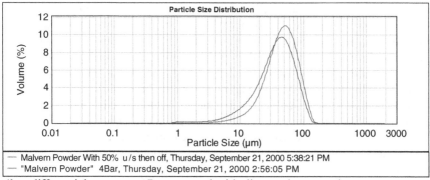

4bar differential pressure, ΔP, compared with dispersed wet result

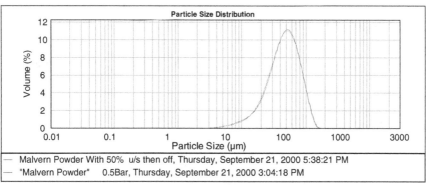

0.5bar ΔP compared with dispersed wet result:

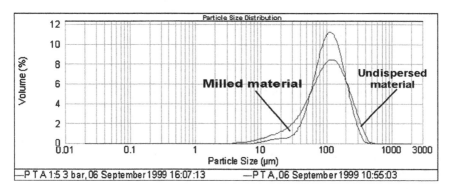

Fig. 6.11 Comparison of wet with dry measurements at 4 and 0.5 bar. Illustration of attrition for another organic material where it is noted that the dry is broadened in comparison to the wet

laser at 632.8 nm, although the situation is slightly more complex in practice, with the relative refractive index playing an important role. We note that the Fraunhofer interpretation of a transparent glass bead standard NBS1003c leads to a false deduction of submicron size fines in the standard. This is due to the passage of light through the particle, which will interfere with that classically scattered from the contour of the particle. Thus a misunderstanding or misinterpretation of the optical properties by assuming them to be infinite (Fraunhofer) or other incorrect values means that the y- (%) and well as the x-axis (size) can be compromised. An inaccurate assessment of the fines in a process can have severe economic repercussions.

At the minimum, during method development, a robustness test based on small, sensible, and systematic (the three S's) changes to the refractive index should be undertaken so that the effect of changes to the optical constants can be assessed. In this way, it can then be judged whether basic measurements of this parameter need to be made – for example by extrapolation of solutions of the material to 100% solids (see Saveyn (2002)).

6.7   Particle Morphology

After size, we consider shape and again this may not be as easy as it first appears.....

6.7.1   Setting the Scene

As stated earlier, size and shape cannot be dealt with as separate issues. Hence we need to have some means of visualising our particles or particulate system in order to get handles on:

- The approximate shape of the particles
- The approximate size
- The state of dispersion
- The nature of the fundamental particles – whether made from crystallites.

This two-dimensional visualisation can be made with microscopy (manual or electron) and made more quantitative with image analysis (at least for the shape aspect of the particles).

The terms used to describe shape in pharmaceutical technology are those common to crystallography and used extensively in other fields (e.g. geology). These terms are encompassed in USP 776 and other definitions (e.g. the old BS 3406 standards):

- Equant – approximately equal length, breadth, and height e.g. a cube or sphere. Such particles may be more robust in terms of attrition than thin particles from which the corners may be fractures.
- Tabular – literally table-like. The thickness of table is less than the length and breadth. This particle will prefer to present the face of the table (length and breadth rather than height) to the visualisation technique. If the particles are subject to flow (air or liquid) then they may prefer to align in a preferred manner

- Plate – thinner than a tabular material
- Flake – yet thinner than a plate!
- Acicular – needle-like particles whose width and thickness will be similar. These can be rounded (e.g. glass fibre) or formed by growth of preferred crystal faces so that they can appear square or rod-like (tetragonal) or hexagonal if ever viewed lengthways. Now we can see why a single descriptor of particle size is impossible. What (single) size is a needle?
- Columnar – column-like. These can be considered to be less of a needle by possessing a larger width and breadth than the acicular system
- Blade or lath – long thin, blade-like particle

Already we can see the qualitative descriptions creeping in! The corollary, thus, is the possible danger (some would say certainty!) of operator subjectivity (Fig. 6.12).

We can also complicate matters by considering groups of particles and the way that they may preferably associate (aggregate or agglomerate):

- Lamellar: stacked plates
- Foliated – stacked sheets like a book. Clay minerals can appear in this form
- Spherulitic: radial clusters of needles forming an overall spherical shape. These can sometimes be induced to fracture into triangular or individual needles
- Cemented – large particles set in a matrix. Commonly encountered in soil science where the term conglomerate is also used.
- Drusy – coated with smaller particles. This can occur as a result of pseudomorphism (a crystal taking the habit of another crystal usually in such a way that a non-preferred or unusual structure is observed) or particle-particle attraction.

Even the definitions of aggregation and agglomeration are confused, with USP considering agglomerate "hard" and aggregate "soft" in comparison to ISO, ASTM and BS. See Nichols et al. (2002), Irani and Callis (1963), ASTM Standard F1877-98 (2003), ISO/DIS 9276-6.

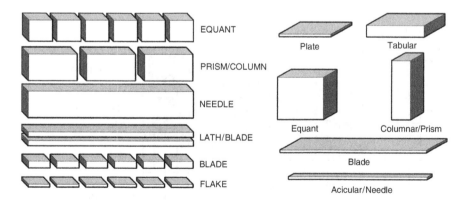

Fig. 6.12 Description of particle shapes as defined by USP (Adapted from (USP General Test <776> 2000) and (Brittain 2001))

We then have to try and make sense of this confusion by placing numbers to quantify the system. Rather like the earlier concepts, we now have a number of diameters to characterize our particles under microscopical or image analysis conditions (seen earlier in Fig. 6.2 and now defined (Brittain 2001)):

- Length – the longest dimension from edge to edge of a particle oriented parallel to the measurement graticule/scale
- Width – the longest dimension of the particle measured at right angles to the length
- Feret's diameter – the distance between imaginary parallel lines tangential to the (randomly oriented) particle and perpendicular to the scale
- Martin's diameter – length of the particle at the point that divides a particle into two equal projected areas

Most imaging techniques connected to a computer will calculate/derive these plus other descriptors:

- Perimeter diameter – diameter of a circle having the same perimeter as the projected outline of the particle
- Projected area diameter – diameter of the circle having the same area as the projected area of the particle in a stable position
- Circularity – see Fig. 6.13:

So we have another set of diameters to get our head around – these again are not necessarily related to other diameters we may have obtained by, say, ensemble techniques where all the particles may have random orientations to the measurement direction.

The Pfizer manual on shape and size gives an excellent format to describe a powder sample under a microscope so I make no apologies for reproducing it in full:

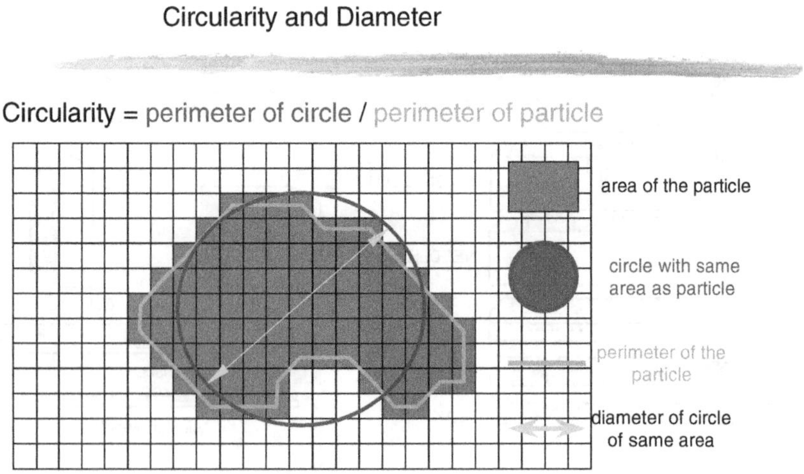

Circularity and Diameter

Circularity = perimeter of circle / perimeter of particle

area of the particle

circle with same area as particle

perimeter of the particle

diameter of circle of same area

Fig. 6.13 Descriptions of circularity (After Wadell (1932))

"The description should include:

- Description of crystal habit
- Average size of particles (note the maximum size)
- Description of association with other crystals
- Any observations about crystallinity
- Photomicrographs (with scale bar and magnification)
- Record of mounting medium used (e.g. silicone oil)
- A note of the microscope used"

The manual gives an excellent example together with a micrograph:

"The powder consists of a mixture of laths, plates, and needles. The laths and needles range in length from about 50 μm to about 1.5 mm with length to breadth aspect ratios of between 3:1 and 15:1. The plates range in size from about 100 μm to 300 μm across. There are a few agglomerates. Between crossed polarisers, the particles show bright interference colours, which extinguish every 90° of rotation and this indicates that the particles are crystalline. The powder was mounted in silicone oil and examined using a Nikon Labophot microscope". The spelling of colour and polariser must indicate a UK origin for this manual!

6.7.2 Microscopy

This is the oldest and the most fundamental measurement of particle morphology.
There are three possible parts to consider of morphology:

- Surface texture or perhaps roughness. This is the short-order structure that we may observe on the surface of the particle usually only with electron microscopy. Although it is not normally considered to be part of the shape of a particle, it will exert considerable influence on such properties as surface area (and therefore dissolution and gas absorption), settling velocity and propensity for particles to adhere to one another and to surfaces. Thus, bulk properties such as flowability may be affected. Fractal analysis is interesting in this area.
- Roundness. This is on a larger scale than surface texture but smaller than the overall shape of the particle. It reflects how angular or otherwise a particle is. Wadell's definition (Wadell 1932) based on Wentworth's earlier definition (Wentworth 1919):

Roundness, R = Average radius of curvature of corners/Radius of largest inscribed circle

This has obvious difficulties in measurement – Wentworth's definition only specified the sharpest corner. Semi-qualitative means ("very angular", "sub-rounded" etc) of assessment are provided in the form of photographs by Powers (1953) that form the basis of the diagrams in Fig. 6.1 of this chapter. It is interesting to note that these were not real particles but photographed models. Folk (Folk 1955) has dealt with assigning class numbers to these ranging from 0 ("perfectly angular") to 6 ("perfectly rounded")

- Sphericity. Again we are back to Wadell (Wadell 1932) who defined sphericity as the ratio between the diameter of the sphere with the same volume of the particle when compared to that of the diameter of the equivalent surface:

$$\psi = (d_v/d_s)^2$$

This is on a bigger scale than roundness and measures the tendency to spherical shape of a particle. Defining three linear dimensions of a particle and making a triaxial ellipsoid assumption (makes a change from spherical!) allows sphericity, φ, to be calculated as follows:

$$\varphi = \sqrt[3]{(d_s d_i/d_l^2)}$$

where d_s, d_i, and d_l are the diameters for the smallest, intermediate and longest dimensions respectively

Microscopes can be manual or electron and we need to explore the limitations of both. The first major obstacle to overcome is that of taking a sample that is representative of the bulk. We may at best be looking at mg of sample under a manual microscope and maybe pg or ng under an electron microscope. We had better be sure that we are looking at something that relates to our problem!

Going back to our example of SiO_2, we may remember that 1 g of a powder containing 1 μm particles would have around 750×10^9 such particles present. Taking 50,000 particles from such a system even on a random basis samples less than seven millionth's of 1% of the sample! And is 1 g all we have of a production lot?

Clearly, if our powder is monodisperse (i.e. every particle the same), then we would only need to take one particle from the bulk for it to be totally representative. Thus the width of the particle size distribution will determine how much sample we need for statistical validity – wider distributions requiring more sample to have adequate numbers of large particles (which dramatically affect the volume or mass distribution) for representative sampling to be assured.

We can calculate the numbers of particles needed in the highest size band for statistical validity at the 1% level as follows:

$$1/100 = 1/n^{0.5}$$

leading to a need for 10,000 particles in the highest size band. The number 10,000 is interesting although it does not provoke a large entry in the Book of Interesting Numbers (Wells 1986). In fact, it provokes no entry at all in the aforementioned volume.10,000 is the same number of images that NBS stated that was needed for statistical validity in image analysis (Dragoo et al. 1987). Clearly the meaning is "particle images":

> "S(tandard error) is proportional to $N^{-1/2}$ where N is the total number of particles measured…This consideration implies that image analysis may require the analysis of on the order of 10000 images to obtain a satisfactory limit of uncertainty" (p. 718, paragraph 1).

An earlier document from Lang (1954) states that 20,000 particles are needed and the average reader is likely to be confused by these recommendations. All this shows that usually we do not count enough particles for statistical validity and thus

Table 6.6 Calculation of weight of sample required for statistical validity at the 1% level (with assumptions as defined). From (Rawle 2002)

D(μm)	Diameter (cm)	Radius (cm)	Density (g/cm3)	Weight in top size fraction (g)	Total weight (g) (=Last column × 100)
1	0.0001	0.00005	2.5	1.31358E-08	1.31358E-06
10	0.001	0.0005	2.5	1.31358E-05	0.001313579
100	0.01	0.005	2.5	0.013135792	1.313579167
1,000	0.1	0.05	2.5	13.13579167	1,313.579167
10,000	1	0.5	2.5	13,135.79167	1,313,579.167
200[a]	0.02	0.01	3.15	0.13240878	13.24
80	0.008	0.004	3.15	0.008474162	0.85
				This is the weight of 10,000 particles	This is where 1% of the particles are in the top size band
				Assuming spheres	

[a]This represents a typical cement

the conclusions that we may draw on *particle size* are likely to be qualitative and not quantitative unless sufficient particles are analyzed. We can also note that 10,000 particles are needed for a standard error of 1% on the mean only but usually we are after a particle size distribution. And it is the last word that is the killer, statistically speaking! Obviously if every particle was the same within a system only one particle would be needed to define that system, so the width of the distribution is important.

Indeed we can crudely calculate (Rawle 2002) the amount of sample required for statistical validity where the top end of the size band is at various points (Table 6.6).

More rigorous theoretical solutions are provided by Masuda and Gotoh (1999) and enhanced by Wedd (2001), but the figures crudely calculated above in the spreadsheet are in the same ballpark. We note that the old maxim that 75 or 100 μm provided the point at which sampling became the predominant error in particle size analysis is easily understood if a sample size of around 1 g is assumed for many analytical techniques.

So we do not put enough sample in! Let us also examine how small we can see with visible light. The limit of resolution is defined as the minimum distance between two points that permits the objective to reveal the existence of both points

Abbé's Theory (for axial illumination):

$$l = \lambda/\text{N.A.}$$

where l=Limit of resolution, λ=wavelength of illumination and N.A.=Numerical Aperture

For oblique illumination:

$$l = \lambda/2\text{N.A.}$$

or, with Rayleigh's equation:

$$l = 1.2\lambda/2\text{N.A.}$$

The last two above are close enough (only 20% different). We need then a definition of Numerical Aperture:

$$N.A. = i \sin (A.A./2)$$

where i is the refractive index of the medium (air or immersion fluid) between the objective and sub-stage condenser and Angular Aperture (A.A.) is the angle between the most divergent rays which can pass through the objective to form an image. This indicates why we can get higher magnifications with an immersion lens.

Those given by the equations are theoretical and represent the best case with well-corrected lenses, proper illumination and high contrast images. N.A. is normally in the range 0.5–1.0 giving a theoretical lower limit of 0.5 μm (the N.A. is usually stated on the lens). Allen states ((Allen 1997), p114) "A more realistic size is 0.8 μm with limited accuracy below 3 μm. BS3406 does not recommend optical microscopy for particles smaller than 3 μm"

Small particles become oversized as their size approaches the limit, as the diffraction disc produced is larger than the particle (equal area: Fraunhofer Approximation, Mie predicts correctly a maximum for the diffraction considering the volume of the particle). Under dark, field illumination can detect particles much smaller than the resolution of light (because they scatter light) – but the apparent size is a disc at the resolution limit of the microscope – so all particles appear to be 0.5 μm in diameter. This is similar state of affairs to that of a condensation nucleus counter (CNC) where all particles appear at around 50 nm when the alcohol has condensed on them. Small discs therefore appear to be oversized as we approach the limits of visible light microscopy (Fig. 6.14).

This can have effect of producing a false maximum just above the theoretical limit for optical microscopy. Quoting Heywood (in Lang (1954)):

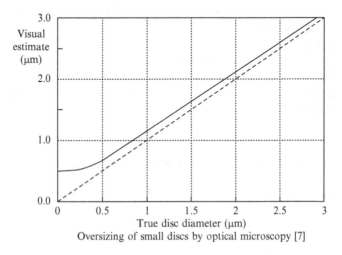

Oversizing of small discs by optical microscopy [7]

Fig. 6.14 Oversizing of small discs by optical microscopy (Taken From Allen, (1997), p. 114)

"Dr Brownowski asked if anyone could comment upon automatic star counting methods: although I cannot say what methods were used for this purpose, there was an interesting similarity between the star and particle counting results when the limit of visibility was approached. Early records of microscopical counts when extended to the extreme limit of resolution, showed a false maximum frequency at a particle size just greater than the lower limit, where it is known that the numerical frequency for normal dusts increases continuously as the particle size decreases. Star counts made by unaided vision about 300 years ago showed a maximum frequency at about fourth magnitude, the limit of such vision being about sixth magnitude. With the aid of telescopes it was easily shown that the numerical frequency of stars increased as the magnitude increased i.e. brightness decreased. The important feature of this comparison was that false effects might be observed if any counting process was worked to the extreme limit of its capabilities."

Fed Stan 209E bases its calculations on particle count increasing exponentially as size decreases. This comment also indicates that the harder we look (e.g. with an electron microscope), then the more we see and the tendency to feel that our particulate system is much smaller than it really is as we are really investigating the dust in the system. It is the equal importance attached to each and every single particle in number counting that is both its disadvantage and advantage. The author has been given a white powder (supposedly of ZnO) and told that it was 0.7 nm in size! It is easy to see how taking any microscopy to its limits can produce a false impression of the material. Incidentally the above sample of ZnO settled to the bottom of a beaker of water in less than one minute....

Once again we must state that microscopy is only a two-dimensional representation of a particle. We must be aware that the preparation conditions can alter the sample and that the material will present a preferred axis to the light.

Refractive index is important from a number of points of view. The higher the relative refractive index between particle and the medium that it is sitting in then the better the contrast and features that are observable. Indeed if the particle's RI and that of the medium it resides in are identical then the particle appears invisible (if this not a contradiction!). Putting the particle in different immersion fluids (e.g. Cargille) allows (by means of the Becke lines) the RI of the particle to be determined.

Notwithstanding all the above comments, visualisation together with the generation of the appropriate micrographs is essential in any particle characterization, but we must be careful with assigning quantitative interpretations to what we see.

Fractal analysis is another possibility; here but the reader should refer to texts by Mandelbrot (1983) and Kaye (1993) for further information.

Finally, we need to consider the magnification that we are using or attempting to use. The trade-off of high magnification is that the depth of field becomes smaller and smaller. The depth of field is the distance within which the image will appear to be sharp and in focus. The amateur photographer will be well aware of the differences in attempting to focus a picture using a 500 mm lens ("high" magnification) as opposed to using a 24 mm (wide-angle) lens. The latter is "low" magnification – in fact, about twice as wide as our normal eye field of vision – usually stated to be 50 mm or so. Thus we could term it 0.5× magnification.

High magnification objective lenses are generally more expensive and need to be brought close to the object to obtain focus. How many of us have crashed such a lens against the microscope slide?! The usual rule is to look at the slide and move the objective as close as possible towards the particle and then gradually move the objective away until focus is achieved.

The mathematical description of depth of field is the distance, D, the objective can be moved such that a point appears to be a circle having a radius no bigger than c:

$$D = 2c/\tan(A.A./2)$$

Thus D depends on the value of c that one is prepared to accept and decreases with decreasing size of the particles. Again a photographic example is useful here. To avoid a long focal range lens becoming too large a mirror can be placed on the front of the lens to reflect the light back thus halving the length of the lens. Such mirror lenses exhibit the feature that out-of-focus spots appear as rings or donuts. The Table below from (Yamate and Stockham 1977) shows the effect on depth of field by increasing the magnification (Table 6.7).

The image or particle can be brought up against various graticules from the simple "ruler" to those with spots to place over the particles in the hope that projected areas can be isolated by the operator. The literature is full of examples of the differences obtained by trained and untrained operators and those between trained operators. And is it easy to rotate the graticule correctly to obtain a Feret's or Martin's diameter?

Last but no means least, we need to consider some aspects of electron microscopy where sample preparation can be vital to correct interpretation. In some cases (e.g. polymers) then increasing the magnification too much will start to boil the sample! Again there is the tendency to observe the "interesting" or unusual rather than the representative. Bearing in mind too that $1 \times 100\,\mu m$ particle has the same mass as one million $\times 1\,\mu m$ particles, then if we happen to spot this one in a million particle, then there is the tendency to ignore it and state that it is not typical (which is true on a number basis!). However, it makes up half of the overall mass of the system – it is the proverbial golden nugget!

Table 6.7 Maximum useful magnification and the eyepiece required for different objectives (From (Yamate and Stockham 1977))

Objective					
Magnification	Focal length	N.A.	Depth of focus	Maximum useful magnification	Eyepiece required
2.5×	56 mm	0.08	50 μm	80	30×
10×	16 mm	0.25	8 μm	250	25×
20×	8 mm	0.5	2 μm	500	25×
43×	4 mm	0.66	1 μm	660	15×
97×	2 mm	1.25	0.4 μm	1,250	10×

N.A Numerical Aperture

The magnification of the objective lens is calculated by dividing the microscope tube length, usually 160 mm, by the focal length

6.7.3   Image Analysis

It is quite clear that the techniques described above can be tedious, long-winded and prone to operator error and subjectivity. Hence the obvious move to speed things up and automate the process of obtaining information. Hopefully with the computer making the judgments then we should have less chance of being prone to vagaries of the operator. Plus we should be able to scan hundreds, thousands, and even tens of thousands of particles in a considerably quicker mode than by hand. We must therefore be getting closer to the statistical validity. However, too with the computer playing a part we have the ability to generate literally hundreds of pieces of information and transformations, so again we must be cautious in sifting the real requirements from those that are ancillary or unnecessary.

There are a number of good texts in the market from the simple and easy e.g. (Allen 1997) and (Washington 1992) to the more advanced e.g. (Russ 1992), (Serra 1993), (Loebl 1985).

The main stages in any image analysis sequence are:

- Image Acquiring
- Object extraction
- Segmentation
- Calculations

A standard microscope will normally have a CCD camera connected to it via a "T" adaptor and the signal passed to a frame grabber board normally residing in a PC. These frame grabber boards advance every three months in terms of speed and features, and this has spawned almost as many home-built packages as those obtained from companies with a known reputation in the field. The image is usually made visible on a monitor and large memory requirements (previously handled by optical discs and similar technology) are required. Those with experience of digital photography will appreciate that even a single modest picture of $2,048 \times 1,024$ picture elements (pixels) takes over 2 MB of memory for black and white only. This will be trebled at a minimum for colour use, although the latter still finds less usage in image analysis. Only in the last 5–10 years has sufficient computer memory become affordable for home use. We have to recall that in the 1970's 1 MB of memory equated to around U\$1 million of cost!

So let us break down the process even further. Once the image has been acquired in some sort of grey scale form, it will need to be digitized (i.e. changed to black and white only: 1's and 0's) in order that particles can be distinguished from the background. Even at this stage, there is discrimination to be had and the threshold setting here will determine what is an edge of a particle (Figs. 6.15 and 6.16).

We then isolate a silhouette of particle allowing the edge to be defined, for example (Fig. 6.17):

Subtract 170 from the image ("thresholding"), Refer all negative numbers to 1 (black) and all positive numbers to 0 (white) – "digitization", remove single pixels – "filtering"):

Fig. 6.15 Thresholding to obtain particle "size" (After diagram by Jean Graf in (Russ 1992))

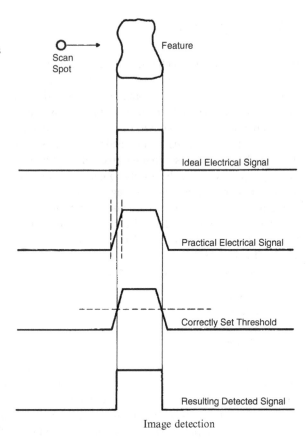

Image detection

Thus for small particles where diffraction limitation is an issue we still have plenty of room to play! We also need to remember that the acquired image may only be a few pixels in size at the so-called lower range of the device we are using. Obviously, we must be aware that shape analysis on something 6 pixels in total size is meaningless. The light that we illuminate the particle with then plays an important role and it will need to be kept constant – lamps warming up will alter the point around the particle at which the thresholding takes effect. If there is too little light then the particle could appear larger. With too much light then a smaller particle could be synthesized.

We will quickly obtain a binary image. This is where many choices can now be made:

- "Holes" in binary images of particles can be filled. A good example of applying this is with glass beads, which can act as a miniature lens and focus light to a bright spot behind. They are not actually donuts but could appear as so if this hole filling routine is not practiced. Now what would be the situation if this were a genuine donut? This is a philosophical question again, as we would have to consider internal and external radii of the "particle".

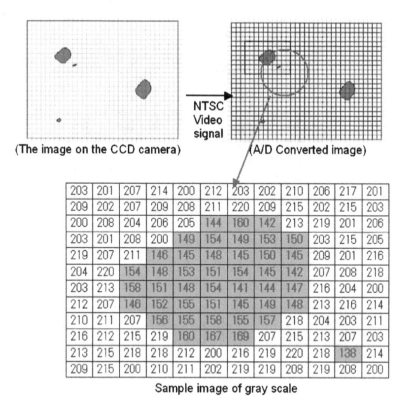

(The image on the CCD camera)   (A/D Converted image)

203	201	207	214	200	212	203	202	210	206	217	201
209	202	207	209	208	211	220	209	215	202	215	203
200	208	204	206	205	144	160	142	213	219	201	206
203	201	208	200	149	154	149	153	150	203	215	205
219	207	211	146	145	148	145	150	145	209	201	216
204	220	154	148	153	151	154	145	142	207	208	218
203	213	158	151	148	154	141	144	147	216	204	200
212	207	146	152	155	151	145	149	148	213	216	214
210	211	207	156	155	158	155	157	218	204	203	211
216	212	215	219	160	167	169	207	215	213	207	203
213	215	218	218	212	200	216	219	220	218	138	214
209	215	200	210	211	202	219	219	208	219	208	200

Sample image of gray scale

Fig. 6.16 Oversimplified (hence incorrect!) digitization. But the principle is crudely the same if a few Fast Fourier Transforms and Laplacian Filtration is included!

- Segmentation. Particles that are touching need to be separated. Ideally, we would want to see only single particles but in practice this is never likely or possible. The technique of erosion and dilation developed by Jean Serra and coworkers at the Ecole des Mines, Fointainbleau, France – see (Wadell 1932) – in the mid-1970's is still the main route to achieve this.

We then end up with the "particles" as binary images on a background. This is when many more calculations can be made for which the computer is well equipped to generate. Many of these calculations will be simply the ratio of two lengths. However, we must note that this digitization process inherent in image analysis has reduced our particle to a silhouette and we therefore lack the textural and morphological information that we saw under the electron microscope.

Different definitions of circularity and convexity etc are to be found in propriety software so the wise user will determine how these were derived.

So what is the type of output analysis that we can obtain? As an example we will look at two different materials dispersed both dry and wet and measured on two different techniques (Wet: FPIA-3000, Sysmex Corporation, Kobe-Osaka, Japan and Dry: Morphologi G2 Malvern Instruments:

0	0	0	0	0	0	0	0	0	0	0	0
0	0	0	0	0	0	0	0	0	0	0	0
0	0	0	0	0	1	1	1	0	0	0	0
0	0	0	0	1	1	1	1	1	0	0	0
0	0	0	1	1	1	1	1	1	0	0	0
0	0	1	1	1	1	1	1	1	0	0	0
0	0	1	1	1	1	1	1	1	0	0	0
0	0	1	1	1	1	1	1	1	0	0	0
0	0	0	1	1	1	1	1	0	0	0	0
0	0	0	0	1	1	1	0	0	0	0	0
0	0	0	0	0	0	0	0	0	0	0	0
0	0	0	0	0	0	0	0	0	0	0	0

Binary image

0	0	0	0	0	0	0	0	0	0	0	0
0	0	0	0	0	0	0	0	0	0	0	0
0	0	0	0	0	1	1	1	0	0	0	0
0	0	0	0	1	0	0	0	1	0	0	0
0	0	0	1	0	0	0	0	1	0	0	0
0	0	1	0	0	0	0	0	1	0	0	0
0	0	1	0	0	0	0	0	1	0	0	0
0	0	1	0	0	0	0	0	1	0	0	0
0	0	0	1	0	0	0	1	0	0	0	0
0	0	0	0	1	1	1	0	0	0	0	0
0	0	0	0	0	0	0	0	0	0	0	0
0	0	0	0	0	0	0	0	0	0	0	0

Fig. 6.17 Edge isolation by considering the pixel in relation to its neighbors

- Wet Images – Glass Fibre + latex. Images (Fig. 6.18).
- Wet measurement – glass fibre + latex. Calculated data and graphs. Note the circularity diagram illustrating how, as a rod is shortened, it becomes more "spherical" (Fig. 6.19).
- Dry measurement (Morphologi G2 Malvern Instruments) (Figs. 6.20–6.22).

In broad terms, then texture and intimate details are provided by microscopy especially electron microscopy. These provide number distributions. Ensemble techniques may measure more sample and a volume or mass distribution may be obtained by a variety of techniques, for example diffraction, sedimentation, sieves etc. Shape distribution and imaging provide a bridge between the two. We need to remember the pitfalls of digitization and thus we may obtain so less or no detail or not a lot, but we still can see the big picture.

Class2 (Detection count : 1438)

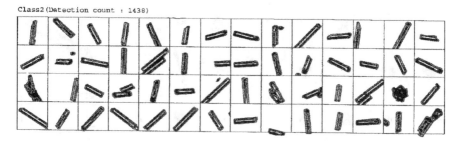

Fig. 6.18 Wet Images – Glass Fibre + latex (FPIA, Sysmex Corporation, Kobe-Osaka, Japan)

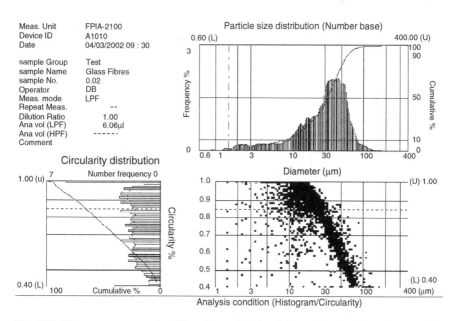

Fig. 6.19 Wet measurement – glass fibre + latex. Calculated data and graphs. Note the circularity diagram illustrating how, as a rod is shortened, it becomes more "spherical"

The relevant ISO committee ISO TC24/SC4 has been working to put together general rules and definitions for imaging within ISO 9276-6: 2008 (under development) and the existing ISO13322-1 and 13322-2 standards. In general terms, we should have expressions that fulfill a number of logical constraints. These shape parameters should be:

- Intuitive – consistent with the way the human mind perceives words such as "circularity"
- Normalized – have values between 0 and 1 making interpretation and data processing easier. This, for example, aspect ratios would be defined as width/length rather than length/width to permit such mathematical manipualtions
- Sensitive to minor deviations – the ability to distinguish small changes

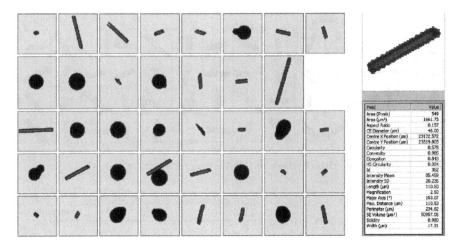

Field	Value
Area (Pixels)	549
Area (μm²)	1661.73
Aspect Ratio	0.157
CE Diameter (μm)	46.00
Centre X Position (μm)	23172.572
Centre Y Position (μm)	23319.803
Circularity	0.578
Convexity	0.995
Elongation	0.843
HS Circularity	0.304
Id	352
Intensity Mean	85.459
Intensity SD	20.235
Length (μm)	110.50
Magnification	2.50
Major Axis (°)	163.07
Max. Distance (μm)	110.53
Perimeter (μm)	234.82
SE Volume (μm³)	50957.06
Solidity	0.980
Width (μm)	17.31

Fig. 6.20 Images of rods and metal spheres and sizing parameters

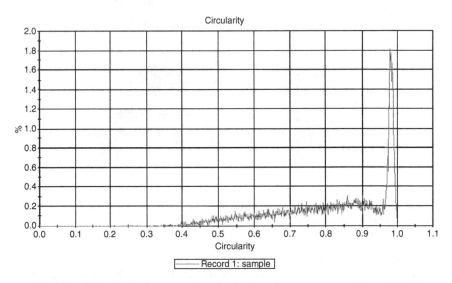

Fig. 6.21 Circularity distribution – note how the circular spheres are separated from the different fibre lengths (obviously the shorter the length in relation to the diameter, then the more circular the particle is

Also, given that numbers can be generated from any other set of numbers, then the quality of any derived set is paramount on the quality of what we first acquire. It's no good processing and massaging poor data with mathematical formulae and definitions. Indeed this is probably true of most instrumentation!

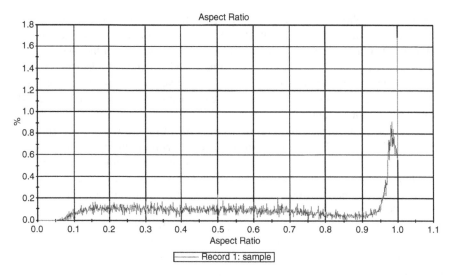

Fig. 6.22 Circularity distribution – note again how we can distinguish the different lengths of fibre from the spheres

6.8 Setting Tolerances and Specifications

We should first consider what is "fit for purpose" and "just good enough" in line with Heywood's earlier statement (Heywood 1966) rather than permitted deviations for deviation's sake. This is actually more difficult to do than it first appears. Relating a drug's efficacy or uptake against the particle size distribution is not a trivial task. However, it is essential for correct interpretation or understanding that this is carried out first and a pass-fail based on "effective for use" needs to be defined. A reasonable specification will then be set rather than the possibility of over-specifying which is sure to lead to further problems.

A probable error balance based on the key parameters (see ISO17025) then needs to be defined.

First, we need to consider the inherent homogeneity of the material. This will then set the minimum mass needed for any particular defined probable error. If we take less than this minimum mass for the analysis, then our permitted tolerances will need to be increased. We then need to specify based on this and typically $1\,\sigma \sim 5\%$ or $+/- 3\sigma \sim 15\%$ is needed here (see Allen, reference (Leschonski 1977, p. 39)) for the median alone. This represents the largest contribution to the *reproducibility* of the analysis (sample-to-sample variation). On top of this, during our method development we will explore the intra-experiment variation (*repeatability*), which may vary due to sample dispersion (perhaps desirable) and sample dissolution (perhaps undesirable). Appropriate parameters can then be set. During this development time, the effect of key parameters (ΔP in the dry dispersion technique, sonication in the wet; refractive index in both) will need quantifying in a *robustness* study. Summing the 3 R's (reproducibility, repeatability, and robustness) will allow a reasoned and argued permitted experimental variation to be set for the material in question. Now, we then need to be aware that moving into the tails of the distribution will involve more statistical variation than that of the median. Thus parameters such as the D_{10} and D_{90} need to have wider permitted tolerances and the same would go for values such as D_{95} and D_{99}. Note that the concept of a D_{100} is stated to be inadmissible in ISO13320 (FDIS) Particle size analysis - Laser diffraction methods - General principles (August 2009).

So how do the international norms cope with this? ISO13320-1 "permits" a variation of 3% in the D_{50} and 5% in the D_{10} and D_{90} with a doubling of these parameters "permitted" under $10\,\mu m$. Thus, this standard assumes that we have dealt with the sampling issues or alternatively that the width of the distribution is narrower than one decade, the recommended width for a verification standard. <USP 429>, which is loosely based around ISO13320-1, "allows" 10% on the D_{50} and 15% on the D_{10} and D_{90} again with a doubling under $10\,\mu m$. Instruments are capable of at least an order of magnitude better than these stated ISO13320-1 values, so these values are practical and sample-specific reflecting the inherent homogeneity (or otherwise) and dispersibility of the material.

Two generalizations may be useful here:

- If we have material larger than around $75\,\mu m$ then representative sampling must be accomplished to bring down any sample-to-sample variations.
- If we have material $<10\,\mu m$ present in the sample, then control of the dispersion stage becomes paramount.

6.9   Assay Development/Validation

In method development in a regulated environment then the stability of the measured results is paramount. In any technique, we will normally examine multiple consecutive measurements to define a precision value under the conditions of the experiment noting that these conditions (say under high dilution) may not exactly replicate the use or storage of the material. A single measurement has perfect precision and no validity. Often the question asked is "Are the results accurate?" Normally this latter question is answered by running appropriate traceable standards with the appropriate methodologies and appropriate calibration (and calibration traceability). The precision is normally defined by the coefficient of variation (CV) also called the relative standard deviation (RSD) for the number of experiments performed.

The three QC parameters that need exploring in a suspension study can be summarized as follows and are known as the three R's:

- Repeatability – repeated measurements on the same aliquot of material looking for changes that take place over short (experiment) or long (e.g. storage) periods of time. In an experiment that only takes a single aliquot (e.g. a sedimentation experiment) we may only have a single measurement and repeatability is not definable here. This also applies to a dry laser diffraction measurement where the exact same group of particles is never seen within the measurement zone.

- Reproducibility – sample-to-sample, operator, technique variation that may indicate a batch homogeneity issue or other variable. A gage study may be implemented here with control and standard samples
- Robustness – how an experimental parameter may affect the stated result. This could be refractive index in a laser diffraction experiment or the effect of ΔP on the pressure-size plot in a dry diffraction result. Typically, this may be the effect of energy input into the system (shear in a rheology experiment, stirring and ultrasound in a wet diffraction measurement). An error balance in a ISO17025 certified laboratory is essential to understand where variability is associated, what limits can be assigned to detection and quantification.

Based on our method development we should be able to find a broad plateau of stability that allows us to define the method (Standard Operating Procedure, SOP) appropriate to the measurement. We need to distinguish this stability need that is vital for QC purposes with that of other dynamic information (e.g. dissolution, dispersion), that may be useful for R& D and formulation purposes.

Validation is the documentation/paperwork exercise that ensures that the integrity of the entire process. The FDA definition is "Establishing documentary evidence which provides a high degree of assurance that a specific process will consistently produce a product meeting its predetermined specification and quality attributes." Thus, we will have a predefined specification and method that we will challenge our product or process with.

It is always sensible to go into this exercise with a "fit for purpose" and a "just good enough" attitude rather than a specification for specification's sake. There is a real danger in regulated industries of over specification when changes to the products properties have no relationship to the efficacy or quality of the drug.

6.10   Rheologial Properties

The rheological (or resistance to flow) properties of a suspension are defined to a great extent by the size, shape, and concentration of the solid or liquid phase present within the liquid system. A simple viscometer or more sophisticated rheometer can be used to examine what happens where shear is applied to any suspension and this is a fundamental characterization test for any slurry or suspension. Note that, in an emulsion, shear forces would have been used to create the system in the first place. In general:

- Increasing the solids content increases the viscosity.
- Decreasing the size of the primary particles in the system tends to increase the viscosity of the system.

Weltmann and Green (1943) provides one example of the above two generalities (Fig. 6.23).

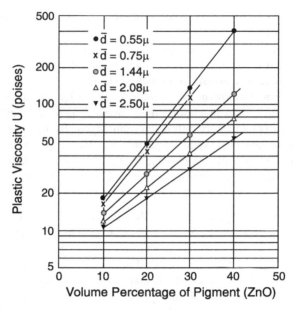

Fig. 6.23 Viscosity of a different sized ZnO pigments at different concentrations (Reference (Weltmann and Green 1943))

Einstein dealt generally with viscosity and concentration in his doctoral thesis and came up with an equation that has been debated, expanded, and used for many years:

$$\eta" = \eta(1 + K\theta)$$

where $\eta"$ = viscosity of the suspension, η = viscosity of the suspending medium and θ = the ratio of the volume of the suspended material to the volume of the medium, K is a constant often stated to be 2.5.

Now this equation, like many, may be reasonably valid at low concentrations (say < 1%) and reasonable particle sizes (say 20–200 μm) where dispersion and suspension issues do not play a significant role. This equation, though, does not deal with the effect of particle shape or distribution and all particles are considered equivalent.

In a particulate system the packing of the material (and hence the manner in which particles will slide over or pass one another) will affect the rheology of the system. This packing will be dependent on the shape of the particles (spheres will flow more easily past each other in comparison to rods). Agglomerates in the system will disrupt the flow or resistance to movement. Conversely, small material packed in between larger material (a wider distribution) may help "lubricate" the system and aid flow. A generic slide can assist the understanding here (Fig. 6.24):

Clearly, the maximum packing fraction depends on the shear rate and the state of aggregation as well as the shape and size distributions. Agglomerates can be dispersed and thus the flow properties may alter in a nonreversible manner.

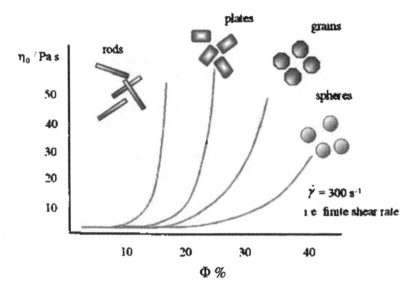

Fig. 6.24 Generic plot of viscosity changes of different loadings of different shaped materials

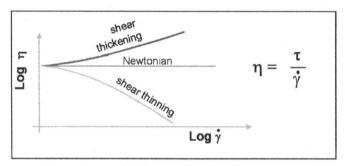

- Newtonian: water, low molecular oils ...
- Shear thinning: polymer melts, emulsions, ceramics...
- Shear thickening: wet sand, PVC plastisols...

Fig. 6.25 Possible scenarios on viscosity behavior as shear is increased on a two-phase system

Typical shear plots suspensions and emulsions can show shear thickening or shear thinning behavior in contrast to a fluid such as water that has Newtonian properties (Fig. 6.25).

Whether a system will shear thin or thicken depends on the order or structure under stationary and shear conditions. Some materials can "unjumble" and flow becomes easier. Conversely increasing the shear in some systems 'bundles up' the particles and flow becomes more difficult.

The charge on the particles within a system affects the state of aggregation and thus changes in pH will almost certainly affect the rheological behavior. This is an introduction to zeta potential and the next section.

6.11 Dispersion Stability

As stated in the introduction, the formation of a stable dispersion is dependent on the steps of wetting, separation, and stabilization. These will be the steps also required in any analytical method designed to generate stable measurements. Generally speaking, the stabilization step may only be required where the inertial forces of larger particles (typically >20 µm) are not sufficient to overcome those of van der Waals.

A typical scenario in which this is seen is with a small clay or TiO_2 where, after relaxation of input sonication energy, there is re-agglomeration of the system manifesting itself in increases in the key size parameters. Often this can be seen most easily by trending the D_{10} (smallest particles in the system). These particles reflect the dispersion stability. This is in contrast to the largest particles in the system (a key marker would be the D_{90}) that is a marker of agglomeration in the system – dispersion efficacy. Note the difference between dispersion stability and dispersion efficacy.

In such cases, we can use either an ionic stabilizer that would absorb on the particle's surface or a steric stabilizer (normally a polymer which would prevent the interaction of the 'main' particles in the system) or a (less-used) combination of the two.

Riddick (1968) first defined the values of zeta potential that define stability in the system. Note that we talk about the system and not simply the charge on the particle. Factors such as pH, ionic concentration, and presence of polyvalent ions play a large role in determining the zeta potential of the particles in the system. Riddick is often misquoted, so we return to the original text bearing in mind that blood cells were the main area of interest (Table 6.8).

Perhaps from not referring to the table directly, many people have made the (wrong!) assumption that a magnitude of zeta potential is greater than 30 mV confers stability on a system. We note that the text refers only to "moderate stability" at this level. Most "natural" biological systems have negative zeta potential at around pH 7. Many inorganics or ores dug from the ground (e.g. galena, PbS) may tend to have positive zeta potential due to metallic ions (Pb^{2+} in the case of galena) being present on the surface – sometimes atmospheric oxygen will bring such ions to the surface in a chemisorption induced segregation process (freshly cleaved galena normally has little or no charge as the Pb^{2+} and S^{2-} ions balance out, but

Table 6.8 Zeta potential and stability (adapted from Riddick (1968))

Stability characteristics	Avg. ZP (in millivolts)
Maximum agglomeration and precipitation	0 to +3
Range of strong agglomeration and precipitation	+5 to −5
Threshold of agglomeration	−10 to −15
Threshold of delicate dispersion	−16 to −30
Moderate stability	−31 to −40
Fairly good stability	−41 to −60
Very good stability	−61 to −80
Extremely good stability	−81 to −100

Fig. 6.26 Photograph of cerium oxide with differing phosphate additions. Note the broad range of concentrations over which stability is conferred

oxygen pulls the Pb^{2+} to the surface). Given that blood cells are around $8\,\mu m$ in equivalent diameter, Riddick's 30 mV stability rule holds roughly for material down to around a micrometer or so. Smaller than this (for example 300 nm TiO_2) then more charge will almost certainly be needed to confer the stability in a system – perhaps around −60 to −70 mV or more in the cited example of TiO_2. In the case of inorganic oxides phosphate (PO_4^{2-}) addition is the normal route to alter the charge. We note though that too much ionic additive is as bad as too little and also that there is usually a wide range of concentration over which stability is conferred in a system. This is best illustrated with a photograph (here a submicron cerium oxide polishing material with varying concentration of sodium hexametaphosphate often called "Calgon") (Fig. 6.26).

Clearly if a range of 0.05–0.5% additive would give stability, we would prefer the lowest concentration on economic (and perhaps environmental) grounds. We also note that too much additive is as bad as none at all. For any ionic added to a system in this manner, there will be an optimum concentration that can be found empirically as above or scientifically as we will later describe.

The zeta potential of such systems can be measured indirectly by measuring the movement (or mobility) of the particle under an electrical field (Field = Volts/Distance).

The conversion of mobility to zeta potential relies on the Smoluchowsky (also spelled Smoluchowski) approximation ($f(\kappa\alpha) = 1.5$) of Henry's equation:

$$U_E = 2\,\varepsilon\,\zeta\,f(\kappa\alpha)/3\eta$$

where U_E is the electrophoretic mobility, ζ is the zeta potential, $f(\kappa\alpha)$ is termed Henry's function and η is the viscosity that the particle experiences in its movement through the fluid. The function $f(\kappa\alpha)$ is made up of two parts – κ is the Debye length and κ^{-1} is used to define the thickness of the electrical double layer. The symbol α refers to the radius of the particle and thus $\kappa\alpha$ measures the ratio of the particle radius to that of the electrical double layer thickness. This double layer thickness can be best defined in an ionic medium as the location of ions and counter ions is easily assumed. The interpretation of the converted mobility to zeta potential measurements in non-aqueous or nonconductive systems is complicated by

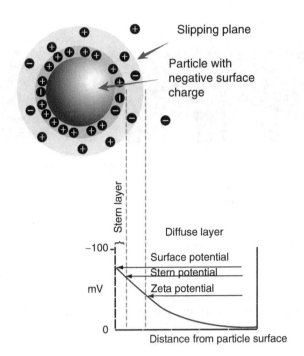

Fig. 6.27 Diagrammatic representation of zeta potential in aqueous medium

Fig. 6.28 Depiction of $f(\kappa\alpha)$

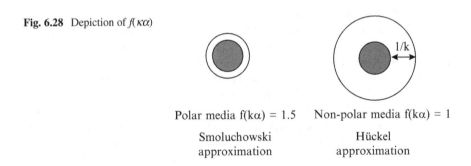

Polar media f(kα) = 1.5   Non-polar media f(kα) = 1

Smoluchowski   Hückel
approximation   approximation

interpretation of the position of the stationary layer in the system. In an aqueous system this can be easily defined as the ions on the surface of the particle have strongly bound counter-ions (Fig. 6.27).

In non-aqueous and non-polar media (of which many more exist than the single aqueous medium!), then it is standard to set the $f(\kappa\alpha)=1.0$, which is the Hückel approximation in Henry's equation (Fig. 6.28).

The validity of this approximation may give interpretation issues as (apparent/calculated) zeta potentials of tens or even hundreds of volts can be obtained in non-polar media. Linking this to a stability issue may pose problems, notwithstanding

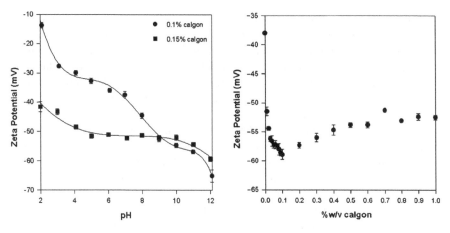

Fig. 6.29 Zeta potential, ionic additives, and pH for TiO_2

the fact that density and size probably play a major part in keeping a particle suspended in a nonpolar solvent.

Perhaps an example is more pertinent. The classic case is that of titanium dioxide manufactured with the objective to maximize scattering of a particle size around 300 nm. The zeta potential against pH can be plotted for different ionic phosphate additives (Fig. 6.29).

Here we can see the rapid absorption of ions to monolayer coverage decreasing the zeta potential markedly followed by a slow rise as multilayers of phosphate shield the inner particle. A minimum in zeta potential (and therefore maximum stability) of around – 60 mV can be seen at around 0.1% Calgon. The robustness to pH change is seen on the plot on the right where the buffering action of the phosphate is readily seen at 0.15% addition. This is the optimum formulation for this system – giving the best robustness to changes in concentration and pH.

6.12 Polymorphism Studies

Polymorphs of a material or compound have the same chemical constitution but different crystal structures. The reason for their importance and detection is twofold:

- Different polymorphs tend to possess different bioactivities and this is relevant both to the efficacy and patentability of the substance
- One polymorph may alter over a period of time to another changing the efficacy of a drug. One needs to know (and "one" is inclusive of the FDA!) if the drug is unstable under standard conditions of storage. We may be able to remedy the situation in certain cases.

Pasteur's discovery of the chirality of tartaric acid (preceded by a laborious tweezers separation) is one manifestation of how identical chemistry can still lead to different crystals forms. Although present in inorganic materials (for example there are three different crystal forms of TiO_2, brookite, rutile, and anatase), it becomes much more important in pharmaceutical chemistry where properties as diverse as tabletability (the ease of making into a tablet), solubility density, and refractive index can be affected.

Crystal structure would normally be evaluated with X-ray diffraction techniques although a few materials do exhibit the obvious crystal differences and often a detailed study of calorific changes on heating with differential scanning calorimetry (DSC) can provide a clue that different forms of a crystal are present. If different crystal forms are obvious (e.g. the monoclinic and orthorhombic forms of paracetemol), then simple microscopy or more sophisticated chemical imaging (size, shape, and chemistry) can be used to differentiate the forms. See Espeau et al (2004).

An estimate is that 30–50% of pharmaceutically important compounds exist in different polymorphic forms (quotation by Boese in article in Chemistry World, (Aldridge 2007)) and thus polymorph screening becomes essential. Within a formulation such as a tablet, then investigation afterwards on a chemical imaging equipment has been able to differentiate from crystal forms.

6.13   Conclusions

So, after this journey where are we? Indeed have we even begun the journey? The measurement and characterization of our suspension system is so vital that it must be tackled in a systematic and logical fashion and there is no substitute for building a firm foundation on the basic principles outlined in this chapter. Careful and systematic experimental work and observation is of considerably more value than any number of Greek symbols in a text. Indeed the latter are probably a crutch for lack of fundamental understanding or "feel" for the system. If this text has gone even a footstep on the never-ending journey of equipping the reader for a lifetime's improvement (Nelligan 2000), then it will have fulfilled its purpose.

References

Aldridge S, 2007. The Shape Shifters Chemistry World, 64–70.
Allen T, 1997. Particle Size Measurement 5th Edition Volume 1 Chapman and Hall.
ASTM E56 Committee on Nanotechnology, 2007. Work item, WK 8705 Standard Practice Guide for Measurement of particle size distribution of nanomaterials in suspension by Photon Correlation Spectroscopy (PCS). Now ASTM Standard E2490–09.
ASTM Standard F1877-98 (Re-approved 2003) Standard Practice for Characterization of particles ASTM, West Conshohocken, PA, U.S.A.
Bendick J, 1947. How much and how many: The story of weights and measures NY: Whittlesey House.

Bleininger AV, 1904. The Manufacture of Hydraulic Cements. Geological Survey of Ohio Fourth Series Bulletin No. 3, Columbus, Ohio.

Brittain HG, 2001. Particle Size Distribution, Part 1 Representations of Particle Shape, Size and Distribution Pharmaceutical Technology, 38–45

Cadle RC, 1965. Particle Size: Theory and Industrial Applications 1st Edition New York: Reinhold Pub Corp.

Davies CN (Editor), 1966. Aerosol Science Academic Press London and New York.

Dragoo AL, Robbins CR, Hsu SM et al, 1987. A critical assessment of requirements for ceramic powder characterization. Advances in Ceramics 21, Ceramic Powder Science, 711–720, The American Ceramic Society Inc.

Espeau P et al, 2004. Polymorphism of paracetamol: Relative stabilities of the monoclinic and orthorhombic phases inferred from topological pressure-temperature and temperature-volume phase diagrams. Journal of Pharmaceutical Sciences 94, 3, 524–539. Published Online: 30 Dec 2004.

Folk RL, 1955. Student operator error in determination of roundness, sphericity and grain size. J. Sedimentary Petrology, 25, 297–301.

Gregg SJ, Sing KSW, 1967. Adsorption Surface Area and Porosity Academic Press.

Hatch T, Choate SP, 1929. Statistical Description of the. Size Properties of Non-Uniform Particulate Substances. J. Franklin Inst. 207, 369–387.

Heywood H, 1947. The scope of particle size analysis and standardization, Institution of Chemical Engineers, Symposium on Particle Size Analysis. 14–24 (February 4th 1947).

Heywood H, 1966. Proc. 1st Particle Size Anal. Conf., (Heffer), 355–359. The quotation appears on Page 355 in the closing address.

Heywood H, 1970. Plenary Lecture "The Origins and Development of Particle Size Analysis" in Particle Size Analysis Conference held at the University of Bradford 9–11 September 1970 Eds. Groves MJ, Wyatt JL The Society for Analytical Chemistry 1972 Published W Heffer & Sons Ltd, Cambridge Pages 1–18. The descriptions of sieving are to be found on pages 4–8 of this article.

Irani RR, C F Callis C F, 1963. Particle Size: Measurement, Interpretation and Application John Wiley & Sons Inc.

ISO13320-1 Particle Size Analysis–Laser Diffraction Methods. Part 1: General Principles ISO Standards Authority (November 1999). Can be downloaded from http://www.iso.ch on payment with credit card.

ISO9276-2:2001. Representation of results of particle size analysis-Part 2: Calculation of average particle sizes/diameters and moments from particle size distributions

ISO9276-6: 2008 Representation of results of particle size analysis-Part 6: Descriptive and quantitative representation of particle shape and morphology.

Joyce Loebl, 1985. Image Analysis: Principles and Practice Joyce Loebl, Marquisway, Team Valley, Gateshead, Tyne and Wear, ISBN 0 9510708 0 0

Kaye BH, 1993. Chaos and Complexity; Discovering the Surprising Patterns of Science and Technology. VCH Weinheim.

Kaye BH, 1998. New Perspectives in Pharmaceutical Technology. Hosakawa Powder Systems' Pharmaceutical technical Seminar, May 19-20 1998.

Lai TYF, Carstensen JT, 1978. Int. J. Pharmaceutics, 1, 1, 33–30.

Lang HR (Editor), 1954. The Physics of Particle Size Analysis Conference arranged by The Institute of Physics & held in the University of Nottingham London: The Institute of Physics, British Journal of Applied Physics Supplement No. 3

Leschonski K, 1977. Proc. Particle Size Anal. Conf., Bradford Anal. Div. Chem. Soc., Bradford, Heyden, 186.

Mandelbrot BB, 1983. The Fractal Geometry of Nature W H Freeman.

Masuda H, Gotoh K, 1999. Study on the sample size required for the estimation of mean particle diameter. Advanced Powder Technol. 10, 2, 159–173.

Muegle RA, Evans HD, 1953. Droplet Size Distributions in Sprays. Ind. Engrg. Chem., 43 (6), 1317–1324.

Nelligan T, 2000. Chris Leslie: Wood, Wire and Wit. Dirty Linen 88, 47–50. The quotation appears on page 49.

Nichols G, Byard S, Bloxham MJ, Botterill J, Dawson NJ, Dennis A, Diart V, North NC, Sherwood JD, 2002 A review of the terms agglomerate and aggregate with a recommendation for nomenclature used in powder and particle characterization. J. Pharm. Sciences, 91, 10, 2103–2109.

Noyes AA, Whitney WR, 1897. The rate of solution of solid substances in their own solutions. J. Am. Chem. Soc., 19, 930–934.

Powers MC, 1953. A new roundness scale for sedimentary particles. J. Sedimentary Petrology, 23, 117–119.

Rawle AF, 1993. The Basic Principles of Particle Size Analysis. Malvern Instruments Technical Paper. Malvern Instruments UK. Downloadable from: http://www.malvern.com

Rawle AF, 2000. Attrition, dispersion and sampling effects in dry and wet particle size analysis using laser diffraction. Paper 0208 14th International Congress of Chemical and Process Engineering "CHISA 2000" 27-31 August 2000, Praha, Czech Republic.

Rawle AF, 2002. Sampling of powders for particle size characterization. Invited Lecture (AME.2-A-02-2002) Session E2 Practical Issues in Ceramic Powder Size Characterization 104th American Ceramic Society Meeting St. Louis (April 29th 2002). See Malvern Application Note MRK456-01 "Sampling for Particle Size Analysis"

Riddick T, 1968. Control of colloid stability through zeta potential with a closing chapter on its relationship to cardiovascular disease. Volume 1 Livingston Pub. Co.

Rose HE, 1954. The measurement of particle size in very fine powders. Chemical Publishing Co. Inc. 212 5th Street New York.

Russ JC, 1992. Computer-Assisted Microscopy: The measurement and analysis of Images. Plenum Press. New York.

Saveyn H, Mermuys D, Thas O, van der Meeren P, 2002. Determination of the Refractive Index of Water-dispersible Granules for Use in Laser Diffraction Experiments. Particle & Particle Systems Characterization 19, 6, 426–432.

Scarlett B, 1985. Measurement of Particle Size and Shape, Some Reflections on the BCR Reference Material Programme Part. Charact., 2, 1–6.

Serra J, 1993. Image Analysis and Mathematical Morphology Volume 1, 4th Edition, Academic Press.

Shiu GK, 1996. Dissolution Methodology: Apparatus and Conditions, Drug Information Journal, 30, 1045–1054.

USP General Test <776>, 2000. USP XXXIV The United States Pharmacopoeial Convention, Rockville, MD), 1965–1967

Wadell H, 1932. Volume, shape, and roundness of rock-particles. Journal of Geology, 40, 443–451.

Washington C, 1992. Particle Size Analysis in Pharmaceutics and other industries. Chapter 10 (215–235), Ellis Horwood, ISBN 0 13 651613 0

Wedd MW, 2001. Procedure for predicting a minimum volume or mass of sample to provide a given size parameter precision. Part. Part. Syst. Charact., 18, 109–113.

Wells D, 1986. The Penguin Dictionary of Curious and Interesting Numbers Penguin Books: London.

Weltmann RN, Green H, 1943. Rheological Properties of Colloidal Solutions, Pigment Suspensions, and Oil Mixtures. Journal of Applied Physics, 14, 11, 569–576.

Wentworth CK, 1919. A laboratory and field study of cobble abrasion. Journal of Geology, 27, 507–521.

Wills BA, 1988. Mineral Processing Technology Mineral Processing technology 4th Edition Pergamon Press. ISBN 0-08-034936-6.

Yamate G, Stockham JD, 1977. Sizing particles using the microscope. Chapter 3 of Stockham JD, Fochtman EG (Eds.) Particle Size Analysis Ann Arbor Science Publishers Inc. ISBN: 0 250 40189 4

Chapter 7
Clinical Trials of Suspension Drug Products

G. Michael Wall and Terry K. Wiernas

Abstract Clinical trials of a pharmaceutical product, regardless of the pharmaceutical form, are generally required by the regulatory agencies around the world as the basis for proof of safety and efficacy in the drug product approval process. Such clinical trials must be conducted in an ethical manner according to a protocol. Some of the most important aspects of clinical trials specific to suspension drug products are concerned with the clinical supplies. Blinding is particularly difficult for suspension drug products since many approved comparators are not suspensions, thus making it very challenging to mask the investigational suspension versus its comparators. Appropriate choices for globally acceptable clinical study design of suspension drug products are discussed.

7.1 Introduction

Clinical trials of a pharmaceutical product, regardless of the pharmaceutical form, are generally required by regulatory agencies around the world as the basis for proof of safety and efficacy in the drug product approval process. Though there are unique regional or national requests, regulatory agencies generally require the same basic elements in the conduct of clinical trials. In this chapter, general requirements for clinical trials will be described briefly and some of the specific challenges of clinical trials for suspension drug products will be discussed.

Clinical trials are used to determine whether a product is of value in the treatment or prevention of a disease or condition by evaluating the safety and effectiveness of the product in human subjects. Numerous authors have reviewed requirements for clinical trials over the years (Machin et al. 2006; Chin and Lee 2008; Friedman et al. 1998; Meinert and Tonascia 1986; Grady and Linet 1990; Pocock 1983). Additionally, a wealth of information is now available on the internet regarding the clinical requirements of the United States Food and Drug Administration (FDA),

G. Michael Wall (✉) and T.K. Wiernas
Alcon Research, Ltd, 6201 South Freeway, Fort Worth, TX, 76134, USA
e-mail: michael.wall@alconlabs.com

A.K. Kulshreshtha et al. (eds.), *Pharmaceutical Suspensions: From Formulation Development to Manufacturing*, DOI 10.1007/978-1-4419-1087-5_7,
© AAPS 2010

European Medicines Agency (EMEA), and the Japanese Pharmaceutical and Medical Devices Evaluation Center (PMDEC). (United States Food and Drug Administration 2009; European Medicines Agency 2009; Japan Pharmaceutical and Medical Devices Evaluation Center 2009).

While regional differences still exist, many regulatory requirements have been standardized via the International Conference on Harmonisation's (ICH) Technical Requirements for Registration of Pharmaceuticals for Human Use (International Conference on Harmonisation 2009). A unique project, ICH brings together the regulatory authorities of Europe, Japan and the United States, as well as experts from the pharmaceutical industry in the three regions to reach consensus on and document the minimal scientific and technical requirements that are acceptable for drug product registration.

In the United States, the FDA regulates clinical studies according to Section 505(i) (drugs and biologics) of the Federal Food, Drug and Cosmetic Act (FD&C Act). The FD&C Act defines drugs as "articles intended for use in the diagnosis, cure, mitigation, treatment, or prevention of disease in man…" and "articles (other than food) intended to affect the structure of any function of the body of man….". Before a new drug can be marketed in the United States, a new drug application (NDA) must be submitted which must contain adequate data on the drug's safety and "substantial evidence" of the product's effectiveness. Under the FD&C Act, it is the FDA that decides what constitutes substantial evidence for each new drug product. During the late 1990s, the FDA's Center for Drug Evaluation and Research made a significant effort to define the concept of substantial evidence and to communicate its expectation to industry. The FDA has provided a detailed guidance document outlining the efficacy data that are necessary to support a new drug approval (Guidance for Industry: Providing Clinical Evidence of Effectiveness for Human Drug and Biological Products, 2009).

7.2   Nonclinical Studies

The comprehensive nonclinical testing program needed to support marketing approval for most new drugs involves years of work and several types of studies in multiple species. Prior to clinical trials, a sponsor must present the data necessary to support the safety of the experimental drug product. An ICH guidance discusses the timing of the various components of nonclinical testing and their relation to the conduct of clinical trials (ICH Note for guidance on nonclinical safety studies for the conduct of human clinical trials and marketing authorization for pharmaceuticals 2008). According to this document, the goals of the nonclinical safety evaluation include "characterization of toxic effects with respect to target organs, dose dependence, relationship to exposure, and potential reversibility. The information is important for the estimation of an initial safety starting dose for the human trials and the identification of parameters for clinical monitoring for potential adverse effects. The nonclinical safety studies, although limited at the beginning of clinical development, should be adequate to characterize the potential toxic effects under the conditions of the supported clinical trial."

Generally speaking, it is necessary to submit nonclinical data at or above the intended human dosage for a duration at least as long as the intended human exposure. The selection of the animal species to be tested is also a critical component which must consider pharmacokinetic parameters in order to demonstrate that the selected animal species are appropriate in order to estimate the potential human toxic effects.

7.3   Ethical Aspects

The most important aspect of a clinical trial is the protection of the safety, welfare, and individual rights of participating human subjects (21CFR Part 50). When product development plans are implemented, the protection of the individual subject should take priority over all the other considerations. For anyone who is involved in the healthcare field, it is important to understand the legal and ethical requirements regarding subject protection. One of the most important ways this is accomplished is for the sponsor to utilize an investigational or institutional review board (IRB) or independent ethics committee (IEC) to oversee the actions of investigators (21CFR Part 56).

Though governed by regulations, these ethics committees operate independently from governments and companies, and are focused on the safety of subjects in a clinical trial. The IRB/IEC must review nonclinical and clinical research, consent forms and other documents on which the trial is based. Clinical trial initiation should occur only after IRB/IEC approval of the appropriate documentation.

A second vehicle for ensuring a safe and reliable clinical trial is to incorporate Good Clinical Practices (GCPs) (ICH Guideline for Good Clinical Practice E6 R1 1996). Globally recognized, GCPs represent ethical and scientific quality standards covering every aspect of a clinical trial, including design, conduct, protection of subjects, monitoring, discontinuation, auditing, inspection, recording, analysis, adverse event handling, and reporting of results that involve the participation of human subjects. Compliance with this quality standard provides public assurance that the rights, safety, and well-being of trial subjects are protected, and are consistent with the principles that have their origin in the Declaration of Helsinki (World Medical Association Declaration of Helsinki 2008), as well as providing reasonable assurance that the clinical trial data are credible. Some aspects of the Declaration of Helsinki, including a prohibition against placebo-controlled trials (which are often required by regulatory agencies for product approval) have resulted in the situation whereby pharmaceutical companies do not always adopt the latest version of the Declaration. Nonetheless, all companies adhere to the basic principles of this Declaration, whose primary aim is to protect the human subjects participating in the clinical studies of new pharmaceutical agents. GCPs require that the study must be scientifically sound and employ appropriate statistical methods to ensure that the results are valid. The study must be conducted according to the protocol with appropriate recording and reporting measures. Only those persons qualified through appropriate training, education and experience can conduct a clinical trial. Investigators chosen by the sponsor are prohibited from having a significant vested interest in the success of a trial or the product of interest.

7.4 Protocol

Every clinical study must have a clinical protocol describing all aspects of the conduct of the trial. The protocol should contain the title, identifying number, all amendments, sponsor contact information, authorized manager, medical expert, qualified investigator and associated laboratories. Also, it should include descriptions of investigational products, summaries of significant findings from previously conducted clinical and nonclinical studies, and summaries of potential product risks and benefits. The study population must be defined, as well as the purpose and objectives of the trial and the trial design. Furthermore, the selection and withdrawal of subjects, treatment of subjects, assessment of efficacy and safety, statistical analysis to be utilized when evaluating the data, aspects of documentation, quality control, ethics, financing and insurance, and publication should also be addressed (ICH Guideline for Good Clinical Practice 1996).

7.5 Clinical Investigators Brochure

For all clinical protocols, a clinical investigator's brochure (CIB) is needed (ICH Guideline for Good Clinical Practice 1996). The CIB includes clinical and nonclinical data on the investigational product that are relevant to the study of the product in human subjects. Its purpose is to provide information in a concise, simple, objective, balanced and nonpromotional format to facilitate investigators' understanding of the rationale for, and their compliance with, many features of the protocol such as the dose, dose frequency/interval, methods of administration, and safety monitoring procedures.

7.6 Clinical Trial Registration

In recent years, the registration of clinical trials on public clinical trial databases (ClinicalTrials.gov 2009) has been required by most international regulatory agencies. Furthermore, many medical journals also require registration of trial data as a prerequisite for publication. For most trials, sponsors must list fundamental information during the conduct of a trial and the results must then be posted to the website in a timely fashion.

7.7   Therapeutic Areas and Route of Administration

Suspensions are used for many therapeutic applications. Routes of administration for suspensions include intranasal, inhalation, oral, parenteral, rectal and topical (including ophthalmic) (Sweetman 2009).

7.8   Well-Controlled Trials

Regulatory agencies typically require well-controlled clinical trials as a basis for the approval of pharmaceutical products. Excerpts of 21CFR314.126 below describe well-controlled trials.

"Adequate and well-controlled investigations provide the primary basis for determining whether there is 'substantial evidence' to support the claims of effectiveness for new drugs. Therefore, the study report should provide sufficient details of study design, conduct, and analysis to allow critical evaluation and a determination of whether the characteristics of an adequate and well-controlled study are present."

In a placebo concurrent controlled trial, "the test drug is compared with an inactive preparation designed to resemble the test drug as far as possible. A placebo-controlled study may include additional treatment groups, such as an active treatment control or a dose-comparison control, and usually includes randomization and blinding of subjects or investigators, or both."

In an active treatment concurrent controlled trial, the test drug is compared with known effective therapy, typically used when "the condition treated is such that administration of placebo or no treatment would be contrary to the interest of the subject. An active treatment study may include additional treatment groups, however, such as a placebo control or a dose-comparison control. Active treatment trials usually include randomization and blinding of subjects or investigators, or both. If the intent of the trial is to show similarity of the test and control drugs, the report of the study should assess the ability of the study to have detected a difference between treatments. Similarity of test drug and active control can mean either that both drugs were effective or that neither was effective. The analysis of the study should explain why the drugs should be considered effective in the study, for example, by reference to results in previous placebo-controlled studies of the active control drug."

A dose-comparison concurrent controlled trial compares at least two doses of the drug. "A dose-comparison study may include additional treatment groups, such as placebo control or active control. Dose-comparison trials usually include randomization and blinding of subjects or investigators, or both."

Where objective measurements of effectiveness are available and placebo effect is negligible, the test drug could be compared with no treatment, referred to as a no-treatment concurrent controlled trial. "No treatment concurrent control trials usually include randomization."

7.9   Clinical Supplies

Many of the issues of clinical trials specific to suspensions involve the clinical supplies used in such studies. The general topic of clinical supplies has been reviewed extensively (Monkhouse and Rhodes 1998).

7.9.1   Content Uniformity

Content uniformity is an important issue with regard to suspension formulations. Though a lack of content uniformity could be a quality issue (e.g., Knight and Luken 1977), variations in content uniformity are commonly due to the settling of the suspension. In clinical trials with suspension drug product, sponsors can conduct studies prior to trials to ensure that merely instructing the subject to shake well, or vigorously, will afford proper resuspension prior to administration. Clinical supplies of suspensions should be labeled such that the patients know to shake the product before dosing. Obviously, suspensions which can be easily resuspended or which do not settle are preferable.

The U.S. FDA requires that nasal sprays and inhalation products, many of which are suspensions, meet strict requirements for spray content uniformity with regard to the number of sprays from the beginning to end of an individual container and over a rigorous range of storage conditions. Also, any recommendations for shaking suspension products should be included, if warranted, on the label (Guidance for Industry, Nasal Spray and Inhalation Solution, Suspension, and Spray Drug Products 2002).

7.9.2   Taste

Taste of suspensions might affect blinding or compliance. Some sponsors intently examine tastes prior to embarking upon clinical trials. Steele et al. (1997) studied the palatability of 22 antimicrobial suspensions by using five independent categories for scoring: appearance, smell, texture, taste, and aftertaste. They found that some pediatric antibiotic suspensions had such unpalatable tastes that they might jeopardize compliance. Antibiotics judged to be so unpalatable as to potentially jeopardize compliance were dicloxacillin, oxacillin, erythromycin/sulfisoxazole, and cefpodoxime. Among the penicillins, amoxicillin and ampicillin were preferred. Within the macrolide class, azithromycin was slightly superior to erythromycin and clarithromycin. Cephalosporins were ranked quite high, the best being loracarbef, cefadroxyl, cefprozil, and cefixime. In a separate study, Sjoqvist et al. (1993) were able to significantly improve the taste of remoxipride by conversion from a solution to an oral suspension with incorporation of sweeteners and flavoring.

7.9.3   Blinding

7.9.3.1   Regulatory Requirements for Blinding

Regulatory agencies require that the clinical trials conducted to support the approval of a drug product must be well-controlled. 21CFR314.126 specifies that adequate measures need to be taken to permit a valid comparison of drug effect including minimization of bias on the part of subjects, observers and analysts of the data. These requirements can be met through "blinding" of the investigational products in a clinical trial. When the identities of the investigational products and placebo are not revealed, the study is referred to as "blinded." The type of blinding used in a clinical trial should be clearly described in the protocol and the eventual clinical study report.

7.9.3.2   Types of Blinding

The FDA guidance documents for various types of clinical trials frequently address blinding. The guidance for clinical trials for the treatment of cutaneous ulcers states that "blinding of the subjects and investigators to the assigned treatment reduces bias." (Guidance for Industry: Chronic Cutaneous Ulcer and Burn Wounds – Developing Products for Treatment 2006). The guidance for clinical trials of antiretroviral drugs states that "blinding reduces biases resulting from differences in management, treatment, or assessment of subjects arising from investigator or subject knowledge of the randomized treatment" (Guidance for Industry, Antiretroviral Dugs Using Plasma HIV RNA Measurements-Clinical Considerations for Accelerated and Traditional Approval 2002). Unsuccessful blinding can cause a clinical trial to fail. For example, Rasmussen et al. (1993) reported that a study of weight loss was compromised by subjects' ability to discern cimetidine from placebo suspension.

As thoroughly discussed by Monkhouse and Rhodes (1998), there are several options regarding the identification of investigational products used in clinical trials. (Table 7.1) To blind a study, all physical aspects of the investigational products should

Table 7.1 Examples of blinding of investigational products used in clinical trials

Trial blinding description	Patient blinded	Investigator blinded	Sponsor blinded
Open-label	No	No	No
Single-blinded (or masked)	Yes	No	No
Double-blinded (or masked)	Yes	Yes	Yes
Observer (or investigator)-blinded (or masked)	Yes	Yes	Usually
Double dummy (in order to blind dissimilar dosage forms, investigational product or placebo are given to each patient, e.g., active suspension + dummy tablet or dummy suspension + active tablet)	Yes	Yes	Usually

be identical. For suspensions, the packaging of both the bottle and the outer containers for the test, reference and placebo products should be identical in appearance to maintain adequate blinding of all subjects and evaluators. Neither the subject nor the investigator should be able to identify the treatment. Regarding physical properties, Monkhouse and Rhodes (1998) clearly describe that blinding should take into account the five senses: sight (aspects of packaging, labeling, markings, size, shape, opacity, color and color changes with time), smell (odor), sound, taste and touch (aspects of route of administration, coatings, isotonicity (osmolality) and viscosity). Also, the dosage intervals and quantity of dosage form administered should be blinded.

Open-Label and Single-Blinded Trial Design

Open-label trials are usually not preferred by the regulatory agencies because of the potential for bias on the part of the subject, investigator or sponsor. An example of an open-label trial design that might be acceptable in some cases includes phase I trials where the sponsor might be evaluating safety only. Like open-label trials, single-blinded trials are also not preferred by regulatory agencies because of the obvious potential for bias.

Double-Blinded Trial Design

By far, the most desirable design for clinical trials for regulatory and medical journal publication acceptance is the double-blinded design. In this case, both the subject and investigator are blinded. Even though the term triple-blinded more clearly delineates the blinding of the sponsor, blinding of the sponsor is usually assumed when a study is referred to as double-blinded. For example, the FDA recommends, that all acute otitis media trials "be double-blinded for study therapy and assessment of outcome unless there is a clearly compelling reason why this cannot be done. Subject reported outcome endpoints are rarely convincing without double-blinding." (Guidance for Industry Acute Bacterial Otitis Media: Developing Drugs for Treatment, Draft Guidance 2008). Even if study supplies cannot be perfectly double-blinded due to physical form, identical packaging can help. For example, in a study of brinzolamide/timolol suspension versus dorzolamide/timolol solution, blinding was achieved by using opaque bottles, thus requiring a sterile transfer of the solution to the same opaque bottles that are used to store the suspension (Manni et al. 2009).

Blending a single-blind phase and a double-blind phase in the same trial is common and acceptable for some types of clinical studies, especially when trying to screen out subjects that might respond to a placebo. For example, the FDA allergic rhinitis guidelines recommend a placebo run-in followed by randomized, double-blinded treatment with the investigational product or placebo (Draft Guidance for Industry, Allergic Rhinitis: Clinical Development Programs for Drug Products 2000). The run-in phase would represent the single-blind portion of the study, while the randomized treatment would represent the double-blinded portion of the trial. Kim et al. (2007) successfully utilized this approach in a suspension nasal spray trial.

Observer-Blinding or Investigator-Blinding Trial Design

Sometimes, double-blinding can be impossible to achieve in suspension drug product trials. One way of handling blinding or masking of a study containing test articles that do not appear identical, e.g., a suspension versus a solution or other dosage form, is to employ observer-blinded, or investigator-blinded, assessment of the subjects. In this way, the investigator who makes clinical assessments is not informed as to the identity of the investigational drug product that the subject is taking by either the subject or other study personnel. The clinical study coordinator is aware of the investigational drug product to which the subject is assigned, and therefore can instruct the subject on correct dosage regimen and other relevant aspects. In these cases, the randomized subject and their caretaker are instructed by the unblinded study coordinator to refrain from discussing any details about the drug they were taking with the investigator or other blinded study staff, unless there is a safety issue. Observer-blinding is acceptable to regulatory agencies in certain situations. As an example, in the FDA guidance for clinical trials on ketoconazole shampoo, it is recommended that in clinical trials where treatment arms cannot be properly masked, assessment by a third party masked evaluator is recommended. (Draft guidance on ketoconazole shampoo 2008).

As an example of observer-blinding, Pichichero et al. (2008) conducted a pharmacodynamic analysis and a clinical trial of amoxicillin sprinkle administered once daily for 7 days compared to penicillin V potassium administered four times daily for 10 days in the treatment of tonsillopharyngitis due to *Streptococcus pyogenes* in children. Saez-Llorens et al. (2005) utilized observer-blinding in a study of subjects dosed once daily with gatifloxacin suspension versus those dosed twice daily with amoxicillin suspension.

Roland et al. (2008) successfully compared effectiveness of an antibiotic suspension versus another antibiotic suspension plus an oral antibiotic. The study coordinator instructed subjects on use, care and administration of the investigational products and the subjects and staff were told not to discuss such with the investigator who was blind to the treatment of the subjects being assessed. Similarly, in a study of an ear drop suspension versus an oral antibiotic suspension, Dohar et al. (2006) also successfully employed observer-masking.

Trials comparing suspension to solution products can be problematic from a blinding perspective if the formulation can be observed in a transparent package. Roland et al. (2003) employed observer-blinding so that the investigational products were dispensed by the study coordinator in opaque containers, and the investigator was not informed of the treatment in conjunction with their assessment of the subject.

In the case of investigational products that utilize commercial packaging that is different, or differing dosage regimens, various approaches have been utilized. Patel et al. (2007) in a study of nasal sprays in containers from different manufacturers, one of which was a suspension, employed the use of aluminum foil pouches that shrouded the appearance of the different nasal spray products, thereby, allowing a double-blinded design. Alternatively, Roland et al. (Roland et al. 2004a, b; Roland et al. 2007) successfully employed observer-masking in a study of different suspension ear drops with dissimilar dosage regimens.

Double-Dummy Trial Design

If a test drug product is a suspension and a comparator is a tablet, the subject obviously knows which treatment they are taking. In such cases, a double-dummy approach could be employed to provide blinding. In the double-dummy design, one group of subjects would receive the test suspension and a placebo tablet while the other group of subjects would receive a placebo suspension and an active tablet (Monkhouse and Rhodes 1998). For example, Lanier et al. (2002) conducted a study of suspension nasal spray, with or without oral antihistamine, and with or without antihistamine eye drops, using a double-dummy study design.

7.10 Conclusion

Clinical trials of a pharmaceutical product, regardless of the pharmaceutical form, are generally required by the regulatory agencies around the world as the basis for proof of safety and efficacy in the drug product approval process. Every such clinical study must have a clinical protocol describing all aspects of the conduct of the trial, as well as a CIB that provides supporting information for evaluation by the investigator and IECs. All clinical trials should be approved by an IEC.

Some of the most important aspects of clinical trials specific to suspension drug products are concerned with the clinical supplies. The investigational product must exhibit content uniformity to ensure appropriate pharmacological effect. Well-controlled trials are typically blinded and there are several designs available for blinding. Blinding is particularly challenging for suspension drug products since many approved comparators cannot be easily modified to resemble the investigational suspension. However, with an appropriate clinical study design, challenges of blinding of suspension drug product trials can be overcome, thereby allowing appropriate assessment of safety and efficacy that is acceptable for registration of suspension drug products around the world.

References

21CFR314.126: Code of Federal Regulations Title 21, Volume 5, Food and drugs, Chapter I – Food and Drug Administration, Dept. Health and Human Services, Subchapter D-Drugs for Human Use, Part 314.126, Adequate and well-controlled studies, 2008, available at: http://www.access-data.fda.gov/scripts/cdrh/cfdocs/cfcfr/CFRSearch.cfm?fr=314.126, accessed June 2, 2009.

21CFR56: Code of Federal Regulations Title 21, Food and drugs, Chapter I – Food and Drug Administration, Dept. Health and Human Services, Subchapter A, General Part 56 Institutional review boards, 2008, available at: http://www.accessdata.fda.gov/scripts/cdrh/cfdocs/cfcfr/CFRSearch.cfm?CFRPart=56&showFR=1, accessed 1 June 2009.

21CFR50: Code of Federal Regulations Title 21, Food and drugs, Chapter I – Food and Drug Administration, Dept. Health and Human Services, Subchapter A, General Part 50 Protection of Human Subjects, 2008, available at http://www.accessdata.fda.gov/scripts/cdrh/cfdocs/cfcfr/CFRSearch.cfm?CFRPart=50, accessed 1 June 2009.

Chin, R. and Lee, B. Y. 2008. Principles and Practice of Clinical Trial Medicine. Oxford, UK: Elsevier, Inc.

ClinicalTrials.gov available at: http://clinicaltrials.gov/, accessed March 26, 2009.

Dohar, J., Giles, W., Roland, P., Bikhazi, N., Carroll, S., Moe, R., Reese, B., Dupre, S., Wall, M., Stroman, D., McLean, C., Crenshaw, K. 2006. Topical ciprofloxacin/dexamethasone superior to oral amoxicillin/clavulanic acid in acute otitis media with otorrhea through tympanostomy tubes. Pediatrics 118(3):e1–e9.

Draft Guidance for Industry, Allergic rhinitis: clinical development programs for drug products, U.S. Department of Health and Human Services, Food and Drug Administration Center for Drug Evaluation and Research, 2000, available at: http://www.fda.gov/downloads/Drugs/GuidanceComplianceRegulatoryInformation/Guidances/ucm071293.pdf, accessed May 25, 2009.

Draft Guidance on Ketoconazole Shampoo, U.S. Department of Health and Human Services, Food and Drug Administration Center for Drug Evaluation and Research, 2008, available at: http://www.fda.gov/cder/downloads/Drugs/GuidanceComplianceRegulatoryInformation/Guidances/ucm08272.pdf, accessed Sept. 25, 2009.

European Medicines Agency (EMEA) available at: http://www.emea.europa.eu/, accessed May 26, 2009.

Friedman, L. M., Furburg, C. D., DeMets, D. L. 1998. Fundamentals of Clinical Trials. New York: Springer Science + Media LLC.

Grady, J. O. and Linet, O. I. 1990. Early Phase Drug Evaluation in Man. Boca Raton: CRC Press.

Guidance for Industry, Providing Clinical Evidence of Effectiveness for Human Drug and Biological Products, U.S. Dept. Health and Human Services, Food and Drug Administration, Center for Drug Evaluation and Research, available at: http://www.fda.gov/cder/guidance/1397fnl.pdf, accessed May 25, 2009.

Guidance for Industry: Providing Clinical Evidence of Effectiveness for Human Drug and Biological Products, U.S. Dept. Health and Human Services, Food and Drug Administration, Center for Drug Evaluation and Research, 2008, available at: http://www.fda.gov/downloads/Drugs/GuidanceComplianceRegulatoryInformation/Guidances/ucm078749.pdf, accessed May 26, 2009.

Guidance for Industry, Nasal Spray and Inhalation Solution, Suspension, and Spray Drug Products – Chemistry, Manufacturing, and Controls Documentation, U.S. Dept. Health and Human Services, Food and Drug Administration, Center for Drug Evaluation and Research, 2002, available at: http://www.fda.gov/cder/guidance/4234fnl.htm#P672_98417, accessed May 26, 2009.

Guidance for Industry: Chronic Cutaneous Ulcer and Burn Wounds – Developing Products for Treatment, U.S. Dept. Health and Human Services, Food and Drug Administration, Center for Drug Evaluation and Research, 5600 Fishers Lane, Rockville, MD, 2006, available at: http://www.fda.gov/cber/gdlns/ulcburn.pdf, accessed May 26, 2009.

Guidance for Industry Acute Bacterial Otitis Media: Developing Drugs for Treatment, Draft Guidance, U.S. Department of Health and Human Services, Food and Drug Administration Center for Drug Evaluation and Research, 2008, available at: http://www.fda.gov/cder/guidance/3892dft.pdf, accessed May 25, 2009.

Draft Guidance on Ketoconazole Shampoo, U.S. Department of Health and Human Services, Food and Drug Administration Center for Drug Evaluation and Research, 2008, available at: http://www.fda.gov/cder/guidance/bioequivalence/recommendations/Ketoconazole_Sham_%2019927_RC7-08.pdf, accessed May 25, 2009.

ICH Guideline for Good Clinical Practice E6 R1, 1996, available at: http://www.ich.org/LOB/media/MEDIA482.pdf, accessed june 1, 2009.

ICH Note for Guidance on Non-clinical Safety Studies for the Conduct of Human Clinical Trials and Marketing Authorization for Pharmaceuticals, CPMP/ICH/286/95, 2008, available at: http://www.emea.europa.eu/pdfs/human/ich/028695endraft.pdf, accessed May 25, 2009.

International Conference on Harmonisation (ICH) of Technical Requirements for Registration of Pharmaceuticals for Human Use. available at: http://www.ich.org/cache/compo/276-254-1.html, accessed May 26, 2009.

Japan Pharmaceutical and Medical Devices Evaluation Center (PMDEC) available at: http://www.nihs.go.jp/pmdec/outline.htm, accessed May 26, 2009.

Kim, K., Quesada, J., Wingertzahn, M. A., Szmaydy-Rikken, N., Darken, P., Shah, T. 2007. Coadministration of intranasal ciclesonide and inhaled fluticasone propionate-salmeterol in perennial allergic rhinitis (PAR). J Allergy Clin Immunol 119(1, Supplement 1):S143.

Knight, P. A. and Lucken, R. N. 1977. Experience with the quality control testing of C. parvum suspension for clinical trial. Dev Biol Stand 38:51–58.

Lanier, B. Q., Abelson, M. B., Berger, W. E., Granet, D. B., D'Arienzo, P. A., Spangler, D. L., Kägi, M. K. 2002. Comparison of the efficacy of combined fluticasone propionate and olopatadine versus combined fluticasone propionate and fexofenadine for the treatment of allergic rhinoconjunctivitis induced by conjunctival allergen challenge. Clin Ther 24(7):1161–1174.

Machin, D., Day, S., Green, S. Eds. 2006. Textbook of Clinical Trials. West Sussex, England: Wilcy.

Manni, G., Denis, P., Chew, P. et al. 2009. The safety and efficacy of brinzolamide 1%/timolol 0.5% fixed combination versus dorzolamide 1%/timolol 0.5% in patients with open-angle glaucoma or ocular hypertension. J Glaucoma 18(4):293–300.

Meinert, C.L. and Tonascia, S. 1986. Clinical Trials Design, Conduct and Analysis. New York: Oxford University Press.

Monkhouse, D. C. and Rhodes, C. T., Eds.1998. Drug Products for Clinical Trials: An International Guide to Formulation, Production, Quality Control. New York: Marcel Dekker, Inc.

Patel, D., Garadi R., Brubaker, M., Conroy, P., Kaji, Y., Crenshaw, K., Whitling, A., Wall, G. M. 2007. Onset and duration of action of nasal sprays in seasonal allergic rhinitis patients: olopatadine hydrochloride versus mometasone furoate monohydrate. Allergy Asthma Proc 28:592–599.

Pichichero, M. E., Casey, J. R., Block, S. L., Guttendorf, R., Flanner, H., Markowitz, D., Clausen, S. 2008. Pharmacodynamic analysis and clinical trial of amoxicillin sprinkle administered once daily for 7 days compared to penicillin V potassium administered four times daily for 10 days in the treatment of tonsilllopharyngitis due to *Streptococcus pyogenes* in children. Antimicrob Agents Chemother 52(7):2512–2520.

Pocock, S. J. 1983. Clinical Trials, a Practical Approach. New York: Wiley.

Rasmussen, M. H., Andersen, R., Breum, L., Gotzsche, P. C., Hilsted, J. 1993. Cimetidine suspension as adjuvant to energy restricted diet in treating obesity. BMJ 306(6885):1093–1096.

Roland, P. S., Anon, J. B., Moe, R. D., Conroy, P. J., Wall, G. M., Dupre, S. J., Krueger, K. A., Potts, S., Hogg, G., Stroman, D. W. 2003. Topical ciprofloxacin/dexamethasone is superior to ciprofloxacin alone in pediatric patients with acute otitis media and otorrhea through tympanostomy tubes. Laryngoscope 113(12):2116–2122.

Roland, P. S., Belcher, B. P., Bettis, R., Makabale, R. L., Conroy, P. J., Wall, G. M., Dupre, S., Potts, S., Hogg, G., Weber, K., The CiproHC Study Group. 2008. A single topical agent is clinically equivalent to the combination of topical and oral antibiotic treatment for otitis externa. Am J Otolaryngol Head Neck Med Surg 29(4):255–261.

Roland, P. S., Kreisler, L.S., Reese, B., Anon, J. B., Lanier, B., Conroy, P. J., Wall, G. M., Dupre, S. J., Potts, S., Hogg, G., Stroman, D. W. 2004b. Topical ciprofloxacin/dexamethasone otic suspension is superior to ofloxacin otic solution in the treatment of children with acute otitis media with otorrhea through tympanostomy tubes. Pediatrics 113(1):e40–e46.

Roland, P. S., Pien, F. D., Schultz, C. C., Henry, D. C., Conroy, P. J., Wall, G. M., Garadi, R., Dupre, S. J., Potts, S. L., Hogg, L. G., Stroman, D. W. 2004a. Efficacy and safety of topical ciprofloxacin/dexamethasone versus neomycin/polymyxin B/hydrocortisone for otitis externa. Curr Med Res Opin 20(8):1175–1183.

Roland, P. S., Younis, R., Wall, G. M. 2007. A comparison of ciprofloxacin/dexamethasone with neomycin/polymyxin/hydrocortisone for otitis externa pain. Adv Ther 24(3):671–675.

Saez-Llorens, X., Rodriguez, A., Arguedas, A., Hamed, K. A., Yang, J., Pierce, P., Echols, R. 2005. Randomized, investigator-blinded, multicenter study of gatifloxacin versus amoxicillin/clavulanate treatment of recurrent and nonresponsive otitis media in children. Pediatr Infect Dis J 24(4):293–300.

Sjoqvist, R., Graffner, C., Ekham, I., Sinclair, W., Woods, J. P. 1993. In vivo validation of the release rate and palatability of remoxipride-modified release suspension. Pharm Res 10(7): 1020–1026.

Steele, R. W., Estrada, B., Begue, R. E., Mirza, A., Travillion, D. A., Thomas, M. P. A. 1997. Double-blind taste comparison of pediatric antibiotic suspensions. Clinical Pediatrics 36(4):193–199.

Sweetman, S. C, Ed. 2009. Martindale: The Complete Drug Reference 36. London: Pharmaceutical Press.

United States Food and Drug Administration (FDA) available at: http://www.fda.gov/, accessed May 26, 2009.

World Medical Association Declaration of Helsinki, Ethical principles for Medical Research Involving Human Subjects, 2008, available at: http://www.wma.net/e/policy/pdf/17c.pdf, accessed June 1, 2009.

Chapter 8
Scale Up and Technology Transfer of Pharmaceutical Suspensions

Yashwant Pathak and Deepak Thassu

Abstract Scaling up of suspension dosage form is considered a challenge as pharmaceutical suspensions are thermodynamically unstable. During the scale-up process, the effect of operational variables on the dispersion process, and physical properties affecting overall attributes need systematic evaluation. Though the scale up process can be studied using dimensional analysis, mathematical modeling, and computer simulation, most of the work in the field of suspension manufacturing still depends on empirical studies and the principles of geometric similarities. Various aspects of the scale up process, technology transfer, and process analytical technology (PAT) applications in suspension manufacturing are discussed.

8.1 Basics Issues of Suspension Formulation

The development of newer drugs and dosage forms is a sign of the contribution of the scientific community toward providing treatment to the people at large for the diseases and discomforts. Most of the newer drugs are poorly water soluble with encouraging in vitro properties, but when it comes to in vivo evaluations, the results are disappointing. The possible reasons are:

1. Poor solubility of the drugs
2. Poor absorption, rapid degradation, and lamination (peptides and proteins) resulting in insufficient concentration
3. Drug distribution to other tissues with high drug toxicities (especially cancer drugs)
4. Fluctuations in plasma levels owing to unpredictable bioavailability

Y. Pathak (✉)

Department of Pharmaceutical Sciences, College of Pharmacy, Sullivan University, Louisville, KY, 40205, USA

e-mail: ypathak@sullivan.edu

D. Thassu

Vice President Pharmaceutical Development, Pharmanova Inc,

Victor, NY 14564

A.K. Kulshreshtha et al. (eds.), *Pharmaceutical Suspensions:* From Formulation Development to Manufacturing, DOI 10.1007/978-1-4419-1087-5_8, © AAPS 2010

The enhancement of oral bioavailability of poorly soluble drugs remains one of the most challenging aspects of drug development (Karanth et al. 2006). Formulation of these drugs in suspensions with desired release profiles is preferred (Caballero and Duran 2000). The dispersion system where one or more dispersed ingredients have therapeutic action is termed a pharmaceutical suspension. These are dispersions of fine suspended in liquid. The solid phase dispersed is termed as fine to coarse particles depending on the application of the pharmaceutical suspensions. Anything below 1 μm is termed as colloidal dispersion and is out of the purview of this discussion. Suspensions are thermodynamically unstable systems even though the ingredients might be chemically stable (Lu 2007). As the solid particle size decreases in the suspension the interfacial energy increases which makes the whole system unstable. This phenomenon further initiates processes of aggregation and particle sedimentation (Nash 1996). Some of the suspension properties, interfacial area associated with suspended particles, polymorphic forms, and growth of large crystals due to Oswald ripening, affects physical stability as well as the bioavailability of suspensions (Lu 2007).

Organoleptic properties of suspensions are getting more and more attention as these products are being developed for pediatric and geriatric use (Gallardo et al. 2000; Sohi et al. 2004). Currently the fast developing area in suspension technology is nano suspensions, where particle size of the dispersed phase is in the nanometer range (Rabinow 2005, 2007; Gruverman 2003, 2004, 2005). Figure 8.1 shows equipment which can be used for manufacturing of nano suspensions. Significant process work would be needed to scale up the nano suspension pharmaceuticals.

Understanding dispersion system technology from the commercial manufacturing perspective with focus on scale up and technical transfer from bench scale (1 l) to large scale (up to 15,000 l) is critical (Table 8.1). Manufacturing and quality control of oral suspensions have presented dilemmas to the industry. There are issues which have led to recalls (FDA Document 2006). These include microbiological, potency, and stability problems. Suspensions being liquid products are prescribed for pediatric and geriatric use; therefore, defective dosage forms can pose greater safety risks (Caballero and Duran 2000).

Assurance of suspension stability is the area that has presented a number of problems. For example, there have been a number of recalls of the vitamins with fluoride oral liquid products because of vitamin degradation. Drugs in the Phenothiazine class, such as Perphenazine, Chlorpromazine, and Promethazine have shown evidence of instability. Suspensions are liquid products and tend to interact with closure systems during stability studies; therefore, studies should be designed to determine whether contact of the drug product with the closure system affects product integrity (FDA Document 1999). Moisture loss (which can cause the remaining contents to become super potent) and microbiological contamination are other problems associated with inadequate closure systems and LDPE containers. Pharmaceutical suspensions can be used for parenteral, topical, or oral applications (Nash 1996). The scaling up and manufacturing of suspensions for topical, oral, or parenteral suspensions is significantly different with requirements being most stringent for the parenteral suspensions (Gallardo et al. 2000).

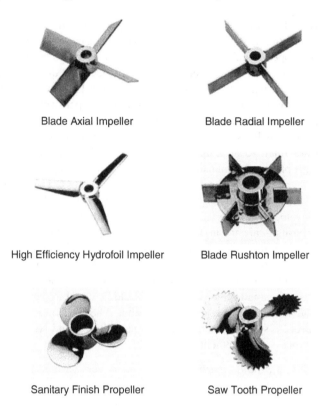

Blade Axial Impeller

Blade Radial Impeller

High Efficiency Hydrofoil Impeller

Blade Rushton Impeller

Sanitary Finish Propeller

Saw Tooth Propeller

Fig. 8.1 Various impellers used in the manufacturing of suspensions

Table 8.1 Production scale up for suspensions

S. no.	Stages of scale up and production	Quantity in liters
1	Laboratory scale	1–15 Liters
2	Phase I study	15–30 Liters
3	Phase II study	1,500 Liters
4	Phase III study	2,000 Liters
5	Validation/production batches	15,000 Liters

8.2   Process Development: Key Issues and Concepts

During the process development of suspensions there is a need for integration of formulator and chemical engineer skills to produce an elegant product with specified properties (Block 2002; Wibowo and Ng 2004). The scale up process should determine the operating conditions applicable to large scale production batches with the goal of obtaining products of the same quality developed on previously optimized laboratory scale experiments (Table 8.1 shows the approximate scale up ranges for phase trials and manufacturing). Scale up process of a suspension, which consists of two immiscible phases, solids and liquids, is a difficult process due to the dynamic behavior of the two individual phases. During early product development, generally, operational variables of powder properties (to be suspended in the liquid phase) are not well established. It is recommended that dimensional analysis, mathematical modeling, and computer simulation (the latter two are less used in pharmaceutical suspension manufacturing) should be performed (Levin 2005). Dimensional analysis (Zlokarnik 2002) is an algebraic treatment of variables affecting a process. The analysis does not result in a numerical equation, but the experimental data are fitted to an empirical process equation that results in scale up (Block 2002). It produces dimensionless numbers and deriving functional relationships among them will completely characterize the process. The analysis can be applied even when the equations governing the process are not known.

Previous development experience is helpful in handling the numerous problems encountered during scale up in formulation development (Newton et al. 1995). Most of the work in the field of dispersing of small particles still depends on the trial and error method and the principles of geometric similarities. The latter describes the interrelationships among system properties on scale up; thus the ratio of some variables in small scale equipment should be equal to that of similar variables in equivalent large scale equipments (Mehta 1988). The next approach, product oriented process synthesis and development (Wibowo and Ng 2004), involves four steps of process development:

1. *Identification of product quality factors*: this covers various factors affecting the properties of the suspension product such as particle size, distribution of the particles, density of the particles, suspending agents, and desired viscosity of the product leading to adequate stability over time and so on.
2. *Product formulation*: to get a stable formulation and palatable products various adjuvants are necessary to be incorporated in product formulation including the flavors colors suspending agents and viscosity builders. The properties and characteristics will decide equipment selection as well as process development.
3. *Flow sheet synthesis*: the process involves the transfer of materials in a sequential pattern depending on the unit operations involved in the manufacturing of the suspensions; hence the flow sheet describes the flow pattern of the material and equipments used.
4. *Product and process evaluation*: finally, in process and product controls, parameters are used to ensure the appropriate desired, stable product.

Specifications to manufacture suspensions include assay, microbial limits, particle size distribution of the suspended drug, viscosity, pH, and dissolution. Viscosity can be important from a processing aspect to minimize segregation (Table 8.2). In few cases, viscosity has been shown to have an impact on bio-equivalency. The pH of suspension formulation affects the effectiveness of preservative systems and may even alter the drug in solution. Particle size distribution is a critical attribute and should be monitored as quality control parameters. Any change in particle size distribution during the shelf life of suspension can have a negative impact on the dissolution kinetics. Particle size, its distribution and change over shelf life is also

Table 8.2 Test methods and check list for suspension performance and stability testing:

S. no.	Test methods for suspensions	Check list for suspension Performance and stability testing
1	Photo microscopic techniques	Appearance
2	The coulter counter	Sedimentation volume and redispersibility
3	Graduated cylinders for sedimentation studies	Sedimentation rate
4	Brookefield Viscometers	Viscosity
5	Specific gravity measurements	pH, density and specific gravity
6	Aging tests	Color, Taste and Odor (organoleptic properties)
7	Zeta potential	Drug content uniformity/Release profiles
8	Aggregation kinetics	Freeze thaw cycling for stability studies

critical for extended release suspensions. Particle size would be a critical parameter for micro and nano suspensions,

8.3 Factors Critical for Process Development

Some of the factors which need to be considered while developing and scaling up a manufacturing process for suspensions are:

1. Increase productivity and selectivity through intensification of intelligent operations and multi scale approach to process control.
2. Design novel equipment based on scientific principles and new production methods; process intensification.
3. Extend chemical engineering methodology to product design and product focused processing.
4. Implement multi scale application of computational chemical engineering modeling and simulation to real life situations from the molecular scale to manufacturing scale (Charpentier 2003).

The major unit operation involved in suspension manufacturing is mixing of solids or suspending the solids in liquid phase. Mixing requirements are dictated by the nature of the compounds being mixed (Joseph 2005). Mostly these are structurally and functionally very complex compounds and there is a need to satisfy the controls imposed by the regulatory bodies on pharmaceutical products (Paul et al. 2004). The majority of mixing operations are carried out in batch or semi continuous modes rather than continuous mixing modes (Fig. 8.2 shows various impellers used for suspension manufacturing). Often multiple tasks are carried out in the same vessel (e.g. chemical reaction, heat and mass transfer, concentration, crystallization) requiring a fine balance in design and operation (Fig. 8.3 shows the typical mixing kettles used for scale up studies, Fig. 8.4 shows a typical continuous manufacturing jacketed kettles, and Fig. 8.5 shows batch production using steel tanks). Poor mixing can result in loss of product yield and purity, excessive crystalline fines or oiled out material and/or

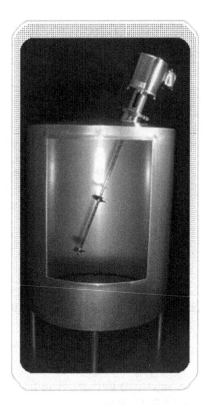

Fig. 8.2 More economical and compact than other sanitary mixers, the modular design can accommodate from 1/8 to 1 HP motors. The mixer is suitable for batch sizes between 20 and 250 l and is currently used in pharmaceutical, food and dairy, and chemical processing applications. It can be used for lab scale to scale up studies

intractable foam (Fig. 8.6 shows the new mixing concept of master plant guarantees for highest product quality and extreme shortening of production time).

Attention needs to be paid to both micro and macro mixing requirements, particularly for suspension manufacturing where competitive chemical reactions and/ or crystallization are involved. Glass-lined mixing equipments are often used in pharmaceutical suspension manufacturing especially in parenteral preparations.

Mixing phenomena and the principles involved in deciding the mixing equipment selection, mixing of viscous and non Newtonian fluids, and transport phenomena involved in suspension and semisolid production is discussed by Block (2002) in detail in his chapter on non-parenteral liquids and semisolids in pharmaceutical process scale up. Dispersion systems often need particle size reduction which can be achieved by compression, impact, attrition or cutting or shear. Equipments for particle size reduction include crushers, grinders, ultra fine grinders, and knife cutters. In suspension production particle size reduction can be an integral part of the processing or one can reduce the particle size of the particles and then

Fig. 8.3 Typical jacketed vessel for suspension manufacture with all the piping and processing alternative, it can be used for continuous or batch processing of suspensions

Fig. 8.4 Typical batch processing of suspensions mixing tanks

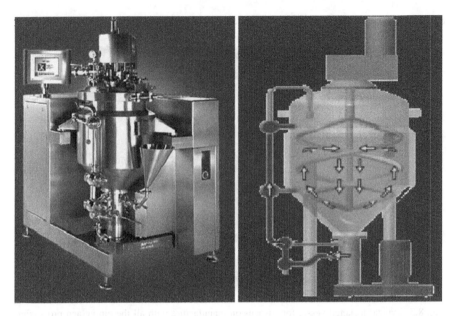

Fig. 8.5 The new mixing concept of master plant MP 2500, fro suspensions and emulsions, with DBI module, pumping, suction, mixing, dispersing and cleaning guarantees for highest product quality and extreme shortening of production time (www.ikausa.com.)

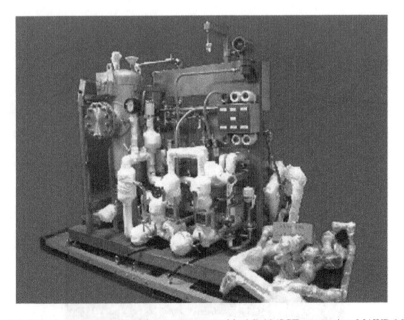

Fig. 8.6 Submicron media mill, high pressure supercritical fluid (SCF) processing. MAWP 6,000 psi @ 300 F. Reduces active pharmaceutical ingredients (API) to submicron size. Designed for class 10,000 area for flexibility and adherence to cGMP. Internal product wetted surfaces have 25–30 micro inch. RA finish. External surfaces have 50 micro inch Ra finish. Entire process is designed to accommodate four (4) skids on wheels. This can be used for Nano-suspensions with dispersed particle size in nanometer range

suspend them in the vehicle. Block (2002) has given in detail the mathematical equations for particle size reduction processing for suspensions as well as emulsions.

Another important unit operation involved in manufacturing of the suspensions, specially the parenteral suspensions, is freeze drying. The purpose of freeze drying is to remove a solvent (usually water) from dissolved or dispersed solids. Freeze drying involves four stages – freezing of the product, sublimation of the solvent under vacuum, applying heat to accelerate the sublimation, and finally condensation to complete the separation process. Though it appears to be simple, freeze drying can often be complicated in practice owing to the different properties of the product. The major factors affecting the efficiency of the freeze drying process are

1. Suspended solvents
2. Rate of freezing
3. Vacuum control
4. Rate of drying/product temperature
5. Residual moisture content
6. Storage temperatures and atmosphere after drying, and
7. Rehydration process.

Freeze drying is an excellent method for providing dry powders for re-suspension offering better stability of the product. The key benefits include retention of morphological, biochemical, and immunological properties; high viability and activity levels of bioactive materials, lower temperature, and oxygen and shear conditions used in the process offer better stability for the drug, retention of structure, surface area and stoichiometric ratios, high yield, long shelf life and reduced storage weight, shipping and handling comfort.

Several patents discuss the freeze drying process and also applying the identification marks on the freeze dried products (US Patent 5,457895, 3,534,440, 2,645,852 and 510,453). These inventions comprise various techniques used to emboss the freeze products with an identifying mark such as the manufacturer's logo, medicinal component strength or other information related to the product. The desired identification mark is first embossed on the base of the container such as a blister packet, and then the liquid suspension is filled in the container and freeze dried therein. These techniques are widely used for parenteral suspensions as well as dry powders to be re-suspended as liquid after addition of water or suitable solvent. Liu (2006) has provided an excellent overview of the freeze drying applications.

8.4 Process Technology: Equipment Design, Rating and Selection and Check List for Suspension Performance and Stability

Mixing scale up should begin with the first laboratory testing, because waiting until the last minute for scaling up will be too late. If the observations are noted properly during the laboratory experiments, it will help significantly during the scaling up.

Equipment used for batch mixing of oral solutions and suspensions is relatively basic. Generally, these products are formulated on weight basis with the batching tank on load cells so that a final Q.S. can be made by weight. Volumetric means, such as using a dip stick or line on a tank, have been found to be inaccurate. Equipment should be of sanitary design. This includes sanitary pumps, valves, flow meters, and other equipment which can be easily sanitized. Ball valves, packing in pumps and pockets in flow meters have been identified as sources of contamination.

In order to facilitate cleaning and sanitization, manufacturing and filling lines should be identified and detailed in drawings and SOPs. In some cases, long delivery lines between manufacturing areas and filling areas have been a source of contamination. Also, SOPs, particularly with regard to time limitations between batches and for cleaning have been found deficient in many manufacturers. Review of cleaning SOPs, including drawings and validation data with regard to cleaning and sanitization is very important.

The design of the batching tank with regard to the location of the bottom discharge valve has also presented problems. Ideally, the bottom discharge valve is flush with the bottom of the tank. In some cases valves, including undesirable ball valves, have been found to be several inches to a foot below the bottom of the tank. In others, the drug or preservative was not completely dissolved and was lying in the "dead leg" below the tank with initial samples being found to be sub potent. For the manufacture of suspensions, valves should be flushed. The batching equipment and transfer lines need to be reviewed and observed constantly.

With regard to transfer lines, they are generally hard piped and easily cleaned and sanitized. In some cases, manufacturers have also used flexible hoses to transfer products. It is not unusual to see flexible hoses lying on the floor, thus significantly increasing the potential for contamination. Such contamination can occur by operators picking up or handling hoses, and possibly even placing them in transfer or batching tanks after they had been lying on the floor. It is also a good practice to store hoses in a way that allows them to drain rather than be coiled which may allow moisture to collect and be a potential source of microbial contamination. Manufacturing areas and operator practices, particularly when flexible hose connection is employed, need to be observed carefully. Another common problem occurs when manifold or common connections are used, especially in water supply, premix, or raw material supply tanks. Such common connections have been shown to be a source of contamination. Optimization of processes is ensured through multi-disciplinary implementation procedures, which match mixing and fluid dynamics to the process chemistry and mass transfer requirements.

Because scale-up is critical to the successful application of mixing technology in industrial processes, a range of geometrically similar reactors 0.2–2.7 m diameter can be used to establish correct scaling procedures. Novel sensors can be used and developed for this purpose. Computational modeling techniques are employed to rationalize experimental programs, describe the fluid dynamics of reactors and extend to a range of industrially relevant applications.

8.4.1   Batch Vs Continuous Pharmaceutical Operations

The manufacturing sector of the pharmaceutical industry is currently undergoing a radical change. The need to improve process and product quality while controlling overall costs is growing. Continuous processing offers a chance to achieve these goals. The implementation of process analytical technology (PAT) tools may further aid the continuous operations, as the tools give real time indication and assurance of the product quality throughout the process. A variety of batch processes are currently being evaluated for continuous production in suspension technology also. At present the batch process is mostly used for the manufacturing of suspensions; hopefully within a few years we may see some continuous manufacturing also.

In most cases, manufacturers will assay samples of the bulk solution or suspension prior to filling. A much greater variability has been found with batches that have been manufactured volumetrically rather than by weight. For example, one manufacturer had to adjust approximately 8% of the batches manufactured after the final Q.S. because of failure to comply with potency specifications. Unfortunately, the manufacturer relied solely on the bulk assay. After readjustment of the potency based on the assay, batches occasionally were found out of specification because of analytical errors. Table 8.2 gives the list of test methods for suspensions and the check list for suspension performance and stability testing (Nash 1996).

8.5   Technology Source and Process Intensification (PI)

The concept of PI was originally pioneered by Colin Ramshaw and his coworkers at ICI. They defined PI as reduction in plant size by at least a factor of 100. PI is a new approach to process and plant design, development, and implementation. Providing a chemical process, like mixing, with the precise environment it needs will result in better products and processes which are cleaner, smaller, and cheaper.

Some of the possibilities of PI which can be adapted to suspension manufacturing are

1. Converting a batch process into a continuous process,
2. Using intensive reactor technologies with high mixing and heat transfer rates (e g. Flex reactors, HEX reactors, www.bhrgroup.com) in place of conventional stirred mixing tanks.
3. Adopting the multidisciplinary approach considering opportunities to improve the process technology underlying the suspension chemistry of the process.
4. Plug and play process technology to provide flexibility in a multi product environment.

Some of the established advantages of PI are reduction in cost, reactor volumes energy usage, operating costs, and higher yield. Very little work has been done and reported applying the process intensification techniques in suspension manufacturing, but then due to competitive prices this approach may have some applications in the pharmaceutical scenario also.

8.6 Process Validation: Planning Coordination, Regulatory Guidelines, Documentation Overview, and Management of Product Process Validation

FDA defines validation as "There shall be written procedures for production and process control designed to assure that the drug products have the identity, strength, quality and purity, they purport or represented to possess." It further says "Establishing documented evidence which provides a high degree of assurance that a specific process will consistently produce a product meeting its predetermined specification and quality attributes" (FDA Guidelines general principles of validation, May 1987). The process validation package should provide

1. Installation Qualification (IQ) verifying that the equipment is installed in a proper manner.
2. Operational Qualification (OQ) verifying the performance of equipment for intended range of applications, and
3. Process Qualification (PQ) verifying the repeatability and consistency of producing the drug product.

Validation is a continuous and evolving process and needs to be given top priority during the production cycle and should be continuously updated. Maintenance, periodic calibrations, and adjustment of the equipments is continuously done, and is always under scrutiny and constant evaluation. There are some maintenance triggers such as

1. Loss of product quality.
2. Replacement of key components in the system.
3. Upgrades of software or hardware.
4. Change in personnel.

These often arise during manufacture and there is always a need to address these triggers preferably with a SOP in place to identify the triggers and solutions.

The FDA guidelines suggest having functional requirements and functional specifications. Functional requirements refer to defining what is to be done; the process should be stated succinctly with little or no reference to actual specific devices. The functional specifications document usually contains:

1. An introduction or purpose.
2. Intended audience.
3. Definitions.
4. Reference documentations.
5. Mechanical, electrical, computer and security requirements and specifications.

Based on process parameters, the range and tolerance of the equipment operation are defined in the functional specifications. There are three important aspects related to the process involved:

1. Identifying the process parameters.
2. Identifying the critical process parameters.
3. Identifying the tolerances.

As with other products, the amount of data needed to support the manufacturing process will vary from product to product. Development (data) should have identified critical phases of the operation, including the predetermined specifications that should be monitored during process validation. For example, for solutions the key aspects that should be addressed during validation include assurance that the drug substance and preservatives are dissolved. Parameters, such as heat and time should be measured. Also, in-process assay of the bulk solution during and/or after compounding according to predetermined limits are also important aspects of process validation. For solutions that are sensitive to oxygen and/or light, dissolved oxygen levels would also be an important test. Again, the development data and the protocol should provide limits.

As discussed, the manufacture of suspensions presents additional problems, particularly in the area of uniformity. Again, development data can address the key compounding and filling steps that assure uniformity. The protocol should provide for the key in-process and finished product tests, along with their specifications. For oral solutions, bio-equivalency studies may not always be needed. However, oral suspensions, with the possible exception of some of the antacids, OTC products, usually require a bio-equivalency or clinical study to demonstrate effectiveness.

8.7 Technology Transfer to Commercial Production: Documentation Package

Technology transfer can be defined as the systematic procedural documentation which results in transfer of the documented knowledge and experience gained during the development and scaling up process for commercialization leading to manufacturing of the final products. It will always involve a minimum of two stations, the sending (transferring) and the receiving station. Technology transfer will include:

1. Documentation transfer.
2. Demonstration of the ability of the transferring station to educate the receiving station properly to perform the desired task.
3. Demonstration of the ability of the receiving station to duplicate the task performed at desired efficiency.
4. Establishing the evaluation process to measure the success.
5. The process at each station that should satisfy all the concerned parties (departments) and the regulatory bodies also.

Technology transfer in the case of suspension manufacturing can happen from the laboratory to process development to manufacture. The manufacturing can involve one location or several locations within the country and also outsourcing to multiple countries. The requirements for different countries are different; if the product is sold in the USA or other countries, the local FDA requirements need to be met, hence the technology transfer package will have take into account all the multiple country FDA requirements.

8.8 Process Analytical Technology in Suspension Manufacturing

The US FDA's process analytical technology initiative encourages the pharmaceutical industry to improve process efficiency by developing a thorough understanding of the manufacturing process and by using this knowledge to monitor what are identified as critical quality parameters. FDA defines the PAT as a system for designing, analyzing, and controlling manufacturing through timely measurements of quality and performance attributes of raw and in process materials and processes with the goal of ensuring the final product quality. The term analytical PAT is used to describe broadly the chemical, physical, microbiological, and mathematical and risk analysis conducted in an integrated manner. The goal of PAT is to enhance understanding and control the manufacturing process which is consistent with our current drug quality system. PAT believes that quality cannot be tested at the product stage; it should be built in to the process or should be by design. There are four dimensions of PAT. When each of these is brought together to optimize the quality, safety, and cost, that can enhance real time understanding of the process for better risk management. Implementation of PAT in suspension development and manufacturing will have a significant effect on the efficiency and quality of production. Figure 8.7 (Radspinner et al. 2005) shows the four dimensions of PAT and their inter relationship.

Some of the advantages of implementing PAT are:

5. Flexibility of manufacturing configuration.
6. Asset utilization through more effective campaigning of batches.
7. Reduction in paper work.
8. Greater productivity and improved quality.
9. Lower regulatory scrutiny.

A summary of the PAT applied to suspension manufacturing is given in Fig. 8.7 (Radspinner et al. 2005).

Currently the batches are processed according to established recipes and slow and wet chemistry based QA/QC tests performed at specified points to determine the success of the manufacturing step (Table 8.3). In order to develop the PAT initiative, an initial investment of both equipments and time is required but at the end the rewards may lead to efficient, intelligently controlled process, faster time to

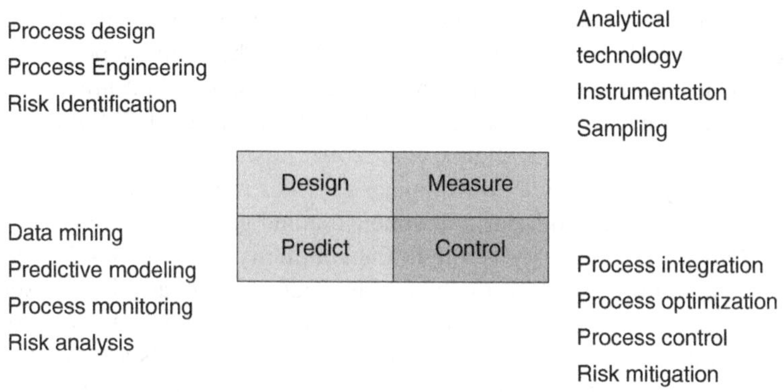

Process design   Analytical
Process Engineering   technology
Risk Identification   Instrumentation
                                                    Sampling

Design	Measure
Predict	Control

Data mining
Predictive modeling   Process integration
Process monitoring   Process optimization
Risk analysis   Process control
                                                    Risk mitigation

Fig. 8.7 Four dimensions of the process analytical technology (Taken with permission from reference Laitinin et al. 2003.)

Table 8.3 Real time monitoring of suspension manufacturing processes

Category	Raw material	Mixing	Formulation, fill and finish
Purpose	Control product quality Ensure product safety	Wet Chemistry analysis	Particle size distribution, Density, pH, Viscosity, Sedimentation volume, Assay, Dissolution studies, freeze thaw cycle, Stability studies over the period of time at Room temperature and accelerated stability studies
Current monitoring Capabilities	Mostly with HPLC and mass analysis for individual ingredients	Use HPLC and other analytical techniques	Based on individual testing for each parameter
Future and desired Monitoring capabilities	Increasing speed of the analytical techniques as well as using one techniques for multiple ingredients to save time and resources	Apply PAT to monitor the mixing efficiency as well as control the particle size distribution	In process PAT application and techniques which can simultaneously determine multiple evaluation parameters

market, and a significant reduction in waste (Kidder et al. 2005; Lewis et al. 2005). The typical steps one may encounter in a risk assessment process incorporating the PAT assessment are (Sekulic 2007).

1. Creation of a flow chart.
2. Identifying the quantity attributes and how these are measured.
3. Identifying and prioritizing process parameters.
4. Experimenting to understand the parameters of evaluation.
5. Prioritizing experiments for risk assessment.
6. Analyzing.PAT decision

Several techniques are proposed for PAT in suspension manufacturing, but Focused Beam Reflectance Measurement (FBRM) is becoming a popular procedure for measuring the particle size in multiphase systems (Laitinin et al. 2003; Barthe and Rousseau 2006; Vaccaro et al. 2006; Saarimaa et al. 2006; Kougoulas et al. 2005; Heath et al. 2002; Blanco et al. 2002; Lasinski et al. 2005). This laser probe offered by Mettler Toledo Company provides a precise, sensitive method for tracking changes in both particle dimension and particle populations in suspension. Part of its popularity is due to the simplicity of the method. A laser is rotated in the multiphase flow, the beam intersects with a particle, and some of the light is scattered back. The total amount of time the detector experiences backscatter multiplied by the rotation speed of the laser yields a measure of the length of the intersected particles, denoted as a chord of the particle. After numerous intersections, a chord length distribution (CLD) can be generated. This CLD will be strongly dependent upon the shape of the particle, with very different distributions generated from spheres, ellipsoids, cubes, and other shapes. Altering the shape of the particle will also change the way CLD is converted into a volume distribution, also known as particle size distribution (PSD). Modeling techniques such as the population balance method can be used to predict how PSDs evolve in time owing to breakage, agglomeration, crystal growth, or nucleation of particles and measurements are necessary to validate those models. Vaccaro et al (2006) have described particle mass distribution (PMD) as another parameter while using FBRM for suspensions. They studied the shape dependent convolution relationships between CLD, PSD and PMD, and based on these convolution relationships, equations relating moments of CLD, PSD and PMD are obtained. Issues relating to the definition of particle size of non spherical objects and its connection is established. Based on the moment relationships particle size was defined for FBRM in terms of CLD equivalent spheres for (non spherical objects) and showed the usefulness of this technique for particles with different shapes.

An interesting application of FBRM which can be used in suspensions is reported by Saarimaa et al 2006. They used the FBRM measurement to continuously assess the aggregation and removal of dissolved and colloidal substances during batch wise dissolved air flotation of process water. They reported the FBRM results to be in agreement with turbidity measurements and chemical analysis by chromatography of pectic acid content. This technique may be adopted for suspensions in process control also.

Kougoulas et al. (2005) have reported the use of FBRM in conjunction with process video imaging (PVI) in a modified mixed suspension mixed product removal (MSMPR) cooling crystallizer. They used this technique to monitor the steady state operation in a modified MPMSR crystallizer. These two techniques show potential for application in suspension manufacturing as PAT.

Blanco et al (2002) have studied the flocculation processes and floc properties using non imaging scanning laser microscope. This methodology allowed them to study the floc stability and resistance to shear forces, reflocculation tendency, and reversibility of the flocs. Furthermore they recommended this technique for studying the optimal dosage of any polymer and the associated flocculation mechanism. Though most of their study is in the paper industry, the techniques can be applied in pharmaceutical suspension PAT also.

Lasinski et al (2005) reported error analysis of the focused beam reflectance measurements when used for suspensions. The assumptions using the FBRM techniques are

1. A particle is stationary (zero velocity) when the laser intersects the particle,
2. The error associated with the intersecting a particle is on multiple occurrences.
3. An error arises in converting a CLD into bi disperse (or multi disperse) PSD.

These errors were determined as a function of independent parameters: the rotation speed of the laser, the volume fraction or number density of particles, the ratio of the particle diameter to the laser diameter, the breadth of the particle size distribution, the velocity profile of the particles, and the preferential orientation of the particles in the flow. To date there has been very little investigation of these errors; their study reported that these errors and assumptions affect the validity and reliability of the FBRM measurements they have made in suspensions formulations. Table 8.4 summarizes the developments and future opportunities for process monitoring and real time quality analytics.

Some of the companies which are supplying the equipments for PAT Brucker Optics, provide techniques like Vibrational Spectroscopy, Fourier Transform Infra red spectroscopy (FTIR), FT-near infra red (FT-NIR) spectroscopy, and Raman Spectroscopy (www.bruckeroptics.com/pat). Malvern instruments provide industrial instruments for particle sizing (www.malvern.co.uk). Niro Inc provides real time process determination equipments (www.niroinc.com) whereas Thermo Electron Corporation supports equipments of FT-NIR analyzers, process tools, and integration and informatics solutions (www.thermo.com) and lastly Mettler Toledo offers equipments for particle system characterization and insitu spectroscopy (www.mt.com/pat). All these equipments need systematic evaluation for their applications in suspension scale up and manufacturing (Shaw 2005).

Table 8.4 Summary of developments and future opportunities for process monitoring and real time quality analytics

Category	In development	Opportunities
Improved processmonitoring methods	Optical sensors for dissolved gasses, pH	Rapid assay for impurities
	NIR for moisture during Lyophilization	Rapid activity assay for preservatives
	Rapid microbiology methods	
	Particle size distribution of the suspensions	
Real time quality analytics	Light scattering assessments of the aggregates and particles	Characterization with modern PAT techniques
	HPLC and mass analysis for the ingredients	Rapid assay techniques

8.9 Enhancing Compliance and Performance in Pharmaceutical Production of Suspensions Using Electronic Production Execution Systems

The major challenges in today's pharmaceutical suspension manufacturing are quality, price, and time to market. These have always been the driving factors for operational excellence in production. Some of the challenges the pharmaceutical industry faces are

1. Increasingly stringent regulations.
2. Increased competition created by globalization.
3. Shrinking drug pipeline and rising reimbursement favoring generic drugs that have added pressure on pharmaceutical manufacturing costs.
4. Regulatory compliance.
5. Time consuming manual data processes.
6. Increased market demand for production flexibility and batch sizes.
7. Lack of real time shop floor data to optimize production and assist in managerial decisions.
8. Limited integration between the shop floor and the extended enterprise.
9. Lack of internal resources to implement and maintain a complex systems architecture.

Pharmaceutical production is a complex process requiring compliance with a wide range of procedures and regulations. Suspensions are liquid products and therefore need to be further carefully controlled; significant amount of data need to be recorded during all the steps of complex lab development, scale up, and later commercial manufacturing. Currently popular manual data management is slow and prone to human error often causing rework, quality cross checks, and batch rejections. FDA is also encouraging the electronic submissions for the NDA's and ANDA's as well as any communications with the authorities. Increasing popularity of unit dose packaging for sterile and non sterile liquid pharmaceuticals further emphasize the necessity of electronic batch record keeping systems. This approach improves compliance, easy review of data and simplifies the quality control process.

In conclusion, there is a need for a change to implement electronic production execution systems for master data management, dispensing management, manufacturing and packaging management, batch review, and finally archiving the batch record. These systems can be adopted at different scale up steps including commercial manufacturing and during post validation phase.

References

Barthe, S. and Rousseau, R.W., 2006, Utilization of focused beam reflectance measurement in the control of crystal size distribution in a batch cooled crystallizer, Chem. Eng. Technol., 29, 206–211.

Blanco, A., Fuente, E., Negro, C. and Tijero, J., 2002, Flocculation monitoring: FBRM as a measurement tool, Can. J. Chem. Eng., 80, 1–7.

Block, L.H., 2002, Non parenteral Liquids and semisolids, In Pharmaceutical scale up, 57–94, Editor; Levin, M., New York, Marcel Dekker Inc.

Caballero, F.G. and Duran, J.L., 2000, Suspension formulation, In Pharmaceutical Emulsions and suspensions, 127–190, New York, Marcel Dekker Inc.

Charpentier, J.C., 2003, Market demand versus technological development: the future of chemical engineering, Int. J. Chem. Reactor Eng., A14.

FDA Document, 1999, Container, closure systems for packaging human drugs and bilogis, chemistry, manufacturing and control documentation, May.

FDA Document, 2006, Guide to inspections oral solutions and suspensions.

Gallardo, V., Ruiz, M. and Delgado, A., 2000, Pharmaceutical suspensions: their applications, 409–464, New York, Marcel Dekker Inc.

Gruverman, I.J., 2003, Breakthrough ultraturbulant reaction thechnology opens frontier for developing life saving nanometer scale suspensions, Drug Deliv. Technol., 3, 52–55

Gruverman, I.J., 2004, A drug delivery breakthrough-nanosuspension formulations for intravenous, oral and transdermal administration of active pharmaceutical ingredients, Drug Deliv. Technol., 4, 58–59.

Gruverman, I.J., 2005, Nanosuspension preparation and formulation, Drug Deliv. Technol., 5, 71–75.

Heath, A.R., Fawell, P.D., Bahri, P.A. and Swift, J.D., 2002, Estimating average particle size by FBRM, Part. Part. Syst. Charact., 19, 84–95.

Joseph, J.A., 2005, Over coming the challenges of blending multiple component dietary supplements, Tablets and Capsule, July, 38–42.

Karanth, H., Shenoy, V.S. and Murthy, R.R., 2006, Industrially feasible alternative approaches in the manufacture of solid dispersions: A technical report, AAPS PharmSciTech, 7, E1–E8.

Kidder, L.H., Lee, E. and Lewis, E.N., 2005, NIR chemical imaging as a process analytical tool, Innovations in Pharmaceutical Technology September, 107–111.

Kougoulas, E., Jones, A.G., Jennings, K.H. and Wood-Kaczmar, M.W., 2005, Use of FBRM and process video imaging in a modified mixed suspension mixed product removal coling crystallizer, 273, 529–534.

Laitinin, N., Antikainin, O. and Yliruusi, J., 2003, Characterization of particle size in bulk pharmaceutical solids using digital image information, AAPS PharmSciTech, 4, 383–387.

Lasinski, M.E., Nere, N.K., Hamilton, R.A., James, B.D. and Curtis, J.S., 2005, Error analysis of FBRM, Poster, presented at Particle technology Forum Poster session, Nov 1st 2005, Cincinnati, OH, no. 287 b.

Lewis, E.N., Schoppelrei, J.W., Lee, E. and Kidder, L.H., 2005, NIR chemical imaging as a process analytical tool. In Process analytical technology, Chapter 7, 187, Editors; Bakeev, K., Blackwell Publishing.

Liu, J., 2006, Physical characterization of pharmaceutical formulations in frozen and freeze dried solid states: Techniques and applications in freeze drying development, Pharm. Dev. Technol., 11, 3–28.

Lu, G.W. 2007, Development of oral suspensions, Am. Pharm. Rev., 10, 35–37.

Levin, M., 2005, How to scale up scientifically: Scale up manufacturing, Pharm. Techol., S4–S12.

Nash, R.A., 1996, Pharmaceutical suspensions, In Pharmaceutical dosage forms: disperse systems, Volume 2, 1–46, Editors; Lieberman, H.A., Rieger, M.M. and Banker, G.S., Marcel Dekker, New York.

Newton, J.M., Chapman, S.R. and Rowe, R.C., 1995, The assessment of scale up performance of the extrusion-spheronization process, Int. J. Pharm., 120, 95–99.

Mehta, A.M., 1988, Scale –up considerations in the fluid bed process for controlled release products, 12, 46–54.

Paul, E.L., Midler, M. and Sun, Y., 2004, Mixing in the fine chemicals and pharmaceutical industries, In Handbook of industrial mixing, Editors; Paul, E.L., Obeng, V.A.A. and Kresta, S.M., New York, John Wiley and Sons.

Rabinow, B., 2005, Pharmacokinetics of drug administered in nanosuspension, Disc. Med., 5, 74–79.

Rabinow, B., 2007, Parenteral nanosuspensions, In Nanoparticulate drug delivery systems, Editors; Thassu, D., Deleers, M. and Pathak, Y.V., Infroma Healthcare, New York.

Radspinner, D., Davies, B. and Kamal, Z., 2005, Process analytical technology and outsourcing-impacts on manufacturing and process knowledge, GOR, 7(4), 55–58.

Saarimaa, V., Sundberg, A. and Holmbom, B., 2006, Monitoring of dissolved air flotation by FBRM, Ind. Eng. Chem. Res., 45, 7256–7263.

Sekulic, S.S., 2007, Is PAT changing the product and process development? Am. Pharm. Rev., 10, 30–34.

Shaw G., 2005, Micrograms to Kilograms: the challenge of scaling, Drug Discovery and Development, July, 1–4.

Sohi, H., Sultana, Y. and Khar, R., 2004, Taste masking technologies in oral pharmaceuticals: Recent developments and approaches, Drug Dev. Ind. Pharm., 30, 429–448.

Vaccaro, A., Sercik, J. and Morbidelli, M., 2006, Modelling focused beam reflectance measurement (FBRM) and its application to sizing of the particles of variable shape, Part. Part. Syst. Charact., 23, 360–373.

Wibowo, C. and Ng, K.M., 2004, Product oriented process synthesis: Creams and pastes, AIChE J., 47, 2746–2767.

Zlokarnik, M., 2002, Dimensional analysis and scale up in theory and industrial application, 1–41, In Pharmaceutical scale up, 57–94, Editor; Levin, M., New York, Marcel Dekker Inc.

Chapter 9
The Science and Regulatory Perspectives of Pharmaceutical Suspensions

Abhay Gupta, Vilayat A. Sayeed, and Mansoor A. Khan

Abstract This chapter details the requirements of the U.S. Food and Drug Administration pertaining to suspension drug products. Extensive information is provided on nomenclature, regulatory filing requirements for new drug applications and abbreviated new drug applications, as well as postapproval changes and supplements. The global move to the common technical document format is also discussed. Finally, the authors present an overview of how quality by design principles can be effectively applied to suspension drug product technology. The views expressed are authors own and do not necessarily reflect the policies of the FDA.

9.1 Introduction

The laws and regulations governing the pharmaceutical industry vary among different countries and regions, but they are all designed to serve the same objective of the protection of public health. These laws are revised as needed based on a scientific understanding and to meet societal health goals. In the United States of America (USA), the Federal Food, Drug, and Cosmetic Act (FD&C Act) is the federal law under which human and veterinary drugs, biological products, medical devices, food and cosmetics are regulated. The Food and Drug Administration (FDA) is given the authority under this act to enforce this law and the law is enforced by the FDA through its eight centers and offices (FDA website).

M.A. Khan (✉)
Division of Product Quality Research, Office of Testing and Research, US Food and Drug Administration, Silver Spring, MD, 20993, USA
mansoor.khan@fda.hhs.gov

V.A. Sayeed
Division of Chemistry, OGD, CDER, US Food and Drug Administration, Rockville, MD

A. Gupta
Division of PQR, Office of Testing and Reserach, CDER, US Food and Drug Administration, Silver Spring, MD

A.K. Kulshreshtha et al. (eds.), *Pharmaceutical Suspensions: From Formulation Development to Manufacturing*, DOI 10.1007/978-1-4419-1087-5_9,
© AAPS 2010

All drug products, prior to be marketed in the USA, are evaluated by the FDA to meet the specifications for identity, strength, quality, purity and potency. FDA's Center for Drug Evaluation and Research (CDER) evaluates New Drug Development and Review, Therapeutic protein and monoclonal antibody products, Generic Drug Review, Over-the-Counter Drug Review and Post Drug Approval Activities, while the Office of Regulatory Affairs (ORA) is entrusted with the inspections to ensure that the regulated products continue to meet the agency's standards. In Europe, the European Medicines Agency (EMEA) is the European Union (EU) body responsible for coordinating the existing scientific resources for the evaluation, supervision and pharmacovigilance of medicinal products. (EMEA website) EMEA was created in 1993 under council regulation (EEC) No 2309/93 and began its activity in 1995. Amendments in regulation (EC) No 141/2000 provided the legal framework and council regulations (EC) No 726/2004 and No 1901/2006 provided new provisions for regulation of medicinal products and pediatric medicines, respectively. In EU, the primary product assessment and marketing authorization is done by the Committee for Medicinal Products for Human Use (CHMP) and the acceptable finding are published by EMEA as a summary of opinion. Upon granting the Marketing Authorization by the European Commission more detailed information is published later as a European Public Assessment Report (EPAR) after deletion of commercially confidential information.

The harmonization of regulatory requirements between the USA, EU, and Japan is ensured by the International Conference on Harmonization (ICH) guidelines. (International Conference on Harmonization website) The ICH effort to harmonize the technical data submission in different jurisdictions, has significantly minimized the variability in the data requirement for medicinal products containing new drugs registered in the USA, EU, and Japan, with the exception of regional requirement. The ICH Steering Committee includes representatives from each of the ICH sponsors and the International Federation of Pharmaceutical Manufacturers Associations (IFPMA), as well as observers from the World Health Organization, the Canadian Health Protection Branch, and the European Free Trade Area.

9.2   CDER Nomenclature for Suspensions

The CDER Nomenclature Standards Committee (NSC) reviews and approves standardized nomenclature monographs and the standard nomenclature is complied in the CDER Data Standards Manual (DSM). DSM nomenclature is either wholly or in part from other nomenclature standards settings bodies, such as the International Conference on Harmonization (ICH), the United States Pharmacopeia (USP), the United States Adopted Names Council (USAN), the American Hospital Formulary Service (AHFS), the Chemical Abstracts Service (CAS), the National Institutes of Standards and Technology (NIST), the International Organization for Standardization (ISO), the American Society for Testing and Materials (ASTM), US Census Bureau, US Postal Service, and the Central Intelligence Agency (CDER Standards Manual website).

The CDER data standards manual defines suspension as a liquid dosage form that contains solid particles dispersed in a liquid vehicle, where a liquid is defined as being pourable, displaying Newtonian or pseudoplastic flow behavior and conforming to its container at room temperature.[1] Suspensions should contain suitable antimicrobial agents to protect against bacteria, yeast, and mold contamination. They should also contain suitable ingredients to prevent caking and solidification of sediments that may settle to the bottom of the container upon standing. Due to the presence of insoluble solid particles dispersed, they should never be injected intravenously or intrathecally. Some suspensions are prepared and ready for use, while others are prepared as solid mixtures intended for constitution just before dispensing with an appropriate vehicle. These specific suspension dosage forms as defined in the CDER standard data manual are listed below:

For Suspension: A product, usually a solid, intended for suspension prior to administration.

For Suspension, Extended Release: A product, usually a solid, intended for suspension prior to administration; once the suspension is administered, the drug will be released at a constant rate over a specified period.

Granule, For Suspension: A small medicinal particle or grain made available in its more stable dry form, to be reconstituted with the solvent just before dispensing to form a suspension; the granules are so prepared to contain not only the medicinal agent, but the colorants, flavorants, and any other desired pharmaceutic ingredient.

Granule, For Suspension, Extended Release: A small medicinal particle or grain made available in its more stable dry form, to be reconstituted with solvent just before dispensing to form a suspension; the extended release system achieves slow release of the drug over an extended period of time and maintains constant drug levels in the blood or target tissue.

Injection, Powder, For Suspension: A sterile preparation intended for reconstitution to form a suspension for parenteral use.

Injection, Powder, For Suspension, Extended Release: A dried preparation intended for reconstitution to form a suspension for parenteral use which has been formulated in a manner to allow at least a reduction in dosing frequency as compared to that drug presented as a conventional dosage form (e.g., as a solution).

Injection, Powder, Lyophilized, For Liposomal Suspension: A sterile freeze dried preparation intended for the reconstitution for parenteral use which has been formulated in a manner that would allow liposomes (a lipid bilayer vesicle usually composed of phospholipids which is used to encapsulate an active drug substance, either within a lipid bilayer or in an aqueous space) to be formed upon reconstitution.

Injection, Powder, Lyophilized, For Suspension: A liquid preparation, intended for parenteral use that contains solids suspended in a suitable fluid medium and conforms in all respects to the requirements for Sterile Suspensions; the medicinal agents intended for the suspension are prepared by lyophilization ("freeze drying"),

---

[1]CDER Data Standards Manual Drug Nomenclature Monographs for Dosage Form http://www.fda.gov/cder/dsm/DRG/drg00201.htm.

a process which involves the removal of water from products in the frozen state at extremely low pressures.

Injection, Powder, Lyophilized, For Suspension, Extended Release: A sterile freeze dried preparation intended for reconstitution for parenteral use which has been formulated in a manner to allow at least a reduction in dosing frequency as compared to that drug presented as a conventional dosage form (e.g., as a solution).

Injection, Suspension: A liquid preparation, suitable for injection, which consists of solid particles dispersed throughout a liquid phase in which the particles are not soluble. It can also consist of an oil phase dispersed throughout an aqueous phase, or vice-versa.

Injection, Suspension, Extended Release: A sterile preparation intended for parenteral use which has been formulated in a manner to allow at least a reduction in dosing frequency as compared to that drug presented as a conventional dosage form (e.g., as a solution or a prompt drug-releasing, conventional solid dosage form).

Injection, Suspension, Liposomal: A liquid preparation, suitable for injection, which consists of an oil phase dispersed throughout an aqueous phase in such a manner that liposomes (a lipid bilayer vesicle usually composed of phospholipids which is used to encapsulate an active drug substance, either within a lipid bilayer or in an aqueous space) are formed.

Injection, Suspension, Sonicated: A liquid preparation, suitable for injection, which consists of solid particles dispersed throughout a liquid phase in which the particles are not soluble. In addition, the product is sonicated while a gas is bubbled through the suspension, and this results in the formation of microspheres by the solid particles.

Powder, For Suspension: An intimate mixture of dry, finely divided drugs and/ or chemicals, which, upon the addition of suitable vehicles, yields a suspension (a liquid preparation containing the solid particles dispersed in the liquid vehicle).

Shampoo, Suspension: A liquid soap or detergent containing one or more solid, insoluble substances dispersed in a liquid vehicle that is used to clean the hair and scalp and is often used as a vehicle for dermatologic agents.

Spray, Suspension: A liquid preparation containing solid particles dispersed in a liquid vehicle and in the form of coarse droplets or as finely divided solids to be applied locally, most usually to the nasal-pharyngeal tract, or topically to the skin.

Suspension, Extended Release: A liquid preparation consisting of solid particles dispersed throughout a liquid phase in which the particles are not soluble; the suspension has been formulated in a manner to allow at least a reduction in dosing frequency as compared to that drug presented as a conventional dosage form (e.g., as a solution or a prompt drug-releasing, conventional solid dosage form).

Suspension/Drops: A suspension which is usually administered in a dropwise fashion.

Tablet, For Suspension: A tablet that forms a suspension when placed in a liquid (formerly referred to as a "dispersible tablet").

The United States Pharmacopeia-National Formulary (USP-NF) defines a few more categories of suspensions. (USP-NF online) These are the Oral Suspensions, which are liquid preparations containing solid particles dispersed in a liquid vehicle, with suitable flavoring agents, intended for oral administration; Topical Suspensions, which are liquid preparations containing solid particles dispersed in a liquid vehicle, intended for application to the skin; Otic Suspensions which are liquid preparations containing micronized particles intended for instillation in the outer ear; and Ophthalmic Suspensions, which are sterile liquid preparations containing solid particles dispersed in a liquid vehicle intended for application to the eye.

9.3 Regulatory Filing in the United States of America

Based on the authority provided under the FD&C Act an application for a suspension drug product may be submitted to the different centers in the FDA as:

1. Over-the-Counter
2. New Drug Application
3. Abbreviated New Drug Application

Over-the-Counter (OTC) drug products are those drugs that can be used safely without the help of a health care practitioner. They are available to consumers without any prescription, enabling them to take control of their own health care in many situations. There are more than 100,000 OTC drug products currently being marketed in the USA. These encompass about 800 significant active ingredients covering more than 80 classes (therapeutic categories) of drugs.

FDA regulates OTC drugs under section 505 and 502 of the FD&C Act and applicable Code of Federal Regulations (21 CFR 330.1) as a New Drug Application or by OTC drug monographs respectively. The Agency has established OTC drug monographs for each class of products covering the acceptable ingredients, doses, formulations, and labeling. These monographs are updated periodically by adding additional ingredients and labeling as needed. Products conforming to a monograph may be marketed without further FDA clearance, while those that do not, must undergo separate review and approval through the "New Drug Approval System" (Office of Nonprescription drugs website).

A *New Drug Application* (NDA) is the regulatory vehicle through which drug sponsors formally propose that the FDA approve a new pharmaceutical for sale and marketing in the USA. The NDA is submitted under section 505(b)(1) or 505(b)(2) of the FD&C Act and applicable Code of Federal Regulations (21 CFR.314). The goal of the NDA is to provide relevant information to permit FDA reviewer to reach the following key decisions: (New Drug Application website).

1. Is the drug safe and effective in its proposed use(s), and do the benefits of the drug outweigh the risks?
2. Is the drug's proposed labeling (package insert) meets the requirements established in the Code of Federal Regulations (21 CFR 201.56 & 201.57) and related guidance's.(FDA guidance documents, labeling)

3. Are the methods used in manufacturing the drug and the controls used to maintain the drug's quality adequate to preserve the drug's identity, strength, quality, and purity?

The NDA must also provide all relevant data and information that a sponsor has collected during the product's research and development to support the safety, efficacy, quality, purity and manufacturability claims. Although the quantity of information and data submitted in NDAs can vary from submission to submission, the components of NDAs are more uniform and are, in part, a function of the nature of the subject drug and the information available to the applicant at the time of submission.

FDA's Office of New Drug Quality Assessment (ONDQA) classifies NDA with a code number from type 1 through 8 that reflects application classification system and provides a way of identifying drug application upon receipt and review process.

1. New Molecular Entity
2. New Active Ingredient (new salt, new noncovalent derivatives, new esters of previously approved drug (not a new molecular entity)
3. New dosage form of Previously Approved Drug (not a new salt or a new molecular entity)
4. New Combination of Two or More Drugs
5. New formulation or New manufacturer of Already Marketed Drug Product
6. New Indication (claim) for Already Marketed Drug
7. Already Marketed Drug Product – No Previously Approved NDA
8. Rx to OTC switch

The review process involves reviewers confirming and assessing the sponsor's conclusions that a drug is safe and effective for its proposed use and has adequate Chemistry Manufacturing and Controls to demonstrate that it would continue to meet the standards of identity, strength, quality, purity and potency through the proposed shelf life as established in the NDA. Upon the acceptance of data at all review and administrative levels, the NDA application is approved and the sponsor can then legally market the NDA drug product in the USA. Detailed information regarding the NDA approval process is available on the FDA website (FDA CDER website).

Approval to manufacture and market generic drug products is obtained by submitting an *Abbreviated New Drug Application* (ANDA) under section 505(j) of the FD&C Act and applicable Code of Federal Regulations (21 CFR part 314, part 320) with the Office of Generic Drugs. The term "abbreviated" implies that the generic drugs are generally not required to include preclinical (animal) and clinical (human) data to establish safety and effectiveness. However, generic drugs are required to be bioequivalent (where applicable) and identical to a brand name drug in active ingredient(s), dosage form, strength, route of administration, quality, performance characteristics and intended use.

Once approved by the FDA, generic drugs can be substituted by a health care provider. The substitution could be from a brand-name drug product to a generic equivalent drug product, from a generic equivalent to a brand-name drug product (Reference listed drug), or from one generic product to another when both are deemed therapeutic equivalent to a brand-name drug product. Brand-name drugs are subject to the same bioequivalence requirements as the generic drugs upon reformulation or other significant changes (e.g., new dosage form). All approved products, both innovator and generic, are listed in FDA's Approved Drug Products with Therapeutic Equivalence Evaluations (Electronic Orange Book website).

In addition to meet the bioequivalence requirements (21 CFR 320.1(e)), the generic product is also required to be pharmaceutical equivalent (21 CFR 320.1(c)) and have adequate Chemistry, Manufacturing and Control (CMC) data to ensure that the generic product is appropriately designed and have methods and controls in place for the manufacture, processing, and packaging of a drug to maintain the identity, strength, quality, purity, and potency of the proposed drug product through its shelf life. (Office of Generic Drugs website) In an effort to streamline the design of bioequivalence studies for specific drug products, FDA has made available product-specific bioequivalence study recommendations on its website, rather than requiring sponsors to request this information from the Agency. Currently the list covers over 225 drug products, mostly solid oral dosage forms, including 15 suspension specific bioequivalence study recommendations.

The technical requirements established by the ICH are used by the FDA in the assessment of the applications submitted in the USA.(Guidance on bioequivalence website) The ICH Safety, Efficacy and Quality documents relating to chemical, pharmaceutical quality assurance, pharmaceutical development, quality risk management and pharmaceutical quality systems in conjunction with the regional requirement can be used by the sponsor in preparing the application as these documents are used in the assessment of the application.

The approval process for a drug application also includes a review of the manufacturer's compliance with the Current Good Manufacturing Practice (CGMP) regulations. Failure to comply may lead to a decision by the FDA not to approve an application to market a drug. FDA inspectors determine whether the firm has the necessary facilities, equipments, and skills to manufacture the drug product for which it had applied for approval. Decisions regarding compliance with CGMP regulations are based upon the inspection of facilities, sample analyses, and compliance history of the firm.

FDA also requires all approved drugs to be manufactured in accordance with the CGMP regulations, otherwise they are considered to be adulterated within the meaning of the FD&C Act, Section 501(a)(2)(B).(21CFR guidance, finished product website) These criteria governing personnel, facilities, equipments, components, methods, and controls used in production, processing, packaging, analysis and holding of product during manufacturing ensure the quality, purity and safety of the pharmaceutical drug product.(Guide to Inspections Dosage Form, website) Due to relative higher number of recalls associated with the oral solutions and suspensions dosage forms, FDA inspections of the manufacturing facilities for these forms warrant special attention (Guide to Inspections, Suspensions, website).

9.4   Suspensions: Requirement to Support Submission in the United States of America

To market a product in the USA an application should be filed with the FDA and must contain sufficient details about the product and process development, identify the elements critical for the performance of the drug product, controls to meet the performance characteristics and supportive data to establish expiration dating. The submission should provide the reviewer with a comprehensive summary of the data and information in the application, including a discussion of the relevance and quantitative aspects of the data supported by the scientific and regulatory justification. The CMC section of the application must address the following elements when appropriate in support of the suspension products.

- Physico-chemical characteristics of the active ingredient
- Selection and control of excipient and their impact and function in the formulation (e.g., electrolytes, viscosity enhancers)
- Regulatory compliance with the proposed level of excipients
- Content uniformity
- Rheological behavior of suspensions
- Flocculation and aggregation behavior of suspension
- Effect of viscosity and sedimentation rate
- Rate of re-suspension
- Effect of temperature and pH
- Dissolution
- Stability (chemical and physical)
- Taste masking
- Packaging, label and dispensing instruction
- CGMP compliance

The physico-chemical characteristics of the active ingredient and the excipients in the formulation do contribute and affect the performance of the drug and provide the necessary or desired functional use or properties to the suspension dosage form. For this reason, a complete understanding of the drug formulations with elements such as chemical composition and function of different excipients, regulatory compliance of excipients for route of administration, and proposed excipients levels based on product therapeutic and target population should be considered.(Guidance, Exploratory IND website) The physical characteristics, particularly the particle morphology/size of the drug substance, are very important for suspensions. Examples of a few of the oral suspensions in which a specific and well defined particle morphology/size control for the drug substance is important include phenytoin suspension, carbamazepine suspension, trimethoprim and sulfamethoxazole suspension, and hydrocortisone suspension.

Content uniformity is another critical element that is of concern with oral suspensions, particularly because of the potential for segregation during manufacture, storage of the bulk, transfer to the filling line, during filling and during product shelf life. Viscosity can be important from a processing aspect to minimize segregation. It is also intended to reduce the rate of settling of particulate matter in the suspension. Depending upon the viscosity, many suspensions may also require continuous or periodic agitation during the filling process. If delivery lines are used between the bulk storage tank and the filling equipment, some segregation may occur, particularly if the product is not viscous. Established procedures and time limits should be in place for such operations to address the potential for segregation or settling as well as other unexpected effects that may be caused by extended holding or stirring. CGMP warrants testing bottles from the beginning, middle and end, including an assay of a sample from the bulk tank, to assure that segregation has not occurred. Such samples should be analyzed individually and not be composited for analysis.

Rheological behavior of suspension is important as it allows a determination of ease of use of a product (e.g., easy of pouring from bottle or injecting through a needle), and the redispersibility of suspension if sedimentation occurs during product storage. The flow properties of a suspension are influenced by the particle shape, size distribution of the suspended particles and on the degree of flocculation. Flocculation also leads to plastic or thixotropic flow properties making it easy to administer and it is well documented that the apparent viscosity of flocculated suspension or suspension prepared with uniform sized particles results in easily redispersible sediments. Thus the effect of particle size, particle–particle interactions, densities of the suspended particles and medium and viscosity of the continuous phase on the sedimentation and aggregation rates should be adequately addressed in the application.

In the oral suspensions, a precise temperature control is especially desirable to control the product quality if temperature is identified as a critical element to maintain product stability and/or inhibit microbial growth. In such a case, it is highly recommended that special attention must be paid to the purified water used in manufacture of the batch, and the temperatures during processing.

In case of suspensions, it is essential to study and discuss the impact of physical instability in addition to the chemical stability of the drug product in the CMC part of the application. Caking is the most common form of physical instability observed in suspensions and is caused by crystal bridging. This phenomenon occurs when simultaneous crystal growth occurs on two or more particles resulting in the formation of crystal-linked particle that leads to the formation of sediment which is difficult to re-disperse. The caking problem is well known for the deflocculated type of suspensions that settles down breaking the electrical double layer of charges.

The particle size distribution of the dispersed systems is another physical attribute that may also change over time leading to an unstable suspension. It may happen due to Ostwald's ripening, polymorphic transformation, or temperature cycling. Ostwald's ripening refers to the growth of larger particles at the expense of smaller particles due to the relative lower solubility of the larger particles. Polymorphic transformation occurs due to difference in the equilibrium solubility of different polymorphs, and leads to precipitation of less soluble polymorph as a result of more soluble polymorph going into solution. This causes an increase in the solids content of the suspension and may lead to physical instability. Since solubility is directly related to temperature for most drug substances, change in storage temperature or temperature cycling can result in change in particle size of the suspended particles. Hence an understanding of the influence of these phenomena on the stability of suspension should be studied and relevant information should be provided in the submission.

For oral suspension drug products, sponsors are required to develop an in-vitro dissolution requirement for release and stability and should establish specifications (limits and test methods) based on the product development knowledge and information from stability batches used to support clinical and/or bioequivalence study. In cases where it is determined that in-vitro dissolution test is not indicative of the product performance, the sponsor must provide developmental data, identify alternate product manufacturing controls and adequately justify the choice of not having an in-vitro dissolution requirement. For oral suspension drug products that are compendia items, compendia methods (e.g. USP-NF) should be used. However, alternative methods can be considered when the compendia methods fail to provide meaningful data with adequate justification.

Though taste is not a regulatory requirement but an adequately designed pharmaceutical product is not only safe and effective; it is also one where patient compliance is woven in the product development. In case of suspension, special attention must be paid in the formulation development process as this can make the difference between product success and product failure. As there are number of technologies available to mask the product taste, it is recommended that the selection of the taste technology must be based on the chemical understanding of the active ingredient, impact on product stability, disease management, and target population and should be adequately supported by product development data in the CMC section.

Selection of packaging material plays an important role in maintaining product quality over its intended shelf life, and thus proper selection of the packaging material that provides adequate protection and no measurable interaction (e.g. extractable, leachables etc.) with the product is critical for product stability. Development studies done to demonstrate that the packaging material used provides adequate protection from the environmental conditions, has no measurable interaction with the product components, and meets with all regulatory requirements must be submitted in the application. The packaging of the drug product should also be labeled according to the current regulatory requirements. In addition, the labels for generic drug products should be the same as the reference listed drug products except for differences as cited in the Code of Federal Regulation (21 CFR 314.94). This portion of an application is generally referred to as the CMC section and includes detailed information about the drug substance and the drug product (21CFR314.50 website).

The CMC section of the application submitted to the FDA should also contain pertinent information about the drug substance and drug product, and provides clarity to general understanding of the data and information in the application, including an understanding of the quantitative aspects of the data to the reviewer.

A full description of the drug substance including its physical and chemical characteristics and stability; the name and address of its manufacturer; the method of synthesis (or isolation) and purification of the drug substance; the process controls used during manufacture and packaging; and such specifications and analytical methods as are necessary to assure the identity, strength, quality, and purity of the drug substance and the bioavailability of the drug products made from the substance, including, for example, specifications relating to stability, sterility, particle size, and crystalline form. The application may provide additionally for the use of alternatives to meet any of these requirements, including alternative sources, process controls, methods, and specifications.

A list of all components used in the manufacture of the drug product (regardless of whether they appear in the drug product); and a statement of the composition of the drug product; a statement of the specifications and analytical methods for each component; the name and address of each facility having a part in the manufacturing and/or control of the drug product; a description of the manufacturing and packaging procedures and in-process controls for the drug product; specifications and analytical methods necessary to assure the identity, strength, quality, purity, and bioavailability of the drug product, including, for example, specifications relating to sterility, dissolution rate, containers and closure systems; and stability data with proposed expiration dating. The application may additionally provide for the use of alternatives to meet any of these requirements, including alternative components, manufacturing and packaging procedures, in-process controls, methods, and specifications.

9.5 Post Approval Changes

If the holder of an NDA or the ANDA intends to make postapproval changes to the CMC section, the applicant must provide a demonstration stating that the pre- and post-change products are equivalent. Other than for editorial changes in previously submitted information (e.g., correction of spelling or typographical errors, reformatting of batch records), an applicant must notify FDA about each change in each condition established in an approved application beyond the variations already provided for in the application. Section 506A of the FD&C provides for reporting categories for reporting these postapproval changes. In April, 2004 FDA (Guidance, Changes to an NDA website), provided guidance to holders of new drug applications (NDAs) and abbreviated new drug applications (ANDAs) who intend to make postapproval changes in accordance with section 506A of the Federal Food, Drug, and Cosmetic Act (the Act) and § 314.70 (21 CFR 314.70). The guidance covers four reporting categories for postapproval changes for drugs other than specified biotechnology and specified synthetic biological products.

Annual Report (AR): The annual submission to the approved application reporting changes that FDA has identified as having minimal potential to adversely affect the identity, strength, quality, purity, or potency of a product as they may relate to the safety or effectiveness of the product.

Change-Being-Effected Supplement (CBE): A submission to an approved application reporting changes that FDA has identified as having moderate potential to adversely affect the identity, strength, quality, purity, or potency of a product as they may relate to the safety or effectiveness of the product. A CBE supplement must be received by FDA before or concurrently with distribution of the product made using the change.

Change-Being-Effected-in-30-Days Supplement (CBE-30): A submission to an approved application reporting changes that FDA has identified as having moderate potential to adversely affect the identity, strength, quality, purity, or potency of a product as they may relate to the safety or effectiveness of the product. A CBE-30 supplement must be received by FDA at least 30 days before distribution of the product made using the change.

Prior Approval Supplement (PAS): A submission to an approved application reporting changes that FDA has identified as having a substantial potential to adversely affect the identity, strength, quality, purity, or potency of a product as they may relate to the safety or effectiveness of the product. A PAS supplement must be received and approved by FDA prior to distribution of the product made using the change.

Alternatively the post approval changes can be managed by the applicant with one or more protocols (i.e., comparability protocols) describing tests, validation studies, and acceptable limits to be achieved to demonstrate the absence of an adverse effect from specified types of changes. A comparability protocol can be used to reduce the reporting category for the specified changes. A proposed comparability protocol should be submitted as a PAS, if not approved as part of the original application (Draft Guidance, Comparability protocols website).

A drug product made with a manufacturing change, whether a major manufacturing change or otherwise, may be distributed only after the holder validates the effects of the change on the identity, strength, quality, purity, and potency of the product as these factors may relate to the safety or effectiveness of the product. For each change, the supplement or annual report must contain information determined by FDA to be appropriate and must include the information developed by the applicant in assessing the effects of the change. The assessment should include a determination that the drug substance intermediates, drug substance, in-process materials, and/or drug product affected by the change conform to the approved specifications. In addition to confirmation that the material affected by manufacturing changes continues to meet its specification, the applicant should perform additional testing, when appropriate, to assess whether the identity, strength, quality, purity, or potency of the product as they may relate to the safety or effectiveness of the product has not been or will be unaffected. If the assessment concludes that a change has affected the identity, strength, quality, purity, or potency of the drug product as established in the approved application, the change should be filed in a PAS, regardless of the recommended general reporting category for the change.

Besides the drug application process and the standard post approval reporting categories cited above, there may be many situations that require a broader understanding or different requirement of the regulatory submissions. As an example, a drug product manufacturing site change except for sterile suspensions can be reported as a CBE-30, if the proposed site has an acceptable CGMP finding. However a change in manufacturing site change for modified oral suspensions may require a bio-equivalence study and it is highly recommended that the sponsor contact the review division prior to making this change for product specific guidance.

Any labeling change based on postmarketing study results, including, but not limited to, labeling changes associated with new indications and usage, clinical studies, revision (expansion or contraction) of population based on data, etc. must be reported as a PAS. A CBE supplement should be submitted for any labeling change that (1) adds or strengthens a contraindication, warning, precaution, or adverse reaction, (2) adds or strengthens a statement about drug abuse, dependence, psychological effect, or overdosage, (3) adds or strengthens an instruction about dosage and administration that is intended to increase the safe use of the product, (4) deletes false, misleading, or unsupported indications for use or claims for effectiveness, or (5) is specifically requested by FDA. These study based PAS changes that are described for labeling would not be applicable for a generic drug product and are only applicable to new drug products.

Labeling with editorial or similar minor changes or with a change in the information concerning the description of the drug product or information about how the drug is supplied that does not involve a change in the dosage strength or dosage form should be described in an annual report.

Multiple related changes involving various combinations of individual changes, where the recommended reporting categories for the individual changes differ, CDER recommends that the filing be in accordance with the most restrictive of those recommended for the individual changes. When the multiple related changes all have the same recommended reporting category, CDER recommends that the filing be in accordance with the reporting category for the individual changes.

9.6 Global Regulatory Environment

An application for an ANDA, NDA or OTC product can be filled in the USA at any time. However, this is not the case in the EU where the application for marketing authorization for medicinal products is evaluated by the EMEA. EMEA routinely publishes submission deadlines and full procedural timetables as a generic calendar. The published timetables identify the submission, start and finish dates of the procedures as well as other interim dates/milestones that occur during the procedure. Timetables are classified under different categories depending on the type of procedure and are generally fixed (currently 11 times in a year) (EMEA Timetables website).

Differences in the interpretation and application of technical guidelines and requirements for product registration sometime lead to duplicate testing of time-consuming and expensive test procedures during the research and development of new medicines to satisfy the requirements of different countries and regions around the world. In an effort to use resources more economically, and to eliminate unnecessary delay in the development and availability of new medicines whilst maintaining safeguards on quality, safety and efficacy, efforts began in 1980 to harmonize requirements with in the European Community. In 1990, the International Conference on Harmonization of Technical Requirements for Registration of Pharmaceuticals for Human Use (ICH) was created bringing together the regulatory authorities of the USA, EU and Japan, and experts from the pharmaceutical industry in these three regions to discuss the scientific and technical aspects of product registration (International Conference on Harmonization website).

One of the ICH goals was to remove redundancy and duplication in the development and review process, such that a single set of data could be generated and submitted in the ICH region to demonstrate the quality, safety and efficacy of a new medicinal product. In addition, ICH has issued a number of other guideline documents for drug applications covering the Safety, Quality, Efficacy, and Multidisciplinary matters. These documents are coded Q, S, E and M, respectively. "Quality" guidelines cover topics that are related to the chemical and pharmaceutical Quality Assurance, e.g., Q1 Stability Testing, Q3 Impurity Testing. "Safety" guidelines cover topics that are related to in vitro and in vivo preclinical studies, e.g., S1 Carcinogenicity Testing, S2 Genotoxicity Testing. "Efficacy" guidelines cover topics that are related to clinical studies in human subject, e.g., E4 Dose Response Studies, Carcinogenicity Testing, E6 Good Clinical Practices. "Multidisciplinary" guidelines cover cross-cutting Topics which do not fit uniquely into one of the above categories, e.g., M2 Electronic Standards for Transmission of Regulatory Information (ESTRI), M4 Common Technical Document (CTD).

This has led to the creation of the ICH Guideline M4, The Common Technical Document (CTD). The CTD provides a harmonized format for new product applications. An Electronic Common Technical Document (eCTD) is also available, allowing for the electronic submission of the CTD from applicant to regulator and to provide a harmonized technical solution to implement the CTD electronically.

CTD is organized into five modules. Module 1 is region specific and contains administrative and prescribing information in the format specified by the relevant regulatory authorities. Modules 2–5 are intended to be common for all regions in a format acceptable to the regulatory authorities. Module 2 contains the CTD summaries. It begins with a general introduction to the drug product, including its pharmacologic class, mode of action, and proposed clinical use, followed by sections on Quality Overall Summary (QOS), Nonclinical Overview, Clinical Overview, Nonclinical Written and Tabulated Summaries and Clinical Summary.

The QOS is the sponsors' summary of critical CMC elements in the CTD. It includes information to provide the Quality reviewer with an overview of Module 3, emphasizing critical key parameters of the product and provides justification in cases where guidelines were not followed. QOS also includes a discussion of key issues that integrates information in the Quality Module of the CTD and supporting information from other Modules of the CTD document.

Module 3 is the Quality section of the CTD. It provides a harmonized structure and format for presenting the CMC information in a registration dossier. It contains information on drug substance and drug product. The drug substance section includes general information about the drug substance (nomenclature, structural and molecular formula, relative molecular mass); its manufacturer, manufacturing process and development history; identity, characterization, quality and control of materials used in the drug substance manufacturing; and tests and acceptance criteria used to control critical steps and intermediates. It also includes specifications, analytical procedures and reference standards used for testing the drug substance; analytical validation information and description of batches and results of batch analyses; description of the container closure systems; and protocol used for and results of stability studies conducted on the drug substance. The drug product section includes a description of the drug product, its composition, dosage form and type of container and closure used. This section also includes information about the pharmaceutical development of the drug product, and identifies and describes the formulation and process attributes (critical parameters) that can influence batch reproducibility, product performance, and drug product quality. Parameters relevant to the performance of the drug product are also discussed. Information about the drug product manufacturer (including contractors), proposed production site(s), batch formula, manufacturing process and critical steps and points at which process controls, intermediate tests, or final product controls are conducted is also provided in this section. Specifications, analytical procedures and reference standards used for testing the drug product; analytical validation information and description of batches and results of batch analyses; and protocol used for and results of stability studies conducted on the drug product are also included in this section.

Nonclinical Study Reports are presented in the Module 4 of the CTD, providing a harmonized structure and format for presenting the nonclinical information on the pharmacology, pharmacokinetics and toxicology in a registration dossier. The pharmacology section discusses and describes studies on the primary and secondary pharmacodynamics, safety evaluation, and pharmacodynamic drug interactions. The pharmacokinetics section describes absorption, distribution, metabolism and excretion of the pharmaceutical and includes information about the analytical methods used for analyzing the biological samples.

Module 5 of the CTD provides a harmonized structure and format for presenting the clinical data and study reports in a registration dossier. It contains reports on the biopharmaceutic studies, studies pertinent to pharmacokinetics using human biomaterials, human PK/PD studies and the efficacy and safety studies.

Detailed information about the structure and content can be found in the following guidances:

Guidance on the Quality section of the CTD (Module 2, QOS, and Module 3) can be found in the guidance for industry M4Q: The CTD – Quality.

Guidance on the Safety section of the CTD (Module 2, the Nonclinical Overview and the Nonclinical Written and Tabulated Summaries, and Module 4) can be found in the guidance for industry M4S: The CTD – Safety.

Guidance on the Efficacy section of the CTD (Module 2, the Clinical Overview and the Clinical Summary, and Module 5) can be found in the guidance for industry M4E: The CTD – Efficacy.

9.7 Quality by Design

In August 2002, FDA announced a significant new initiative, Pharmaceutical CGMPs for the twenty-first Century, to enhance and modernize the regulation of pharmaceutical manufacturing and product quality.[2] This initiative aims to apply scientific and engineering principles in regulatory decision-making, establish specifications based on product and process understanding, and to evaluate manufacturing processes thereby improving the efficiency and effectiveness of both manufacturing and regulatory decision-making. It encourages voluntary development and implementation of innovative approaches in pharmaceutical development, manufacturing, and quality assurance. An important element of this initiative is the *quality by design* (QbD). Quality by design means designing and developing a product and associated manufacturing processes to ensure that the product consistently attains a predefined quality at the end of the manufacturing process by understanding the influence of formulation and manufacturing process on the quality of the drug product. It also provides a sound framework for the transfer of product knowledge and process understanding from drug development to the commercial manufacturing processes and for postdevelopment changes and optimization. A good QbD approach to product development includes a systematic evaluation of product attributes, understanding and refining of the formulation and manufacturing processes, application of quality risk management principles to establish appropriate control strategies, and use of knowledge management and formal design of experiments to generate and refine the design space throughout the lifecycle of the product.

Design space is the multidimensional combination and interaction of input variables (e.g., material attributes) and process parameters that have been demonstrated to provide assurance of quality (ICH Q8 and Annex to ICH Q8 websites) The linkage between the process inputs (input variables and process parameters) and the critical quality attributes (CQA) may also be described in the design space. Analysis of historical data may also provide the basis for establishing the design space. The design space may be defined in terms of ranges of input variables or parameters, or through more complex mathematical relationships. Individual design spaces could be established for one or more unit operations and are often simpler to develop. A single design space could also be established spanning multiple operations providing more operational flexibility. For the design space to be applicable to multiple operational scales, it should be described in terms

---

[2] Pharmaceutical cGMPs for the twenty-first Century – A Risk-Based Approach. Final Report – Fall 2004 http://www.fda.gov/cder/gmp/gmp2004/GMP_finalreport2004.htm.

of relevant scale-independent parameters, which may also prove helpful for technology transfer or site changes. Design space could lead to helpful information about potential failure modes.

Working within the design space is not considered a change. Movement out of the design space is considered to be a change and would normally initiate a regulatory postapproval change process. An input variable or process parameter need not be included in the design space if it has no effect on delivering CQAs when the input variable or parameter is varied over the full potential range of operation. If possible, all variables that affect the process and product quality should be included in the design space. An explanation should be provided in the application to describe what variables were considered, how they affect the process and product quality, and which parameters were included or excluded in the design space.

Variables and their ranges within which consistent quality can be achieved can be identified by the use of various quality risk management tools and use of an effective pharmaceutical quality system. (ICH Q9, ICH Q10 websites) Quality risk management is the systematic process for the assessment, control, communication, and review of risks to the quality of the drug product across the product lifecycle. It supports a scientific and practical approach to decision making and provides documented, transparent, and reproducible methods to manage the risk based on current knowledge about assessing the probability, severity, and, sometimes, detectability of the risk. An effective quality risk management approach can ensure the high quality of the drug product to the patient by providing a proactive means to identify and control the potential quality issues during development and manufacturing. Pharmaceutical quality system is a management system to direct and control a pharmaceutical company by establishing, implementing and maintaining a set of processes that provides products with consistent quality. It includes: developing and using effective monitoring and control systems for process performance and product quality; and identifying and implementing appropriate product quality improvements, process improvements, variability reduction, innovations, and pharmaceutical quality system enhancements. It is achieved by managing knowledge from development through the commercial life of the product up to and including product discontinuation; and through the use of quality risk management tools to identify and control the potential risks to quality throughout the product lifecycle. Use of quality risk management and pharmaceutical quality system can improve the decision making if any quality problem arises, facilitate better and more informed decisions, provide regulators with greater assurance of a company's ability to deal with potential risks, and affect the extent and level of direct regulatory oversight. A greater understanding of the product and its manufacturing process can create a basis for more flexible regulatory approaches.

A sponsor wishing to incorporate the QbD principles may utilize prior knowledge, initial experimental data and/or risk assessment tools, e.g., Ishikawa (fishbone) diagram (Fig. 9.1) or Failure Mode Effects Analysis (FMEA), to identify all potential variables that may be critical to the quality of the finished dosage form. A screening design may then be used to exclude variables not found to be critical to the quality of the finished dosage form. Formal design of experiments or other

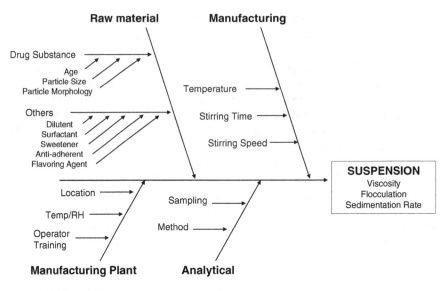

Fig. 9.1 Ishikawa diagram

experimental approaches may then be used to evaluate the impact of the remaining variables on the CQA, to gain greater understanding of the manufacturing process, and to develop appropriate control strategies. For a suspension dosage form, the initial screening may, for example, identify the rheological behavior, viscosity, sedimentation rate and stability of the suspension as the CQA and the particle size of the drug, concentrations of the surfactant and the suspending agent, vessel temperature and stirring speed during manufacturing as the critical input variables. A full factorial design or a response surface design may then be used to study these critical variables and to identify the potential interactions and individual impact on the CQA. The information gained could be used to generate the design space allowing the sponsor ability to adjust the processing conditions (e.g., stirring time or vessel temperature) in response to a change in the properties of incoming raw material (e.g., particle size of the drug substance). Appropriate scale-up studies will then demonstrate that the data is consistently reproducible.

9.8 Conclusion

It is the duty and responsibility of the national governments to ensure that the drugs available to its citizen are not only safe and effective, but also properly manufactured and are packaged appropriately with easily understood labeling. The regulatory considerations discussed in this chapter were focused towards suspension dosage forms for oral drug delivery, although the principles and procedures discussed may be applicable to suspensions dosage forms which have been administered by other routes (e.g., parenteral) and other pharmaceutical dosage forms (e.g., tablets). However, additional regulations may be applicable in these cases, e.g., ANDA fillings in the USA for the suspensions for parenteral drug delivery are required to have same chemical composition as that in the reference listed drug products. It is for this reason that all new drug products must be reviewed and approved by the national regulatory authorities before entering the market. Mechanism exist through which changes may be made to already approved drug products, but authorities must be notified of all such changes, and in some cases, prior approval may be required.

References

21CFR211. Current Good Manufacturing Practice for Finished Pharmaceuticals. http://www. gpoaccess.gov/cfr/

21CFR314.50. Applications for FDA Approval to Market a New Drug. http://www.gpoaccess. gov/cfr/

Annex to Q8: Pharmaceutical Development. http://www.ich.org

CDER Data Standards Manual. http://www.fda.gov/cder/dsm/

Draft Guidance for Industry: Comparability Protocols – Chemistry, Manufacturing, and Controls Information. http://www.fda.gov/cder/guidance/5427dft.pdf

Electronic Orange Book. http://www.fda.gov/cder/ob/default.htm

EMEA Human Medicines – Procedural timetables/Submission dates. http://www.emea.europa.eu/ htms/human/submission/submission.htm

European Medicines Agency. http://www.emea.europa.eu

FDA CDER. http://www.fda.gov/cder/

FDA Guidance Documents: Labeling. http://www.fda.gov/cder/guidance/index.htm#Labeling

Federal Food, Drug, and Cosmetic Act. United States Code (U.S.C.) Title 21, Chapter 9. http:// www.access.gpo.gov/uscode/title21/chapter9_.html

Guidance for Industry: Changes to an Approved NDA or ANDA. http://www.fda.gov/cder/ guidance/3516fnl.pdf

Guidance for Industry, Investigators, and Reviewers: Exploratory IND Studies. http://www.fda. gov/cder/guidance/7086fnl.htm

Guidance for Industry: Individual Product Bioequivalence Recommendations. http://www.fda. gov/cder/guidance/bioequivalence/

Guide to Inspections of Dosage Form Drug Manufacturer's – CGMPR'S. http://www.fda.gov/ora/ inspect_ref/igs/dose.html

Guide to Inspections Oral Solutions and Suspensions. http://www.fda.gov/ora/inspect_ref/igs/oral.html

ICH Q8: Pharmaceutical Development. http://www.ich.org.

ICH Q9: Quality Risk Management. http://www.ich.org.

ICH Q10: Pharmaceutical Quality System. http://www.ich.org

International Conference on Harmonization of Technical Requirements for Registration of Pharmaceuticals for Human Use. http://www.ich.org

New Drug Application. http://www.fda.gov/cder/handbook/ndabox.htm

Office of Generic Drugs. http://www.fda.gov/cder/ogd/

Office of Nonprescription Products. http://www.fda.gov/cder/offices/otc/default.htm

Suspensions. United States Pharmacopeia 30 – National Formulary 25 Supplement 2. http://www. uspnf.com

U.S. Food and Drug Administration. http://www.fda.gov

Chapter 10
Solid Nanosuspensions: The Emerging Technology and Pharmaceutical Applications as Nanomedicine

Sudhir Verma and Diane Burgess

Abstract One of the foremost reasons for such a widespread interest in nanosuspensions as drug delivery systems is their ability to provide formulations of poorly soluble drugs with increased saturation solubility and higher dissolution rates. Nanosuspensions manifest these properties because of their small size and high surface area. The unique properties of the nanosuspensions along with their fast development times, low production costs, ease of manufacture and scale-up as well as their ability to extend the lifecycle of off-patent drugs have resulted in rapid commercialization. Nanosuspensions provide us with a useful tool for formulating an ever-increasing number of poorly water soluble drugs in pharmaceutical development programs.

10.1 Introduction

A better understanding and superior control of materials at the nanoscale level has led to rapid advances in the field of nanotechnology (Farokhzad and Langer 2006; Wagner et al. 2006; Caruthers et al. 2007). These advances have also made rapid inroads into one of the most challenging areas of medicine, i.e., drug delivery. It is not that the use of nanoscale materials is something absolutely new to this area, as Doxil, the first liposomal product, was marketed in the US as early as in 1995. However, greater awareness and the recent successes of nanotechnology-based products have spurred research in this area. Nanotechnology has made a significant impact on drug delivery, including targeting of drug nanoparticles to specific cancer tissues; delivery of drug nanoparticles directly to the brain; use of nanoparticles as imaging agents for diagnosis of cancer and cardiovascular diseases; biodegradable

D. Burgess (✉) and S. Verma
School of Pharmacy, 69 North Eagleville Road, Unit 3092, University of Connecticut,
Storrs, CT, 06269, USA
e-mail: d.burgess@uconn.edu

A.K. Kulshreshtha et al. (eds.), *Pharmaceutical Suspensions: From Formulation
Development to Manufacturing*, DOI 10.1007/978-1-4419-1087-5_10,
© AAPS 2010

nanoparticles for controlled drug delivery; pulmonary delivery of nanoparticles; gene delivery using nanoparticles; and the use of quantum dots in elucidating gene trafficking through cellular pathways. One of the most direct applications of nano-technology is the production of solid nanosuspensions (for purposes of enhanced solubility and improved bioavailability). This application was quickly embraced by the pharmaceutical industry due to the huge potential for commercial success. In addition to improvement in product performance, reformulation as a nanosus-pension can extend product patent life.

10.2 Types of Nanosuspensions

Pharmaceutical nanosuspensions can be classified into four main categories: (a) solid or crystalline drug nanosuspensions, which usually consist of crystalline drug particles, in the nano-size range, stabilized with the help of surfactants or polymers; (b) polymeric-coated drug nanosuspensions or polymeric nanoparticles, where the drug is coated or encapsulated within a polymer matrix or capsule, such as poly (lactic-co-glycolic acid) (PLGA), poly (butyl cyanoacrylate) (PBCA), etc., (Bhardwaj et al. 2005); (c) solid lipid nanoparticles, where the drug is entrapped in a lipid matrix (Mehnert and Mader 2001); and, (d) liposomes, which are spherical phospholipid bilayer vesicles containing the drug dissolved or dispersed in either the inner aqueous compartment or the lipid bilayer (Patil and Burgess 2005). Crystalline or solid nanosuspensions stand out from other nanosuspension formu-lations in terms of the minimal requirement for excipients, high drug loading, ease of manufacture and scale-up and other advantages that will be discussed in detail in this chapter.

10.3 Definition

The US Food & Drug Administration (FDA) has not yet established a formal definition of nanotechnology. However, the US National Nanotechnology Initiative (NNI) (a federal R&D program established to coordinate the multi-agency efforts in nanoscale science, engineering, and technology, in which FDA and 22 other federal agencies participate) defines "nanotechnology" only if all of the following are met:

1. Research and technology development at the atomic, molecular or macromo-lecular levels, in the length scale of approximately 1–100 nm.
2. Creating and using structures, devices, and systems that have novel properties and functions because of their small and/or intermediate size.
3. Ability to control or manipulate on the atomic scale.

In the field of pharmaceutical sciences, especially drug delivery, the benefits of small particle size can be observed well above 100 nm. Most of the commercialized products and published literature in this field show unique and advantageous properties with drug particles in the size range of 100–1,000 nm. For example, in a pharmacokinetic study in dogs, danazol (a poorly soluble gonadotropin inhibitor with an aqueous solubility of 0.001 mg/mL at 25°C (Connors and Elder 2004)) showed higher bioavailability (C_{max} and AUC 3.01 and 16.5 µg/mL, respectively) when formulated as a nanosuspension with a mean particle size distribution of 169 nm, than when formulated as a conventional suspension (Danocrine) with mean particle size of 10 µm (C_{max} and AUC 0.20 and 1.0 µg/mL, respectively). The absolute bioavailability of Danocrine was 5.2%, whereas that of the danazol nanosuspension was 82.3 % (Liversidge and Cundy 1995). Consequently, pharmaceutical nanosuspensions can be defined as sub-micron colloidal dispersions of discrete drug particles, ranging from 100 to 1,000 nm, stabilized with polymers or surfactants or a mixture of both (Kipp 2004; Patravale et al. 2004; Rabinow 2004a; Wagner et al. 2006). These have unique properties of high surface area and small size that allows for such desirable qualities as increased bioavailability, altered disposition as well as drug targeting. The dispersion medium for nanosuspensions is generally aqueous but it can also be hydroalcoholic or non-aqueous (Muller et al. 2001, 2002; Keck and Muller 2006).

10.4 Advantages/Disadvantages

10.4.1 Increased Saturation Solubility and Dissolution Rate

One of the foremost reasons for such a widespread interest in nanosuspensions as drug delivery systems is their ability to provide formulations of poorly soluble drugs with increased saturation solubility and higher dissolution rates. Nanosuspensions manifest these unique properties because of their small size and high surface area. The kinetics of drug particle dissolution can best be described using the Noyes Whitney equation

$$\frac{dC}{dt} = \frac{DS}{Vh}(C_s - C)$$

where D is the diffusion coefficient of the solute in solution, S is the surface area of the exposed solid, h is the thickness of the diffusion layer, C_s is the saturation solubility of the solute particle, C is the concentration of the solute in the bulk solution at time t, V is the volume of solution, and dC/dt is the dissolution rate (Martin 2001). The increased dissolution rate observed with nanosuspensions can be explained by the high surface area of the drug particles present in them, as surface area is directly proportional to the rate of dissolution. Dissolution rate is also inversely proportional to the thickness of the diffusion layer. So the question arises as to how does a smaller particle size affect this parameter? It has been experimentally proved in a number of studies that the smaller the particle size, the smaller the diffusion layer thickness and the higher the dissolution rate (Bisrat and Nystrom 1988; Hintz and Johnson 1989; Wang and Flanagan 2002; Tinke et al. 2005). The relationship between the boundary layer thickness and the particle size can be best understood using the Prandtl equation:

$$h_H = k \left(\sqrt{\frac{L}{V}} \right)$$

in which h_H is the hydrodynamic boundary layer thickness (diffusion layer thickness of the Noyes Whitney equation), V is the relative velocity of the flowing liquid against a flat surface, k is a constant, and L is the length of the surface in the direction of flow. It describes the effect of surface curvature on boundary layer thickness – the greater the surface curvature, the thinner the boundary layer. Small particles have larger surface curvature as compared to large particles and therefore boundary layer thickness is relatively small in the case of nanoparticles, since drug diffusion from the particle surface to the bulk is rapid (Patravale et al. 2004).

In addition to affecting boundary layer thickness, the higher surface curvature of small particles also increases the saturation solubility (C_s) of the drug, thus increasing the ($C_s - C$) term in the Noyes Whitney equation. The inter relationship between curvature, interfacial free energy and solubility of the particles is given by the Gibbs–Thomson effect or Kelvin effect (Adamson 1997; Hiemenz and Rajagopalan 1997; Borm et al. 2006) According to this theory, smaller particles have higher curvature or their surface is more convex which is not favorable energetically. Therefore these particles have higher dissolution pressure and show higher equilibrium solubility. The Gibbs–Thompson effect for spherical solids is described by Ostwald–Freundlich equation:

$$\ln \left(\frac{S}{S_0} \right) = \frac{2v\gamma}{rRT} = \frac{2M\gamma}{\rho rRT}$$

where S is drug solubility at temperature T, S_0 is the solubility if $r=\infty$, M is the molecular weight of the compound, v is the molar volume, γ is the interfacial tension, and ρ is the density of the drug compound. It predicts that the larger particles have lower solubility than the smaller particles. Though this effect is not substantial at particle sizes above 2 µm, as we go below 2 µm, the effect of particle size on solubility becomes much more pronounced (Hiemenz and Rajagopalan 1997; Borm et al. 2006). Increased saturation solubility increases the concentration gradient of the dissolved drug between the drug particle surface and the bulk solution, which results in increased dissolution velocity due to increased drug diffusion from the surface into the bulk solution, as per Fick's first law.

$$J = -D \left(\frac{dC}{dx} \right)$$

where *J* is the flux, *D* is the diffusion coefficient, *C* is the concentration, and *x* is the distance perpendicular to the drug surface (Martin 2001).

10.4.2   *Superior Clinical Performance*

A direct consequence of the increase in saturation solubility and fast dissolution rates of pharmaceutical nanoparticles is their improved pharmacokinetics in terms of: increased rate and extent of absorption; rapid onset of action (Yin et al. 2005; Jinno et al. 2006); reduced variations in fed and fasted state (Wu et al. 2004); reduced gastric irritancy (Liversidge and Conzentino 1995); and better clinical efficacy (Kayser et al. 2003).

10.4.3   *Formulation Benefits*

Nanosuspensions provide us with a unique tool for formulation of drugs that fall under class II and IV categories of the biopharmaceutical classification. Nanosuspensions are the formulations of choice for high dose compounds that show poor solubility both in oil (high melting point) and water (high log P) (Rabinow 2004a). In addition, suspension formulations can achieve high drug loading compared to solutions, which is very helpful especially for intramuscular, subcutaneous and intraocular applications where the total volume administered is a constraint. Since harsh chemicals and solvents are usually avoided during nanoparticle processing, the risk of toxicity from these sources is typically not a concern. In some cases higher drug stability is achieved when formulated as nanosuspensions (Moschwitzer et al. 2004).

10.5   Approved and other Developmental Products

The unique properties of the nanosuspensions along with their fast development times, low production costs, ease of manufacture and scale up as well as their ability to extend the lifecycle of off-patent drugs have resulted in rapid commercialization. The popularity of nanosuspensions can be judged by the fact that in the last decade, six products based on crystalline nanosuspensions have been marketed in the US with many more in development. Table 10.1 is a comprehensive list of the products that have been marketed and/or are in various stages of clinical development (Rabinow 2004a; Wagner et al. 2006).

Table 10.1 Marketed and developmental formulations based on solid nanosuspensions

Drug	Indication	Route	Status	Company
Emend (aprepitant)	Anti-emetic	Oral	Marketed	Merck/Elan
Rapamune (sirolimus)	Immuno-suppresant	Oral	Marketed	Wyeth/Elan
Megace ES (Megesterol acetate)	Eating disorders	Oral	Marketed	Par/Elan
Tricor (fenofibrate)	Lipid regulation	Oral	Marketed	Abbott/Elan
Trigilde (fenofibrate)	Lipid regulation	Oral	Marketed	Sciele Pharma/Skyepharma
Abraxane (paclitaxel)	Anti-cancer	IV	Marketed	Abraxis Bioscience/AstraZeneca
Paliperidone palmitate	Schizophrenia	IM	Phase III	J&J/Elan
NPI 32101	Atopic dermatitis	Topical	Phase II	Nucryst
Panzem NCD	Glioblastoma	Oral	Phase II	Entremed/Elan
BioVant (calcium phosphate)	Vaccine adjuvant	IM	Phase I	BioSante
Undisclosed multiple	Anti-infectiveAnti-cancer	Oral/IV	Preclinical to Phase II	Baxter
Cytokine inhibitor	Crohn's disease	Oral	Phase II	Cytokine PharmaSciences /Elan
Diagnostic agent	Imaging agent	IV	Phase I/II	Photogen/Elan
Thymectacin	Anti-cancer	IV	Phase I/II	Newbiotics/Elan
Busulfan	Anti-cancer	Intrathecal	Phase I	Supergen/SkyePharma
Budesonide	Asthma	Pulmonary	Phase I	Sheffield/Elan
Silver	Eczema, atopic dermatitis	Topical	Phase I	Nucryst
Insulin	Diabetes	Oral	Phase I	BioSante

10.6   Formulation

10.6.1   Theory

Preparation of nanosuspensions involves formation of a large number of very small particles with enormous surface area in a dispersion medium which is usually aqueous.

The high surface area associated with these tiny particles results in thermodynamically unstable high energy systems. Because like molecules tend to remain together and repel unlike molecules there is an increase in the interfacial tension of the system which in turn raises the total free energy. This increase in free energy is given by the following equation:

$$\partial G = \partial A \gamma$$

in which ∂G is the increase in Gibbs free energy, ∂A is the increase in surface area, and γ is the interfacial tension at the solid/liquid interface (Hiemenz and Rajagopalan 1997; Martin 2001). To achieve stability, the system tends to decrease the surface area and hence interfacial tension, by promoting agglomeration and crystal growth. Therefore, the addition of stabilizing agents, which reduce interfacial tension and prevent agglomeration, is indispensable to any successful formulation of stable nanosuspensions. Maximum stability is achieved when these agents are present during the process of particle formation (in bottom up methods) or during particle size reduction (top down methods). Stabilizing agents that are used include phospholipids, polymers or surfactants (both ionic and nonionic) or a combination of these (Peters et al. 1999; Kayser 2000; Trotta et al. 2001, 2003; El-Shabouri 2002; Jacobs and Muller 2002).

Recently, there were reports in the literature about research being carried out to prepare specialized stabilizers for nanosuspensions. For example, Lee et al. (2005) synthesized amino acid block copolymers based on various combinations of lysine, phenylalanine, leucine and alanine via ring opening polymerization. Various compositional and morphological effects of block copolymers were investigated in this study. It was determined that copolymers containing lysine and phenylalanine or lysine and leucine, were able to form stable nanosuspensions of naproxen, with particle size of 200–300 nm, by wet milling. However, copolymers of lysine and alanine were unable to stabilize drug nanoparticles. The hydrophobicity of the copolymers was determined to be of more critical importance than their morphology, in their ability to achieve stable nanoparticles.

There are two mechanisms by which colloidal particles can be stabilized in nanosuspensions:

1. Electrostatic repulsion
2. Steric stabilization

Stability due to electrostatic repulsion can be explained by the classic theory on lyophobic colloid stability given by Derjaguin, Landau, Verwey and Overbeek, also known as the DLVO theory (Hiemenz and Rajagopalan 1997; Martin 2001). According to this theory, there are two major forces acting on colloidal particles in a dispersion medium: electrostatic repulsive forces due to overlap of electrical double layers and van der Waals attractive forces. The electrostatic double layer arises as a consequence of the charge at the solid–liquid interface. When an immiscible solid, liquid or gas is immersed in a liquid, a surface charge may be acquired by: (1) preferential adsorption of ions; (2) dissociation of ionizable surface groups (3) isomorphic substitutions; (4) adsorption of polyelectrolytes; and, (5) accumulation of electrons on a surface. To maintain electrical neutrality of the system as a whole, counter ions present in the media are attracted toward the surface to form a double layer of ions: a tightly bound first layer of ions, also known as the Stern layer; and a diffuse layer of ions, also called the Gouy or Gouy –Chapman layer. When particles in a medium approach close enough to allow overlap of their individual double layers, repulsion occurs and they move away from one another thus preventing aggregation. It is very difficult to express electrostatic repulsion quantitatively due to ever-changing dynamics of the system when two particles approach. A simplified quantitative expression of the repulsion between two spherical particles of equal radius can be given by the Derjaguin approximation:

$$V_R = 64a\pi n_\infty \kappa^{-2} kT\gamma^2 \exp(-\kappa H)$$

where V_R is the potential energy of repulsion, n_∞ is the bulk concentration of the ions, a is the particle radius, k is the Boltzmann constant, κ is the reciprocal of the thickness of the double layer, H is the distance between the two particles, T is the absolute temperature, and γ is given by

$$\gamma = [\exp(ze\psi / 2kT) - 1] / [\exp(ze\psi / 2kT) + 1]$$

in which ψ is the potential associated with the double diffuse layer. The following assumptions have been made in the above equations: potential is small and constant, diffuse double layer is thick and particles are far apart so that there is minimal overlap of their double layers (Hiemenz and Rajagopalan 1997; Cherng-ju 2004).

The attractive forces between the dispersed particles are van der Waals forces which arise due to dipole or induced-dipole interactions within the atoms of the material. The energy of attraction between two spheres of equal radius a, and separated by a distance H (for $a >> H$) is given by

$$V_A = \frac{-Aa}{12H}$$

where A is the Hamaker constant.

The total energy of interaction (V_T) between the colloidal particles of a nanosuspension can be given by

$$V_T = V_R + V_A$$

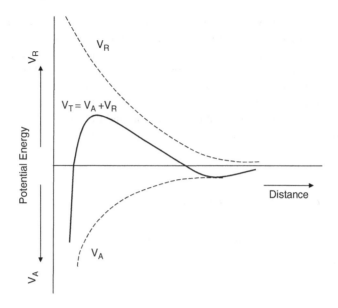

Fig. 10.1 Potential energy diagram of two interacting nanoparticles

The interactions between these two forces are often shown in the potential energy diagram (Fig. 10.1). At very small and large distances, attractive forces, which vary linearly with the distance, are dominant resulting in a primary minimum and secondary minimum, respectively. At intermediate distances, repulsive forces, which decay exponentially, are dominant resulting in net repulsion between the particles of the dispersion thus preventing aggregation or flocculation.

Another mechanism by which stability can be achieved is steric stabilization. Polymers usually provide stability via this mechanism. When polymers are added to the nanosuspensions in high concentration, they get adsorbed onto the surfaces of newly formed particles of the hydrophobic drug with their hydrophobic parts attached to the particle surface and their hydrophilic chains extending into the aqueous environment. These long polymeric chains that extend into the water prevent the two particles from coming very close to each other due to steric effects, also known as volume restriction effects. In addition, when two such particles approach each other, the hydrated polymer chains are compressed and water is squeezed out resulting in a deficiency of water in the region between the two approaching particles. This results in an osmotic effect, with water rushing into the depleted region and thus moving the particles apart from one another (Burgess 2005).

Electrostatic stabilization is more susceptible to the ionic strength of the dispersion medium. High concentrations of ions in the dispersion medium lead to the screening of the surface charge, which decreases the thickness of the diffuse double layer. The depleted double layer makes the dispersed particles more susceptible to aggregation at the primary minimum and hence destabilization. On the other hand, the hydration of the polymers is more susceptible to variation in temperature and hence suspensions stabilized only sterically are more prone to destabilization by temperature fluctuations. Therefore, a combination of both ionic surfactants and nonionic surfactants/polymers can be beneficial in optimizing stability of colloidal dispersions. In addition, the presence of a polymeric stabilizer along with an ionic stabilizer reduces the self repulsion between the ionic surfactant molecules which facilitates close packing of the stabilizer layer around the particle. A close packed layer is usually much more efficient in preventing agglomeration.

10.6.2   Methods of Manufacture

Nanosuspensions can be manufactured by different processes:

- Top down processes
- Bottom up processes and
- Combination processes

10.6.2.1   Top Down Process

The top down approach consists of breaking down bigger particles into smaller particles by various milling techniques. Particle size reduction can be achieved by different methods namely: media milling, high pressure homogenization and microfluidization. No harsh solvents are used in these processes and high drug loading can be achieved. However, these are high energy processes in which a lot of heat is generated, and therefore, unless cooling accessories are employed thermolabile materials are difficult to process by these methods (Rasenack and Muller 2004).

10.6.2.1.1   Media Milling/Nanocrystals®

Media milling or Nanocrystals® is the patented technology of Elan Drug Delivery Systems. It was first developed by Liversidge et al. in 1991. This technology was used successfully to market the first nanosuspension product, Rapamune. The process involves charging the milling chamber as shown in Fig. 10.2 with water, milling media, drug and stabilizer and rotating the milling shaft at high speed. High shear forces are generated in the milling chamber due to the impact of the milling media, and attrition between the particles and the milling media causes the particles to fracture along weak points. The milling media generally consists of glass, zirconium oxide or highly crosslinked polystyrene resin beads (Merisko-Liversidge et al. 2003). One of the main disadvantages of this technique is contamination of the product with undesirable particles from the media and equipment parts as a result of erosion.

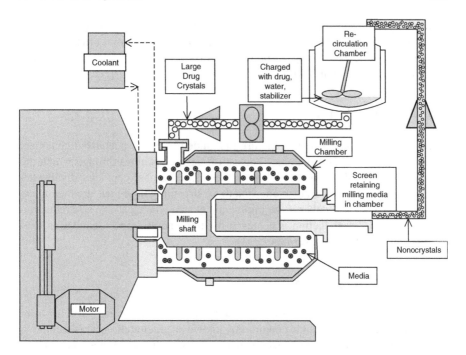

Fig. 10.2 The Media Milling Process is shown in a schematic representation. The milling chamber charged with polymeric media is the active component of the mill. The mill can be operated in a batch or re-circulation mode. A crude slurry consisting of drug, water and stabilizer is fed into the milling chamber and processed into a nanocrystalline dispersion. The typical residence time required to generate a nanometer-sized dispersion with a mean diameter <200 nm is 30–60 min. (Adapted from Merisko-Liversidge et al. 2003, with permission)

10.6.2.1.2   High Pressure Homogenization

High pressure homogenization was first developed and patented by R.H. Muller in the early nineties (Müller et al. 1996). This technology, now owned by Sykepharma LLC, is commonly known as Dissocubes®. It involves passage of a macro suspension of the drug through a small aperture under high pressure. The high velocity of the suspension in the small aperture reduces the pressure tremendously, resulting in the formation of bubbles as per Bernoulli's law. When the suspension emerges from the narrow aperture there is a drop in velocity and an increase in pressure to the atmospheric pressure. This causes bubbles to implode and generate high energy shock waves which are mainly responsible for particle size reduction. Turbulent flow and high shear stress also contribute to particle fracture (Phipps 1971; Schultz et al. 2004). Various poorly soluble drugs such as buparvaquone (Muller and Jacobs 2002), budesonide (Jacobs and Muller 2002), and clofazimine (Peters et al. 2000) RMKP22 (Grau et al. 2000) have been successfully processed into nanosuspensions using this technique. The particle size achieved depends on the hardness of the drug crystal itself, the number of cycles and the applied pressure (Muller et al. 2001). The main disadvantage of this technique is that the initial drug should be micronized before processing to prevent blockage of the narrow orifice during operation (Patravale et al. 2004).

10.6.2.1.3 Microfluidization

In this technique, suspension of the drug is guided through specialized chambers under high pressure. The chambers consist of narrow openings through which the suspension is forced to pass at high velocities. The chamber geometry divides the suspension into two different streams and these streaming liquid jets are then made to impinge against each other in the impingement area where particle size reduction occurs. Particle size reduction is achieved by high energy impact, cavitation, and shear forces Fig. 10.3 (Gruverman 2003).

10.6.2.2 Bottom Up Process

The bottom up approach refers to the building up of the nano-sized particles from their molecular solutions, and is commonly known as precipitation. This technique is useful when the active pharmaceutical agent can be dissolved in a non aqueous water miscible solvent. Bottom up processes can be carried out at ambient temperatures, and therefore heat sensitive materials can be processed easily. There are different variations of this approach, such as (1) solvent–antisolvent, (2) supercritical fluid process, (3) emulsion-solvent evaporation, and, (4) spray drying.

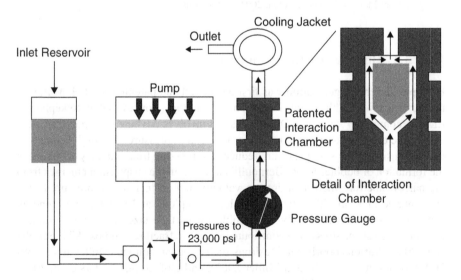

Fig. 10.3 Basic schematic of single pump microfluidizer and interaction chamber (Adapted from Gruverman 2003, with permission)

10.6.2.2.1   Solvent Anti-solvent

One of the earliest applications of this method for the formulation of poorly soluble drugs is the Hydrosol technology that was patented in 1988 (List and Sucker
1988, 1995). Hydrosols consist of the finely precipitated colloidal drug particles
in an aqueous medium. In this technique, the drug is dissolved in a water miscible
organic solvent such as ethanol. The organic solution is then poured slowly into a
vessel containing a large amount of water. The stabilizers can be either added to
the organic solution of the drug or they can be present in the aqueous phase. Water
constitutes the bulk of the dispersion, almost 95% of the formulation. These dispersions are highly susceptible to particle growth and are stable only for short
periods of time. Freeze drying or spray drying is usually employed immediately
after precipitation to maintain the particle size (Gassmann et al. 1994). Nanoparticles
can   also   be   produced   by   controlled   addition   of   the   antisolvent into the solution of a drug in an organic solvent in the presence of a stabilizer (Rogers et al. 2004). In such procedures, the rate of addition of the anti
solvent, mixing speed, ratio of the solvent to anti solvent, concentration of the
stabilizer, and solubility of the drug itself play an important role in the successful
formation of nano-sized particles (Rasenack and Muller 2002; Schwarzer and
Peukert 2002). Different variations of the solvent-antisolvent process have been
investigated for the preparation of nanoparticles, such as evaporative precipitation
in aqueous solution (Sarkari et al. 2002). In this method, a heated solution of the
drug in water immiscible solvent is atomized in a heated aqueous solution containing suitable stabilizer with the help of a fine nozzle. Fine particles coated with
stabilizers are obtained due to rapid evaporation of the organic solvent. Other
methods of salting out such as change in temperature and pH (Pozarnsky and
Matijevic 1997) can also be used to prepare ultra fine drug particles to improve
their dissolution and hence bioavailability.

10.6.2.2.2   Supercritical Fluid Process

As described above, nano-sized particles can be produced by addition of anti-solvent, evaporation of solvent, or changes in temperature and pH etc. However, the
above methods are based only on mixing, highly efficient mixers and controlled
conditions are required to propagate the changes efficiently throughout the bulk to
produce reproducible batches of nanoparticles. Therefore, there is a need for a solvent or anti-solvent that has liquid properties yet can diffuse or mix rapidly like a
gas, so that the change can be affected instantaneously. This makes supercritical
fluids, which have high diffusivities and unique solvent properties, ideal candidates
for the preparation of nanoparticles. Carbon dioxide is the most extensively used
supercritical fluid (SCF). It has all the desired properties of a suitable pharmaceutical solvent such as nontoxicity, low inflammability, environmentally clean, chemical inertness, cheap, abundant, easy to remove and low critical conditions (pressure

and temperature). Another advantage of supercritical process is the smooth surface morphology and low surface energy of the nanoparticles prepared as compared to other techniques such as micronization (Hooton et al. 2004). On the basis of how the SCF is used, this methodology of particle engineering can be classified into two broad categories (Texter 2001):

1. Supercritical fluid as a solvent (RESS / RESOLV)

In this method, SCF CO_2 is used as a solvent for insoluble drugs. Solubility of drugs in the SCF CO_2 is highly dependent on the SCF pressure. In RESS or Rapid Expansion of Supercritical Solutions, the drug is first solubilized in SCF CO_2 at high pressure in a chamber. The solution is then pumped into an expansion chamber through a nozzle to allow for the rapid expansion of SCF CO_2. Rapid expansion of the SCF causes the drug solubility to decrease rapidly, resulting in a high degree of supersaturation which leads to the formation of nanoparticles. The parameters affecting the size and shape of the particles include: temperature, pressure drop, distance of impact, and design of the atomization assembly, etc. Usually, the particle size obtained with this process ranges from 1 to 5 μm, due to aggregation of the fine particles initially generated during the process. However, Pathak et al. (2006) improved this process by atomizing the SCF solution of the drug into liquid solvent containing stabilizers to prepare particles below 100 nm in size. They named their method Rapid Expansion of Supercritical Solutions into Liquid Solvent or RESOLV. Precipitation of the particles in the presence of solvents containing stabilizers prevented the aggregation of primary particles formed due to evaporation of SCF in CO_2 (Pathak et al. 2007). Figure 10.4 shows ibuprofen nanoparticles produced by the RESOLV process (Pathak et al. 2006).

2. Supercritical fluid as an antisolvent (GAS/SAS/ASES/SEDS)

In this process, SCF CO_2 is used as an anti solvent because of its ability to preferentially solubilize non polar, small molecule solvents compared to the larger drug molecules. In these processes, SCF CO_2 is used to decrease the solubilizing power of the organic solvents in which the drug is dissolved. This can be achieved in three different ways:

(a) In the Gas Anti Solvent (GAS) or the Supercritical Anti Solvent (SAS) method, SCF CO_2 is directly introduced into a solution of the drug. Dissolution of the solvent into SCF supersaturates the solution with respect to the drug, resulting in the precipitation of fine drug particles.
(b) In Aerosol Solvent Extraction System or ASES, a drug solution is sprayed into a vessel containing SCF CO_2, which results in the rapid extraction of the solvent and subsequent precipitation.
(c) In Solution Enhanced Dispersion by Supercritical Fluids or SEDS, the drug solution is atomized with the help of SCF CO_2 through a specially designed nozzle that maximizes interaction.

A detailed compilation of the literature and patents covering different processes based on SCF can be found in the review by Jung and Perrut (2001).

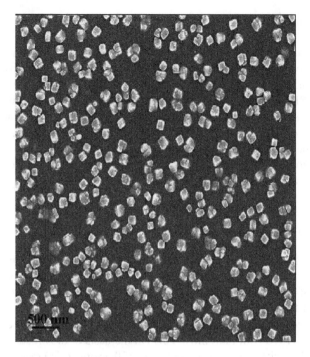

Fig. 10.4 An SEM image of the PVP (3 mg/mL)-protected ibuprofen nanoparticles obtained with a high ibuprofen concentration (1.5 mg/mL in CO_2) in RESOLV (40°C, 200 bar) (Adapted from Pathak et al. 2006, with permission)

10.6.2.2.3   Emulsion-Solvent Evaporation

Emulsions have a long history for being used as microreactors for the production of nanosize inorganic particles such has titanium dioxide (Chhabra et al. 1995; Mori et al. 2001), copper (Qi et al. 1997), silica (Arriagada and Osseoasare 1995), calcium carbonate (Hirai et al. 1997), etc. This procedure allows for the preparation of monodisperse particles free from aggregates. The particle size distribution depends on the size of the stabilized emulsion droplets, and thus can be controlled easily (Shchukin and Sukhorukov 2004). Due to these advantages, emulsions are being used as templates for the preparation of drug nanosuspensions (Trotta et al. 2001; Trotta et al. 2003; Kocbek et al. 2006). Two types of emulsions can be used for the preparation of the drug nanosuspensions. The first type of emulsion consists of the drug solubilized in a volatile organic solvent, such as methylene chloride, which is then emulsified with the help of suitable stabilizers to produce stable nanoemulsions. Removal of the solvent from the nanodroplets under reduced pressure produces drug nanosuspensions (Tan and Nakajima 2005). The second type involves the use of solvents such as butyl lactate, ethyl acetate and benzyl alcohol that are partially miscible with water. In this method, solvent containing the dissolved drug is emulsified in water to produce an O/W water emulsion. This emulsion is then diluted with excess water which leads to the formation of drug nanoparticles.

One of the main disadvantages of these methods is the need to remove excess stabilizer and solvent by suitable techniques such as dialysis, ultrafiltration, diafiltration, and tangential flow filtration, etc (Dalwadi et al. 2005). Shekunov et al. (2006) recently proposed a new method for the preparation of nanosuspensions by emulsification. In this method, which is known as supercritical fluid extraction of emulsions (SFEE), cholesterol acetate nanosuspensions with mean volume weighted diameter of less than 500 nm were prepared by extraction of the dispersed phase solvent (ethyl acetate) by spraying the emulsion into a chamber containing SCF CO_2. The level of ethyl acetate in the final suspension was less than 54 ppm for all conditions studied. Another advantage of SFEE over SAS is the morphology of crystal formed as shown in Fig. 10.5.

10.6.2.2.4   Spray Drying

Spray drying has been used for a number of applications in the pharmaceutical industry such as conversion of crystalline substances into amorphous materials, preparation of solid dispersions, coating, drying of solids, etc. Usually the particle size obtained by spray drying lies in the micrometer range. However, Yin et al. (2005) reported the preparation of nanocrystalline dispersions of a research compound BMS-347070 with Pluronic F-127. On the basis of the broadening of peaks in powder X-Ray diffraction, the particle size of the spray dried material was determined to be in the range of 40–60 nm. Another variation of this technique is known as spray freezing. In this technique the active pharmaceutical ingredient (API) is dissolved in an organic solvent and sprayed through a nozzle beneath the surface of liquid nitrogen. Spraying in such conditions results in immediate freezing of the atomized droplets which are then lyophilized to remove the solvents and obtain the drug nanoparticles (Hu et al. 2002; Rogers et al. 2002).

10.6.2.3   Combination Techniques

Rapid precipitation of drug particles obtained using the bottom up processe often results in particles which have dendritic morphology due to rapid crystal growth in one direction. These needle-like crystals are more fragile than normal crystals and hence can be easily size reduced. The rapid precipitation results in the formation of crystals with more imperfections and crystal defects, compared with their traditional counterparts that have been separated by slow crystallization techniques. These imperfections and impurities result in many weak planes where the particles can be readily broken down. Therefore, it is prudent to apply size reduction techniques to these newly formed particles to create nanosuspensions. Two such technologies that have been recently patented are:

Baxter Nanoedge® which consists of homogenization of freshly prepared particles with a piston gap homogenizer to prepare stable drug nanosuspensions (Rabinow 2004b) (Fig. 10.6). Immediate homogenization of freshly created particles helps

Fig. 10.5 (**a**) Morphology of griseofulvin crystals produced using SFEE method in comparison with (**b**) crystals produced using supercritical anti-solvent precipitation under the same conditions of temperature, pressure and solution flow rate (Adapted from Shekunov 2006, with permission)

not only in achieving smaller particle size but also improves surface coverage by the stabilizer. Particles formed by rapid precipitation are usually amorphous and tend to crystallize with time. This additional homogenization step also provides energy for transformations which helps in the formation of stable nanosuspensions.

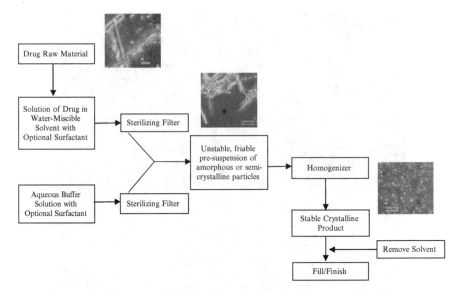

Fig. 10.6 Schematic of Baxter Nanoedge technology (Adapted from Rabinow 2004b, with permission)

Microfluidization Reaction Technology (MRT) is an amalgamation of the "bottom up" and "bottom down" approaches. In MRT pressurized solutions of the drug and its anti solvent are pumped through a coaxial feed system within a Microfluidizer® reaction chamber, where streams of these liquids collide with each other at supersonic speeds Fig. 10.7 (Gruverman 2003). The velocities of the streams can go up to 300 m/s. Extreme turbulent flow conditions produced by the impinging of jets of liquids helps in creating an ideal mixing environment which is very conducive for the homogenous and rapid nucleation with little time for crystal growth. The intense mixing also minimizes local variations in pH or concentration improving uniformity of composition, crystal size and phase purity. In this system different mixing regimens, macro-, meso-, or micro-mixing, can be achieved by precise control of the feed rate and mixing location. Particle size reduction of the newly formed particles is achieved in the interaction chamber due to high shear forces and cavitation. This approach allows for a greater ability to control the growth rates of nanoparticles to produce uniform, optimally sized nanoparticles in a more efficient, cost-effective manner.

10.6.3 Characterization of Nanosuspensions

Physicochemical and biological characterization forms an integral part of dosage form development with application to method optimization, process control, formulation aesthetics, and most importantly, stability. In the case of nanosuspensions,

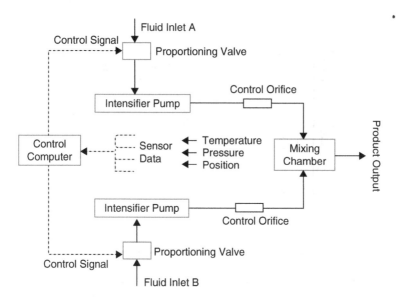

Fig. 10.7 Schematic of microfluidics reactor technology (Adapted from Gruverman 2003, with permission)

the biological performance depends on the physical state of the particles, and so, an understanding of the biophysical properties is of critical importance. Depending on the intended route of administration, many physicochemical characterization studies are necessary for suspension formulations such as: assay, determination of degradation products, particle size analysis, viscosity, sterility, pyrogenicity, pH, syringebility, moisture content (if stored as a solid for reconstitution), preservative content, etc. This chapter focuses only on the physicochemical properties of particular relevance to nanosuspensions. These include mean particle size, particle size distribution, surface charge, morphology, crystallinity and dissolution.

Particle size forms the crux of the whole multi-billion dollar nanotechnology industry and is the central issue of many scientific discussions (Barth and Flippen 1995; Haskell 1998; Burgess et al. 2004; Shekunov et al. 2007; Snorek et al. 2007). Accurate determination of mean particle size and particle size distribution is one of the most important characterization tests as the unique characteristics shown by nanosuspensions are due to their size. Particle size determination techniques can be broadly classified into three categories: ensemble, counting, and separation techniques. Ensemble methods (such as laser diffraction and photon correlation spectroscopy) are those techniques that assess some bulk property of the entire distribution and, with the help of appropriate mathematical modeling, convert that property into a particle size distribution. Direct counting includes techniques such as light obscuration, optical, and electron microscopy where individual particles are counted to construct a particle size distribution. Separation techniques involve classification of particles into different size ranges based on their behavior (e.g. ultracentrifugation, hydrodynamic chromatography, etc.). Table 10.2 gives a list of the

Table 10.2 List of particle size distribution techniques

Technology	Optimal particle shape	Size range (μm)		Distribution		Sample	
		Min	Max	Width	Modality	Dispersion	Concentration
Acoustic spectroscopy	O⬚	0.01	10	⋏		Wet	High
Chord-length measurement	O⬚	1	10,000	⋏⬤	⋀⋀	Wet	Low high
Disc centrifuge	O⬚▱▱	0.005	100	⋏⬤	⋀⋀	Wet	Low
Dynamic image analysis	O⬚▱	0.05	3,500	⋏⬤	⋀⋀	Wet and dry	Low
Elliptically polarized light scattering	O⬚▱	0.05	10	⋏		Wet	Low
Electrical sensing zone	O⬚▱	0.4	1,600	⋏⬤	⋀⋀	Wet	Low
Hydrodynamic chromatography	O⬚▱∿	0.01	50	⋏⬤	⋀⋀	Wet	Low
Laser diffraction	O⬚	0.01	>5,000	⋏⬤		Wet and dry	Low
Light obscuration	O⬚	0.5	5,000	⋏⬤	⋀⋀	Wet	Low
Photon correlation spectroscopy	O⬚▱	0.003	3	⋏⬤		Wet	Low
Polarization intensity differential scattering	O⬚	0.04	0.4	⋏⬤		Wet	Low
Sieve analysis	O⬚	5	10,000	⋏⬤		Wet and dry	High
Scanning electron microscopy	O⬚▱∿	0.001	5	⋏⬤		Wet and dry	Low
Static image analysis (optical microscopy)	O⬚▱∿	1	10,000	⋏⬤		Wet and dry	Low
Time of flight	O⬚▱	0.3	500	⋏⬤	⋀⋀	Wet and dry	Low

O, spherical; ⬚, blocky; ▯, acicular; ▱, platy, tabular, bladed; ∿, fibours.

techniques, along with their applicable size range, that can be used for particle size determination (Snorek et al. 2007)

Though ensemble techniques are usually the methods of choice because of their convenience, accuracy and speed, they fair poorly at the extremes of their limits. The mathematical modeling techniques used in these methods are predisposed to exclude those particles that lie in the extremes of the instrument limit or which represents a small proportion of the distribution from the calculations giving rise to skewed distributions (Frantzen et al. 2003). Provided good sampling techniques are used, counting methods provide the actual distribution of the system. However, they tend to be slow and time consuming and therefore are not suitable for routine analysis. Separation techniques only give the range of sizes and not the actual distribution. In separation techniques, it should be ensured that particle separation occurs primarily due to particle size and not due to any interaction between the particles and the parts of the separating technique. As a consequence of all these limitations, it is advisable to use a combination of techniques to obtain a complete picture of the particle size distribution (Burgess et al. 2004).

Surface properties, such as charge and surface roughness, play an important role not only in the physical stability of nanosuspensions, but they also have a bearing on the *in vivo* functioning of the formulation. A direct correlation has been shown in several studies between opsonization, phagocytosis, and the charge on the particle (Devine et al. 1994; Roser et al. 1998). Negatively charged particles activate the complement system at much lower concentrations than neutral or positively charged particles. Surface charge density also affects the clearance of the particles by reticulo-endothelial system, the higher the charge density, the faster the clearance rate (Juliano and Stamp 1975). Electrophoretic or electoacoustic techniques can be used to determine surface charge. As the particle size is decreased, the surface roughness starts to play a dominant role in interactions between the colloidal particles. Depending on the ratio of the particle size to roughness, surface heterogeneities can affect both electrostatic and van der Waals forces. When this ratio is high, adhesion forces decrease by an order of magnitude (Walz 1998). Surface roughness of nanoparticles can be determined by atomic force microscopy (Young et al. 2002).

Information about bulk morphology and molecular arrangement, i.e., crystalline vs. amorphous of the API, is a key component for designing a stable formulation. Conversion of crystalline forms to amorphous forms is often encountered during milling procedures. Different polymorphic forms may be generated during precipitation techniques, depending on the solvents employed. Process parameters may also effect the arrangement of molecules in the nanoparticles. For example, when fluconazole was dissolved in acetone and processed via supercritical fluid antisolvent (SAS) technique, anhydrate form I was precipitated at 40°C and form II was obtained at 80°C. The anhydrate form II can also be obtained at 40°C if ethanol is used as a solvent instead of acetone (Park et al. 2007). These changes can affect both the performance and stability of the formulation, since different polymorphs have different solubility and chemical stability at a given temperature. Changes in the crystal structure and generation of solvates or amorphous material during processing or stability can be investigated by techniques such as X-ray diffraction and differential scanning calorimetery. Scanning electron microscopy and atomic force microscopy can be used to determine the shape and structural characteristics of the particles (Shi et al. 2003).

Dissolution testing of formulations based on solid nanosuspensions is very challenging. Conventional dissolution techniques of dispersed systems involve sample and separate methods, where the dosage form is added to the dissolution media and the amount released or dissolved is determined by sampling, filtering out the dispersed system, and analyzing using a combination of chromatographic and spectroscopic techniques. Separation of the unbound or molecular drug from the bound or nanoparticulate drug becomes really difficult for particles in the nanometer range. Most of the available filters can easily separate particles above 200 nm, but as size is decreased excessive pressure is required for filtration, and blockage of filter pores is a big problem. Slippage and dissolution of particles can occur when excessive pressures are used during filtration. Other separation methods such as dialysis membranes or ultracentrifugation can be used. However, these methods suffer from long processing times and may not be suitable for characterizing the fast dissolution rates of nanosuspension. The answer to these problems may lie in developing alternate methods that do not require separation or those that monitor other properties of the colloidal system such as turbidity.

The use of fiber optic UV probes is gaining immense popularity for the in situ determination of chemical compounds due to its ease of use, no sampling requirement, and ability to almost continually collect data. In this technique, data acquisition is fast and data can theoretically be collected at an interval of as small as 5 s, which provides a large number of data points. Any discrepancies due to scattering can be removed by second derivative analysis of the UV scans. These advantages can be exploited for characterizing nanosuspension products with high dissolution rates (Bijlani et al. 2007). Recently, a turbidimetric method has also been patented and used for the evaluation of solubility and dissolution characteristics of nanosuspensions (Tucker 2004; Crisp et al. 2007). In this method, change in turbidity with nanoparticle dissolution is monitored to quantify drug release.

10.7 Applications

Nanosuspensions have gained importance in the past several years as a result of their formulation and biopharmaceutical benefits, as detailed below. In addition, nanosuspensions can be administered via all major routes, such as oral (Liversidge and Conzentino 1995; Liversidge and Cundy 1995; Vergote et al. 2001; Kayser et al. 2003; White et al. 2003; Yin et al. 2005), intramuscular, intravenous (Scholer et al. 2001; Moschwitzer et al. 2004), intraocular, and pulmonary (Jacobs and Muller 2002).

10.7.1 Formulation Benefits of Nanosuspensions

Nanosuspensions are used as a means to formulate insoluble drugs, and compared to other methods (e.g. organic solvents, micellar solutions, complexation with water soluble carriers such as cyclodextrins, emulsions and liposomes), nanosuspensions have the major advantage of high drug loading, whereas other formulation methods used for insoluble drugs generally have high excipient to drug ratio. For example, two marketed liposomal formulations of doxorubicin, Myocet® and Doxil®, have drug:lipid ratios of 0.27:1 and 0.4:1, respectively (Allison 2007). In comparison, solid nanosuspensions can achieve drug:stabilizer ratios as high as 5:1 (Pace et al. 1999) which makes them very attractive dosage forms for many applications (for example, parenteral and ophthalmic applications (Jarvinen et al. 1995) where low volumes are desired).

Another major advantage of nanosuspensions is that they can provide higher chemical stability compared to other methods to formulate insoluble drugs, since the drug is in the solid state rather than the solution state. For example, omeprazole, a proton pump inhibitor, degrades in acidic environment and is currently available as a powder to be reconstituted immediately before administration, with a shelf life of 4 h at room temperature. Recently, a nanosuspension formulation of omeprazole has been reported that is stable up to 1 month at 4°C (Moschwitzer et al. 2004). This is explained due to the crystalline nature of the drug and the presence of the stabilizer layer on the surface of the particles that reduces the overall rate of degradation.

10.7.2 Biopharmaceutical Benefits of Nanosuspensions

Nanosuspensions have been widely used to increase the rate and extent of absorption of poorly water-soluble drugs. This increased bioavailability (Kondo et al. 1993a, b; Liversidge and Conzentino 1995; Liversidge and Cundy 1995; Yin et al. 2005; Kesisoglou et al. 2007) occurs primarily due to increased dissolution velocity (Vergote et al. 2001; Hu et al. 2002; Rasenack and Muller 2002; Kayser et al. 2003; Hecq et al. 2005, 2006; Jinno et al. 2006; Kocbek et al. 2006; Zhang et al. 2007), but can also be a result of increased permeability (Jia et al. 2002) and enhanced particulate uptake (Desai et al. 1996). For example, nanocrystalline cilostazol, a BCS Class II drug with an aqueous solubility of $3\,\mu g/mL$ at 25°C and a median particle size of 220 nm, dissolved immediately in water. On the other hand, jet milled cilostazol (median particle size, $2.4\,\mu m$) and hammer milled cilostazol (median particle size, $13\,\mu m$) only dissolved approximately 60% and 18%, respectively in 5 min (Jinno et al. 2006) (Fig. 10.8).

In a bioavailability study conducted in dogs in fasted state, nanocrystalline cilostazol gave an AUC which was 7 times higher than that produced by hammer milled cilostazol. Approximate absolute bioavailability was as high as 86% with

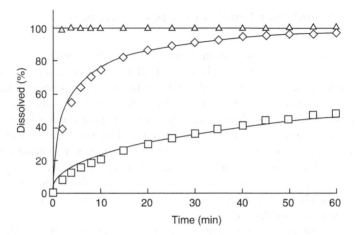

Fig. 10.8 *In vitro* dissolution profiles of cilostazol from the suspensions in water at 37°C. Dissolution study was performed at 50 rpm following USP Apparatus 2. About 5 mg of cilostazol was applied in 900 mL water. Results are expressed as the mean with the bar showing S.D. values of six experiments and simulated curves (*solid lines*). Keys: ▲, NanoCrystal® spray-dried powder; ■, jet-milled crystal; ■, hammer-milled crystal (Adapted from Jinno et al. 2006, with permission)

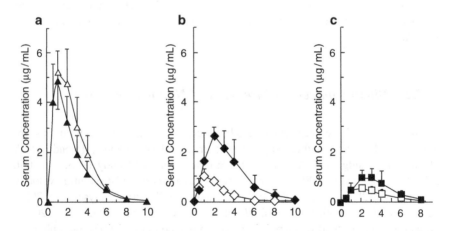

Fig. 10.9 Serum concentration–time profiles of cilostazol after oral administration of the suspensions at a dose of 100 mg/body in beagle dogs. Results are expressed as the mean with the bar showing S.D. values of four experiments. Keys: (**a**) △, NanoCrystal® suspension (fasted); ▲, NanoCrystal® suspension (fed); (**b**) ◊, jet-milled suspension (fasted); ◆, jet-milled suspension (fed); (**c**) □, hammer-milled suspension (fasted); ■, hammer-milled suspension (fed) (Adapted from Jinno et al. 2006, with permission)

nanocrystalline cilostazol. Also administration of cilostazol in the nanocrystalline form helped in reducing the effect of food on its absorption, giving a more consistent serum concentration profile both in fed and fasted conditions (Fig. 10.9).

In another *ex vivo* study, it was found that the permeability rate and the total amount permeated of the investigational drug 301029 through a Caco-2 monolayer were about four times higher for a nanoparticle formulation with a particle size of 280 nm than for a microparticle formulation with a particle size of 7 μm (Jia et al. 2002) (Fig. 10.10). This permeability data correlated well with pharmacokinetic data obtained in rats, where the nanoparticle formulation showed a T_{max} of 1 h against 4 h shown by the microparticle formulation. Also, a fourfold increase in the AUC of the drug was obtained with nanonized 301029 when compared to micronized formulation indicating that nanosizing helped in both the rate and extent of absorption. In addition to the increased dissolution and permeability, superior adhesion properties of the submicron particles in the gastro intestinal mucosa also assist in increasing the absorption and enhancing the local action of water insoluble drugs (Lamprecht et al. 2001; Jacobs et al. 2001; Kayser 2001; Muller and Jacobs 2002).

Nanosuspensions also help in reducing the side effects of formulations of water insoluble drugs. In the case of BCS class II drugs, dissolution is the rate-limiting step, and administration of such compounds as nanosuspensions often results in increased dissolution and absorption rate, as indicated by the reduced T_{max}. This decreases the residence time in the gut, which is beneficial in case of irritant drugs such as naproxen. Liversidge and Conzentino (1995) demonstrated that 270 nm naproxen particles caused less gastric irritation when compared to 20–30 μm particles.

Nanosuspensions are not only helpful in increasing the oral bioavailability of drugs, but they can also prove useful in i.v. administration of water insoluble compounds (Scholer et al. 2001). They are a stable and effective alternative to other

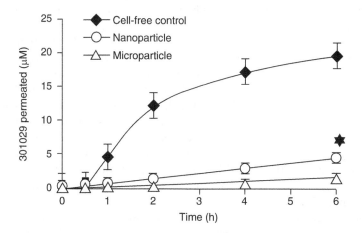

Fig. 10.10 Cellular permeability of regular microparticle 301029 and nanoparticle 301029 across Caco-2 monolayer. The microparticle 301029 or nanoparticle 301029 was added to the apical side of the cells and the permeated 301029 was determined from the basolateral side by the LC/MS method. The data represents the mean±SD (n=6). *p<0.05 compared with regular microparticle 301029 by one-way analysis of variance (ANOVA) (Adapted from Jia et al. 2002, with permission)

challenging i.v. formulations such as liposomes (Peters et al. 2000). I.V. administration of itraconazole as nanosized particles increased the efficacy or pharmacological effectiveness of the drug due to altered disposition (White et al. 2003; Yeh et al. 2005; Rabinow et al. 2007). Administration of itraconazole as a nanosuspension with an average particle size of 581 nm resulted in sequestration of the nanoparticles in the monocyte phagocytic system (MPS) and this then acted as a drug depot from which the drug was released slowly over a long period of time. Engulfment of the nanoparticles in MPS resulted in a reduced C_{max}, and increased $t_{1/2}$ as compared to the i.v. solution. Also due to sequestration a much higher dosage level can be administered safely as compared to an i.v. solution. Abraxane is a nanosuspension formulation of paclitaxel containing amorphous particles of paclitaxel 130 nm in size, that have been stabilized with the help of albumin for i.v. administration. The conventional i.v. formulation of paclitaxel, Taxol, is a clear solution of the drug in cremophor and ethanol. As a consequence of these co-solvents, i.v. administration of Taxol is generally associated with pain on injection. These co-solvents are also partly responsible for the "taxane associated toxicity," especially the fluid retention and hypersensitivity that usually occurs following the administration of Taxol (Weiss et al. 1990; Harries et al. 2005; Pinder and Ibrahim 2006). No harsh chemicals or co-solvents are employed in the nanosuspension formulation (Abraxane) and hence it is free from these side effects (Harries et al. 2005; Green et al. 2006; Stinchcombe 2007). Abraxane also produces a prolonged effect due to its particulate nature and the higher doses can be administered without causing toxicity.

Drug targeting to specific tissues or organs can also be achieved by utilizing nanosuspension based formulations. Usually on i.v. administration, if the particles do not dissolve quickly, then the body tries to remove these foreign particulates by opsonization and phagocytosis (Owens Iii and Peppas 2006). In the process of opsonization, the particles are coated by the opsonin (such as immunoglobulin G) which makes them susceptible to phagocytosis. Phagocytosis is a clearance mechanism of the body in which the foreign objects are engulfed by macrophages and destroyed or broken down into parts that can be removed easily by the body. If the particle or the object cannot be metabolized, then they are sequestered into the organs of MPS. This sequestration of the nanoparticles is very helpful in early stages of diseases, such as AIDS, in which the MPS system acts as a carrier or breeding ground for the HIV virus. In one such *in vitro* study it was demonstrated that indinavir nanoparticles were actively taken by up human monocyte-derived macrophages and afforded significant anti-retroviral activities (Dou et al. 2007). A similar study with the anti-leishmania drug aphidicolin also showed promising results against the Leishmania parasite that also resides and multiplies within the macrophages (Kayser 2000).

Besides targeting the MPS system, drug nanosuspensions can also be used to target tumors due to the enhanced permeation and retention effect (EPR effect) (Wu et al. 1993). Tumors have lower plasma clearance and leaky vasculature which results in filtering and accumulation of these circulating nanoparticles into the tumor tissues. For tumor targeting, it is necessary to prevent uptake of particles by the MPS systems, which is usually achieved by coating the particles with hydrophilic polymers such as PEG's which delay the adsorption of opsonin proteins and subsequent phagocytosis (Storm et al. 1995).

Delivery of drugs to the eye is another area where nanosuspensions can be very useful. The amount of the drug that can be administered in the eye using a solution dosage form is limited by the volume of the cul-de-sac, rapid turnover of the lachrymal fluid, and loss due to drainage via the naso-lachrymal duct. Suspensions are preferred over solutions in ocular drug delivery since the drug particles act as a reservoir from which the drug is constantly depleted, thus providing a higher bioavailability. The bioavailability of the drug from the suspension depends on the particle size of the drug and also on the saturation solubility and dissolution rate of the drug (Schoenwald and Stewart 1980; Hui and Robinson 1986). Nanosuspensions of sub-micron drug particles with high surface area that have high dissolution rates are an ideal dosage form for ocular delivery. The irritation potential of nanosuspensions is very low due to their small particle size, compared to normal suspensions which have particle sizes in the range of micrometers. Figure 10.11 compares the rise in intraocular pressure (IOP) as a measure of bioavailability of prednisolone from a nanosuspension, a conventional suspension and a solution. Here, it can be clearly seen that the nanosuspension formulation with a particle size of 211 nm gave the maximum intensity and duration of effect (Kassem et al. 2007).

Particle size plays a crucial role in the delivery of drugs through the respiratory tract. It affects particle deposition, adhesion, dissolution, and clearance mechanisms, thus affecting the overall clinical performance (local action and/or bioavailability, toxicity and side effects) of the drug. Large particles, greater than 5 μm, generally get deposited in the upper respiratory tract, while particles below 5 μm are able to get transported deep into the lungs and alveoli (Chow et al. 2007). Nanosuspensions usually

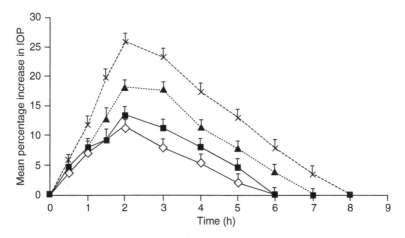

Fig. 10.11 Effect of drug particle size on mean percentage increase in IOP for normotensive Albino rabbits receiving 50 μL of 0.1% prednisolone solution or suspensions of different mean particle diameters particle diameter 211 nm (X), particle diameter 1.626 μm (▲), particle diameter 4.0 μm (■), solution (♦) (Adapted from Kassem et al. 2007, with permission)

have no or only a tiny fraction of particles above 5 μm, therefore they give superior performance compared to conventional aerosol suspensions that have particle sizes ranging from 1 to 5 μm. Other advantages of nanosuspensions are fast dissolution rates and higher saturation solubility which helps in better control of the disease condition. Another major advantage of nanosuspensions is that they have better size homogeneity than conventional suspensions and thus, minimum loss of drug occurs in the upper respiratory tract. These advantages have been confirmed in an Andersen cascade impactor study in which ultrasonic nebulization of Nanocrystal™ beclomethasone dipropionate produced 56–72% of the emitted dose as the respirable fraction, whereas the commercial formulation, Vanceril®, produced only 36% of the emitted dose as the respirable fraction. In addition, deposition in the induction port of the cascade impactor, which is a measure of the loss of the drug due to deposition in throat, was 9–10% in case of nanosuspensions, while it was as high as 53% in case of the commercial formulation (Ostrander et al. 1999). Nanosuspensions for pulmonary administration can be prepared using surfactants such as lecithin and Span 85 which are suitable for administration into the lungs (Jacobs and Muller 2002).

10.8   Conclusions/Future Directions

Nanosuspensions provide us with a useful tool for formulating an ever increasing number of poorly water soluble drugs that are coming out of the discovery programs. This technology has not only helped in improving the dissolution and bioavailability of water insoluble drugs, but also holds promise in other aspects of drug delivery (e.g. drug targeting, reducing toxicity, altering pharmacokinetics, pulmonary delivery, ocular delivery and patient compliance). Nanosuspensions can be prepared by a variety of methods such as media milling, high pressure homogenization, microfluidization, precipitation using supercritical fluid (RESS, GAS, ASES, SEDS, SFEE, etc.), spray drying and microfluidizer reaction technology. Most of the commercially available nanosuspension products are prepared by media milling. Media milling is a straight forward method and does not require any preconditioning steps. The simple instrumentation and proven clinical success has resulted in rapid commercialization of products based on this method. Although there are several nanosuspension based products in the market, there is a lack of knowledge in the areas of nanosuspension stability and characterization. For example, till now there is no scientific rationale for selection of stabilizer and the mechanism(s) of stabilization is not fully understood. There are no standard methods for characterization of nanosuspension release, and differentiation between small particles and molecular drug can be challenging. Size characterization also necessitates careful interpretation because direct observation (for example microscopy), which is inherently exhausting, becomes very difficult, and ensemble techniques (which measure some property of the particles and mathematically transform that into particle size distribution) are only as good as their mathematical algorithms used.

Particle size distributions obtained from these ensemble techniques are often incomparable, and also, they do not furnish any information about morphology or state of the particles (single/agglomerated) in the suspension. An improved understanding of formulation factors along with their influence on the biopharmaceutical effects of nanosuspensions will further advance this important area of drug delivery.

References

Adamson, A. W. (1997). Physical chemistry of surfaces. New York, Wiley

Allison, S. D. (2007). "Liposomal drug delivery." Journal of Infusion Nursing **30**(2): 89–95.

Arriagada, F. J. and K. Osseoasare (1995). "Synthesis of nanosize silica in aerosol of reverse microemulsions." Journal of Colloid and Interface Science **170**(1): 8–17.

Barth, H. G. and R. B. Flippen (1995). "Particle-size analysis." Analytical Chemistry **67**(12): R257–R272.

Bhardwaj, V., S. Hariharan, et al. (2005). "Pharmaceutical aspects of polymeric nanoparticles for oral drug delivery." Journal of Biomedical Nanotechnology **1**: 235–258.

Bijlani, V., D. Yuonayel, et al. (2007). "Monitoring ibuprofen release from multiparticulates: in situ fiber-optic technique versus the HPLC method: a technical note." AAPS PharmSciTech **8**(3).

Bisrat, M. and C. Nystrom (1988). "Physicochemical aspects of drug release. VIII. The relation between particle size and surface specific dissolution rate in agitated suspensions." International Journal of Pharmaceutics **47**(1–3): 223–231

Borm, P., F. C. Klaessig, et al. (2006). "Research strategies for safety evaluation of nanomaterials, Part V: role of dissolution in biological fate and effects of nanoscale particles." Toxicological Sciences **90**(1): 23–32.

Burgess, D. J. (2005). Injectable dispersed systems. Boca Raton, Taylor & Francis Group

Burgess, D. J., E. Duffy, et al. (2004). "Particle size analysis: AAPS workshop report, cosponsored by the food and drug administration and the united states pharmacopeia." AAPS Journal **6**(3).

Caruthers, S. D., S. A. Wickline, et al. (2007). "Nanotechnological applications in medicine." Current Opinion in Biotechnology **18**(1): 26–30.

Cherng-ju, K. (2004). Advanced pharmaceutics: physicochemical principles. Boca Raton, CRC Press LLC

Chhabra, V., V. Pillai, et al. (1995). "Synthesis, characterization, and properties of microemulsion-mediated nanophase TiO_2 particles." Langmuir **11**(9): 3307–3311.

Chow, A. H. L., H. H. Y. Tong, et al. (2007). "Particle engineering for pulmonary drug delivery." Pharmaceutical Research **24**(3): 411–437.

Connors, R. D. and E. J. Elder (2004). "Using a portfolio of particle growth technologies to enable delivery of drugs with poor water solubility." Drug Delivery Technology **4**(8): 78–83.

Crisp, M. T., C. J. Tucker, et al. (2007). "Turbidimetric measurement and prediction of dissolution rates of poorly soluble drug nanocrystals." Journal of Controlled Release **117**(3): 351–359.

Dalwadi, G., H. A. E. Benson, et al. (2005). "Comparison of diafiltration and tangential flow filtration for purification of nanoparticle suspensions." Pharmaceutical Research **22**(12): 2152–2162.

Desai, M. P., V. Labhasetwar, et al. (1996). "Gastrointestinal uptake of biodegradable microparticles: effect of particle size." Pharmaceutical Research **13**(12): 1838–1845.

Devine, D. V., K. Wong, et al. (1994). "Liposome-complement interactions in rat serum – implications for liposome survival studies." Biochimica et Biophysica Acta-Biomembranes **1191**(1): 43–51.

Dou, H., J. Morehead, et al. (2007). "Laboratory investigations for the morphologic, pharmacokinetic, and anti-retroviral properties of indinavir nanoparticles in human monocyte-derived macrophages." Virology **358**(1): 148–158.

El-Shabouri, M. H. (2002). "Nanoparticles for improving the dissolution and oral bioavailability of spironolactone, a poorly-soluble drug." STP Pharma Sciences **12**(2): 97–101.

Farokhzad, O. C. and R. Langer (2006). "Nanomedicine: developing smarter therapeutic and diagnostic modalities." Advanced Drug Delivery Reviews **58**(14): 1456–1459.

Frantzen, C. B., L. Ingebrigtsen, et al. (2003). "Assessing the accuracy of routine photon correlation spectroscopy analysis of heterogeneous size distributions." AAPS PharmSciTech **4**(3): E36

Gassmann, P., M. List, et al. (1994). "Hydrosols – alternatives for the parenteral application of poorly water soluble drugs." European Journal of Pharmaceutics and Biopharmaceutics **40**(2): 64–72.

Grau, M. J., O. Kayser, et al. (2000). "Nanosuspensions of poorly soluble drugs - reproducibility of small scale production." International Journal of Pharmaceutics **196**(2): 155–159.

Green, M. R., G. M. Manikhas, et al. (2006). "Abraxane®, a novel Cremophor®-free, albumin-bound particle form of paclitaxel for the treatment of advanced non-small-cell lung cancer." Annals of Oncology **17**(8): 1263–1268.

Gruverman, I. J. (2003). "Breakthrough ultraturbulent reaction technology opens frontier for developing life-saving nanometer-scale suspensions & dispersions." Drug Delivery Technology **3**(1): 52.

Harries, M., P. Ellis, et al. (2005). "Nanoparticle albumin-bound paclitaxel for metastatic breast cancer." Journal of Clinical Oncology **23**(31): 7768–7771.

Haskell, R. J. (1998). "Characterization of submicron systems via optical methods." Journal of Pharmaceutical Sciences **87**(2): 125–129.

Hecq, J., M. Deleers, et al. (2005). "Preparation and characterization of nanocrystals for solubility and dissolution rate enhancement of nifedipine." International Journal of Pharmaceutics **299**(1–2): 167–177.

Hecq, J., M. Deleers, et al. (2006). "Preparation and *in vitro/in vivo* evaluation of nano-sized crystals for dissolution rate enhancement of ucb-35440-3, a highly dosed poorly water-soluble weak base." European Journal of Pharmaceutics and Biopharmaceutics **64**(3): 360–368.

Hiemenz, P. C. and R. Rajagopalan (1997). Principles of colloid and surface chemistry. New York, Marcel Dekker Inc

Hintz, R. J. and K. C. Johnson (1989). "The effect of particle size distribution on dissolution rate and oral absorption." International Journal of Pharmaceutics **51**(1): 9–17.

Hirai, T., S. Hariguchi, et al. (1997). "Biomimetic synthesis of calcium carbonate particles in a pseudovesicular double emulsion." Langmuir **13**(25): 6650–6653.

Hooton, J. C., C. S. German, et al. (2004). "An atomic force microscopy study of the effect of nanoscale contact geometry and surface chemistry on the adhesion of pharmaceutical particles." Pharmaceutical Research **21**(6): 953–961.

Hu, J. H., T. L. Rogers, et al. (2002). "Improvement of dissolution rates of poorly water soluble APIs using novel spray freezing into liquid technology." Pharmaceutical Research **19**(9): 1278–1284.

Hui, H. W. and J. R. Robinson (1986). "Effect of particle dissolution rate on ocular drug bioavailability." Journal of Pharmaceutical Sciences **75**(3): 280–287.

Jacobs, C., O. Kayser, et al. (2001). "Production and characterisation of mucoadhesive nano-suspensions for the formulation of bupravaquone." International Journal of Pharmaceutics **214**(1–2): 3–7.

Jacobs, C. and R. H. Muller (2002). "Production and characterization of a budesonide nanosus-pension for pulmonary administration." Pharmaceutical Research **19**(2): 189–194.

Jarvinen, K., T. Jarvinen, et al. (1995). "Ocular absorption following topical delivery." Advanced Drug Delivery Reviews **16**(1): 3–19.

Jia, L., H. Wong, et al. (2002). "Effect of nanonization on absorption of 301029: *Ex vivo* and *in vivo* pharmacokinetic correlations determined by liquid chromatography/mass spectrometry." Pharmaceutical Research **19**(8): 1091–1096.

Jinno, J., N. Kamada, et al. (2006). "Effect of particle size reduction on dissolution and oral absorption of a poorly water-soluble drug, cilostazol, in beagle dogs." Journal of Controlled Release **111**(1–2): 56–64.

Juliano, R. L. and D. Stamp (1975). "The effect of particle size and charge on the clearance rates of liposomes and liposome encapsulated drugs." Biochemical and Biophysical Research Communications **63**(3): 651–658.

Jung, J. and M. Perrut (2001). "Particle design using supercritical fluids: Literature and patent survey." Journal of Supercritical Fluids **20**(3): 179–219.

Kassem, M. A., A. A. Abdel Rahman, et al. (2007). "Nanosuspension as an ophthalmic delivery system for certain glucocorticoid drugs." International Journal of Pharmaceutics **340**(1–2): 126–133.

Kayser, O. (2000). "Nanosuspensions for the formulation of aphidicolin to improve drug targeting effects against Leishmania infected macrophages." International Journal of Pharmaceutics **196**(2): 253–256.

Kayser, O. (2001). "A new approach for targeting to *Cryptosporidium parvum* using mucoadhesive nanosuspensions: research and applications." International Journal of Pharmaceutics **214**(1–2): 83–85.

Kayser, O., C. Olbrich, et al. (2003). "Formulation of amphotericin B as nanosuspension for oral administration." International Journal of Pharmaceutics **254**(1): 73–75.

Keck, C. M. and R. H. Muller (2006). "Drug nanocrystals of poorly soluble drugs produced by high pressure homogenisation." European Journal of Pharmaceutics and Biopharmaceutics **62**(1): 3–16.

Kesisoglou, F., S. Panmai, et al. (2007). "Application of nanoparticles in oral delivery of immediate release formulations." Current Nanoscience **3**(2): 183–190.

Kipp, J. E. (2004). "The role of solid nanoparticle technology in the parenteral delivery of poorly water-soluble drugs." International Journal of Pharmaceutics **284**(1–2): 109–122.

Kocbek, P., S. Baumgartner, et al. (2006). "Preparation and evaluation of nanosuspensions for enhancing the dissolution of poorly soluble drugs." International Journal of Pharmaceutics **312**(1–2): 179–186.

Kondo, N., T. Iwao, et al. (1993a). "Pharmacokinetics of a micronized, poorly water-soluble drug, HO-221, in experimental animals." Biological and Pharmaceutical Bulletin **16**(8): 796–800.

Kondo, N., T. Iwao, et al. (1993b). "Improved oral absorption of a poorly water-soluble drug, HO-221, by wet-bead milling producing particles in submicron region." Chemical and Pharmaceutical Bulletin **41**(4): 737–740.

Lamprecht, A., U. Schäfer, et al. (2001). "Size-dependent bioadhesion of micro- and nanoparticulate carriers to the inflamed colonic mucosa." Pharmaceutical Research 18(6): 788–793.

Lee, J., S. J. Lee, et al. (2005). "Amphiphilic amino acid copolymers as stabilizers for the preparation of nanocrystal dispersion." European Journal of Pharmaceutical Sciences **24**(5): 441–449.

List, M. and H. Sucker (1995). "Hydrosols of pharmacologically active agents and their pharmaceutical compositions comprising them." US Patent 5389382.

List, M. L. and H. B. Sucker (1988). "Pharmaceutical colloidal hydrosols for injection." GB Patent 2200048.

Liversidge, G. C., K. C. Cundy, et al. (1991). "Surface modified drug nanoparticles." US Patent 5145684.

Liversidge, G. G. and P. Conzentino (1995). "Drug particle-size reduction for decreasing gastric irritancy and enhancing absorption of naproxen in rats." International Journal of Pharmaceutics **125**(2): 309–313.

Liversidge, G. G. and K. C. Cundy (1995). "Particle-size reduction for improvement of oral bioavailability of hydrophobic drugs .1. Absolute oral bioavailability of nanocrystalline danazol in beagle dogs." International Journal of Pharmaceutics **125**(1): 91–97.

Martin, A. (2001). Physical pharmacy. Philadelphia, Lippincott Williams & Wilkins.

Mehnert, W. and K. Mader (2001). "Solid lipid nanoparticles: production, characterization and applications." Advanced Drug Delivery Reviews 47(2–3): 165–196.

Merisko-Liversidge, E., G. G. Liversidge, et al. (2003). "Nanosizing: a formulation approach for poorly-water-soluble compounds." European Journal of Pharmaceutical Sciences 18(2): 113–120.

Mori, Y., Y. Okastu, et al. (2001). "Titanium dioxide nanoparticles produced in water-in-oil emulsion." Journal of Nanoparticle Research 3(2): 219–225.

Moschwitzer, J., G. Achleitner, et al. (2004). "Development of an intravenously injectable chemically stable aqueous omeprazole formulation using nanosuspension technology." European Journal of Pharmaceutics and Biopharmaceutics 58(3): 615–619.

Müller, R. H., R. Becker, et al. (1996). "Pharmaceutical nanosuspensions for medicament administration as systems with increased saturation solubility and speed of dissolution." US Patent 5858410.

Muller, R. H. and C. Jacobs (2002). "Buparvaquone mucoadhesive nanosuspension: preparation, optimisation and long-term stability." International Journal of Pharmaceutics 237(1–2): 151–161.

Muller, R. H., C. Jacobs, et al. (2001). "Nanosuspensions as particulate drug formulations in therapy: Rationale for development and what we can expect for the future." Advanced Drug Delivery Reviews 47(1): 3–19.

Muller, R. H., K. Krause, et al. (2002). "Method for controlled production of ultrafine microparticles and nanoparticles. EP Patent 1194123 (A2).

Ostrander, K. D., H. W. Bosch, et al. (1999). "An in-vitro assessment of a NanoCrystal (TM) beclomethasone dipropionate colloidal dispersion via ultrasonic nebulization." European Journal of Pharmaceutics and Biopharmaceutics 48(3): 207–215.

Owens Iii, D. E. and N. A. Peppas (2006). "Opsonization, biodistribution, and pharmacokinetics of polymeric nanoparticles." International Journal of Pharmaceutics 307(1): 93–102.

Pace, S. N., G. W. Pace, et al. (1999). "Novel injectable formulations of insoluble drugs." Pharmaceutical Technology 23(3): 116–134.

Park, H. J., M. S. Kim, et al. (2007). "Recrystallization of fluconazole using the supercritical antisolvent (SAS) process." International Journal of Pharmaceutics 328(2): 152–160.

Pathak, P., M. J. Meziam, et al. (2006). "Formation and stabilization of ibuprofen nanoparticles in supercritical fluid processing." Journal of Supercritical Fluids 37(3): 279–286.

Pathak, P., M. J. Meziani, et al. (2007). "Supercritical fluid processing of drug nanoparticles in stable suspension." Journal of Nanoscience and Nanotechnology 7(7): 2542–2545.

Patil, S. D. and D. J. Burgess (2005). Injectable dispersed systems. Boca Raton, Taylor & Francis Group.

Patravale, V. B., A. A. Date, et al. (2004). "Nanosuspensions: a promising drug delivery strategy." Journal of Pharmacy and Pharmacology 56(7): 827–840.

Peters, K., S. Leitzke, et al. (2000). "Preparation of a clofazimine nanosuspension for intravenous use and evaluation of its therapeutic efficacy in murine Mycobacterium avium infection." Journal of Antimicrobial Chemotherapy 45(1): 77–83.

Peters, K., R. H. Muller, et al. (1999). "An investigation into the distribution of lecithins in nanosuspension systems using low frequency dielectric spectroscopy." International Journal of Pharmaceutics 184(1): 53–61.

Phipps, L. W. (1971). "Mechanism of oil droplet fragmentation in high pressure homogenizers." Nature 233(5322): 617–619.

Pinder, M. C. and N. K. Ibrahim (2006). "Nanoparticle albumin-bound paclitaxel for treatment of metastatic breast cancer." Drugs of Today 42(9): 599–604.

Pozarnsky, G. A. and E. Matijevic (1997). "Preparation of monodisperse colloids of biologically active compounds .1. Naproxen." Colloids and Surfaces. A, Physicochemical and Engineering Aspects 125(1): 47–52.

Qi, L. M., J. M. Ma, et al. (1997). "Synthesis of copper nanoparticles in nonionic water-in-oil microemulsions." Journal of Colloid and Interface Science 186(2): 498–500.

Rabinow, B., J. Kipp, et al. (2007). "Itraconazole IV nanosuspension enhances efficacy through altered pharmacokinetics in the rat." International Journal of Pharmaceutics **339**(1–2): 251–260.

Rabinow, B. E. (2004a). "Nanosuspensions in drug delivery." Nature Reviews Drug Discovery **3**(9): 785–796.

Rabinow, B. (2004b). "Nanoedge drug delivery solves the problems of insoluble injectable drugs." Supplement to Scrip World Pharmaceutical News, October, 13–16.

Rasenack, N. and B. W. Muller (2002). "Dissolution rate enhancement by in situ micronization of poorly water-soluble drugs." Pharmaceutical Research **19**(12): 1894–1900.

Rasenack, N. and B. W. Muller (2004). "Micron-size drug particles: Common and novel micronization techniques." Pharmaceutical Development and Technology **9**(1): 1–13.

Rogers, T. L., I. B. Gillespie, et al. (2004). "Development and characterization of a scalable controlled precipitation process to enhance the dissolution of poorly water-soluble drugs." Pharmaceutical Research **21**(11): 2048–2057.

Rogers, T. L., J. H. Hu, et al. (2002). "A novel particle engineering technology: spray-freezing into liquid." International Journal of Pharmaceutics **242**(1–2): 93–100.

Roser, M., D. Fischer, et al. (1998). "Surface-modified biodegradable albumin nano- and microspheres. II: effect of surface charges on *in vitro* phagocytosis and biodistribution in rats." European Journal of Pharmaceutics and Biopharmaceutics **46**(3): 255–263.

Sarkari, M., J. Brown, et al. (2002). "Enhanced drug dissolution using evaporative precipitation into aqueous solution." International Journal of Pharmaceutics **243**(1–2): 17–31.

Schoenwald, R. D. and P. Stewart (1980). "Effect of particle size on ophthalmic bioavailability of dexamethasone suspensions in rabbits." Journal of Pharmaceutical Sciences **69**(4): 391–394.

Scholer, N., K. Krause, et al. (2001). "Atovaquone nanosuspensions show excellent therapeutic effect in a new murine model of reactivated toxoplasmosis." Antimicrobial Agents and Chemotherapy **45**(6): 1771–1779.

Schultz, S., G. K. Wagner, et al. (2004). "High-pressure homogenization as a process for emulsion formation." Chemical Engineering & Technology **27**(4): 361–368.

Schwarzer, H. C. and W. Peukert (2002). "Experimental investigation into the influence of mixing on nanoparticle precipitation." Chemical Engineering & Technology **25**(6): 657–661.

Shchukin, D. G. and G. B. Sukhorukov (2004). "Nanoparticle synthesis in engineered organic nanoscale reactors." Advanced Materials **16**(8): 671–682.

Shekunov, B. Y., P. Chattopadhayay, et al. (2006). "Nanoparticles of poorly water-soluble drugs prepared by supercritical fluid extraction of emulsions." Pharmaceutical Research **23**(1): 196–204.

Shekunov, B. Y., P. Chattopadhyay, et al. (2007). "Particle size analysis in pharmaceutics: principles, methods and applications." Pharmaceutical Research **24**(2): 203–227.

Shi, H. G., L. Farber, et al. (2003). "Characterization of crystalline drug nanoparticles using atomic force microscopy and complementary techniques." Pharmaceutical Research **20**(3): 479–484.

Snorek, S. M., J. F. Bauer, et al. (2007). "PQRI recommendations on particle-size analysis of drug substances used in oral dosage forms." Journal of Pharmaceutical Sciences **96**(6): 1451–1467.

Stinchcombe, T. E. (2007). "Nanoparticle albumin-bound paclitaxel: a novel Cremphor-EL®-free formulation of paclitaxel." Nanomedicine **2**(4): 415–423.

Storm, G., S. O. Belliot, et al. (1995). "Surface modification of nanoparticles to oppose uptake by the mononuclear phagocyte system." Advanced Drug Delivery Reviews **17**(1): 31–48.

Tan, C. P. and M. Nakajima (2005). "Beta-Carotene nanodispersions: preparation, characterization and stability evaluation." Food Chemistry **92**(4): 661–671.

Texter, J. (2001). "Precipitation and condensation of organic particles." Journal of Dispersion Science and Technology **22**(6): 499–527.

Tinke, A. P., K. Vanhoutte, et al. (2005). "A new approach in the prediction of the dissolution behavior of suspended particles by means of their particle size distribution." Journal of Pharmaceutical and Biomedical Analysis **39**(5): 900–907.

Trotta, M., M. Gallarate, et al. (2003). "Preparation of griseofulvin nanoparticles from water-dilutable microemulsions." International Journal of Pharmaceutics 254(2): 235–242.

Trotta, M., M. Gallarate, et al. (2001). "Emulsions containing partially water-miscible solvents for the preparation of drug nanosuspensions." Journal of Controlled Release 76(1-2): 119–128.

Tucker, C. J. (2004). Real time monitoring of small particle dissolution by way of light scattering US Patent 6750966.

Vergote, G. J., C. Vervaet, et al. (2001). "An oral controlled release matrix pellet formulation containing nanocrystalline ketoprofen." International Journal of Pharmaceutics 219(1–2): 81–87.

Wagner, V., A. Dullaart, et al. (2006). "The emerging nanomedicine landscape." Nature Biotechnology 24(10): 1211–1217.

Walz, J. Y. (1998). "The effect of surface heterogeneities on colloidal forces." Advances in Colloid and Interface Science 74(1–3): 119–168.

Wang, J. Z. and D. R. Flanagan (2002). "General solution for diffusion-controlled dissolution of spherical particles. 2. Evaluation of experimental data." Journal of Pharmaceutical Sciences 91(2): 534–542.

Weiss, R. B., R. C. Donehower, et al. (1990). "Hypersensitivity reactions from taxol." Journal of Clinical Oncology 8(7): 1263–1268.

White, R. D., J. Wong, et al. (2003). "Pre-clinical evaluation of itraconazole nanosuspension for intravenous injection." Toxicological Sciences 72: 51–51.

Wu, N. Z., D. Da, et al. (1993). "Increased microvascular permeability contributes to preferential accumulation of stealth liposomes in tumor tissue." Cancer Research 53(16): 3765–3770.

Wu, Y., A. Loper, et al. (2004). "The role of biopharmaceutics in the development of a clinical nanoparticle formulation of MK-0869: a Beagle dog model predicts improved bioavailability and diminished food effect on absorption in human." International Journal of Pharmaceutics 285(1–2): 135–146.

Yeh, T. K., Z. Lu, et al. (2005). "Formulating paclitaxel in nanoparticles alters its disposition." Pharmaceutical Research 22(6): 867–874.

Yin, S. X., M. Franchini, et al. (2005). "Bioavailability enhancement of a COX-2 inhibitor, BMS-347070, from a nanocrystalline dispersion prepared by spray-drying." Journal of Pharmaceutical Sciences 94(7): 1598–1607.

Young, P. M., D. Cocconi, et al. (2002). "Characterization of a surface modified dry powder inhalation carrier prepared by "particle smoothing"." Journal of Pharmacy and Pharmacology 54(10): 1339–1344.

Zhang, D. R., T. W. Tan, et al. (2007). "Preparation of azithromycin nanosuspensions by high pressure homogenization and its physicochemical characteristics studies." Drug Development and Industrial Pharmacy 33(5): 569–575.

Index

CPSIA information can be obtained at www.ICGtesting.com
Printed in the USA
LVOW05*1652280115

424709LV00001BA/2/P